HANDBUCH DER UROLOGIE

ENCYCLOPEDIA OF UROLOGY

ENCYCLOPÉDIE D'UROLOGIE

HERAUSGEGEBEN VON · EDITED BY
PUBLIÉE SOUS LA DIRECTION DE

C. E. ALKEN
HOMBURG (SAAR)

V. W. DIX
LONDON

H. M. WEYRAUCH
SAN FRANCISCO

E. WILDBOLZ
BERN

XI/1

SPRINGER-VERLAG BERLIN HEIDELBERG GMBH 1967

TUMOURS I

BY

ARTHUR JACOBS · ERIC RICHES

ORGANIC DISEASES

BY

LESLIE N. PYRAH

WITH 141 FIGURES

SPRINGER-VERLAG BERLIN HEIDELBERG GMBH 1967

ISBN 978-3-642-46087-6   ISBN 978-3-642-46085-2 (eBook)
DOI 10.1007/978-3-642-46085-2

Library of Cougress Catalog Card Number 58-4788

Titel-Nr, 5886

Contents

Tumours of Bladder. By Arthur Jacobs. With co-operation of G. W. Blomfield, K. M. Girdwood, J. M. Scott and T. Symington. With 47 Figures.

Contributors to Volume XI/1

ARTHUR JACOBS, F.R.C.S. (Glas.), F.R.C.S. (Ed.), 93, Kelvin Court, Glasgow W 2 / Great Britain

Professor LESLIE N. PYRAH, Ch. M., F.R.C.S. Dr., 29, Park Square, Leeds 1 / Great Britain

Sir ERIC RICHES, M. S., F.R.C.S., 22, Weymouth Street, London W 1 / Great Britain

Tumours of the Kidney and Ureter

By

SIR ERIC RICHES

With 68 Figures

A. Introduction

Renal and ureteric tumours form a relatively small but highly important part of urological practice. Their behaviour appears capricious, for whilst some are extremely malignant others pursue a benign course, although nearly all have malignant potentialities. Although early diagnosis is vitally necessary and is often possible by the use of modern methods, it does not always lead to cure because some of these tumours have an intrinsic malignancy which at present defies treatment. In the past the nomenclature has been confused; the current tendency to simpler classification, which will be followed here, has led to some clarification of the picture but there still remain some types of such infrequent occurrence that their behaviour cannot be foretold.

B. Aetiology

I. General factors

The ultimate cause of renal tumours is not known and their occasional association with congenital anomalies, cysts, or infection appears to be coincidental. Adenocarcinoma may be found in a horse-shoe kidney (SCHOON-OVER 1953) and it may occur in a polycystic kidney (JOHNSON 1953) (Fig. 1). It is sometimes associated

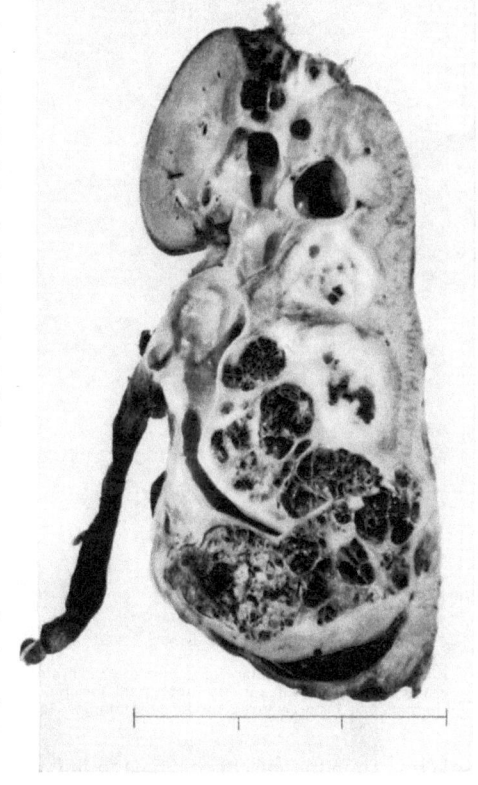

Fig. 1. Polycystic kidney containing a grade 1 adenocarcinoma. From a woman of 39 (scale in inches)

with a solitary cyst, and WALSH (1951) stated that of 500 cases of solitary cyst reported in the literature 7 per cent had an associated malignant tumour of some type.

This figure is disputed by some workers who consider that many of the reported cases are examples of cystic degeneration of a tumour (EMMETT, LEVINE and WOOLNER, 1963).

Infection of a kidney containing a tumour is not infrequent; infection of the pelvis may lead to squamous metaplasia and be followed by squamous cell carcinoma. Tuberculous infection was reported in seven cases of renal adenocarcinoma by Neibling and Walters (1948) and we have since seen it in one case (Riches 1954) (Fig. 2) but the clinical history pointed to the fact that the neoplasm preceded the tuberculosis. Lucké (1938) has produced evidence that renal adenocarcinoma in the leopard frog is caused by a virus.

II. Carcinogenic substances

That carcinogenic substances excreted in the urine can produce malignant papillary tumours in the bladder is generally accepted (Scott and Boyd 1953, McDonald and Lund 1954). The occurrence of papillary tumours of the renal pelvis in dye workers after long exposure has been reported by Macalpine (1947) who pointed out that a concomitant hydronephrosis with resulting stagnation of urine in the kidney was an associated factor more likely to produce tumours.

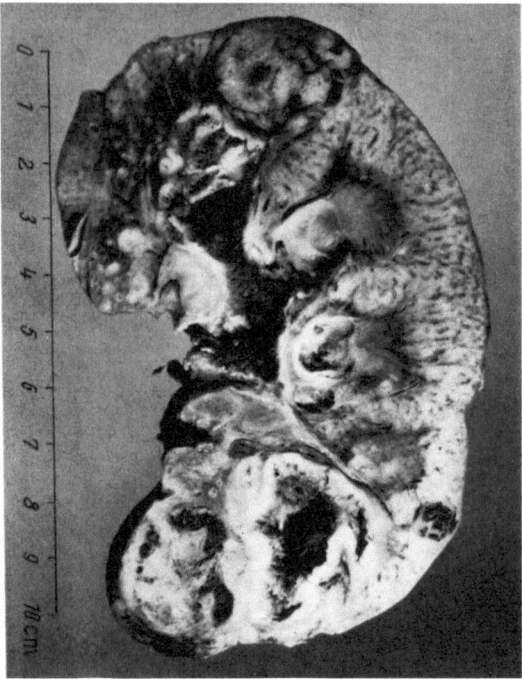

Fig. 2. Tuberculous kidney of a man of 63 containing deposits of grade 1 adenocarcinoma in the upper and lower poles. The other kidney had been removed for adenocarcinoma 14 years earlier

The substances which can produce papillary tumours of the pelvis and ureter as well as in the bladder (the Urothelium) are metabolites of certain aromatic amines of which the most potent is β-naphthylamine. They can enter the body by inhalation, ingestion or absorption through the skin and this constitutes an occupational risk for workers in the chemical, rubber and cable industries. Carcinoma of the urothelium is recognised as an industrial cancer in these occupations; that it occurs less frequently in the renal pelvis than in the bladder is because there is less time for the carcinogens to act where there is more rapid evacuation of the urine. It is characteristic of such growths that they appear many years after exposure to the agent.

Endogenous metabolites may be concerned in the production of renal tumours. Kerr and his colleagues (1963) reported a case of renal adenocarcinoma associated with an increased excretion of carcinogenic *ortho*-aminophenols in the urine from the affected kidney.

Renal tumours were produced in rats by Zollinger (1953) after repeated injections of lead phosphate and Boyland et al. (1962) found that 1 per cent of lead acetate added to the diet of Wistar rats for a year caused adenocarcinoma of the kidney.

III. Hormone induction

The possibility of a hormonal factor in the production of renal tumours is suggested by the two to one preponderance of males over females with carcinoma of the kidney. Experimentally KIRKMAN and BACON (1950) have produced malignant renal tumours in male hamsters by treatment with oestrogens.

IV. Ionizing radiation

Exposure of rats and mice to X-rays has produced renal tumours (KOLETSKY and GUSTAFSON, 1955; ROSEN et al., 1961). Carcinoma of the kidney has followed the use of radioactive thorotrast for pyelography in man.

PAVONE-MACALUSO (1963) has given a comprehensive account of the experimental production of renal tumours.

V. Malignant change in benign tumours

Whilst there is no proof that a benign adenoma can become a malignant adenocarcinoma there is a considerable probability that this does occur. NEW-COMB (1937) after a study of 248 benign nodules in 1172 autopsies concluded that some Grawitz tumours originated in adenomata. THACKRAY (1951) pointed out the greater difficulty in distinguishing histologically between a benign adenoma and a low grade carcinoma than between the higher grades of carcinoma; the subsequent behaviour in respect of metastases may be the final criterion. BUCKLEY (1940) described a case in which malignant transformation was occurring in a large tubular adenoma and quoted EWING's (1940) opinion that such transitions were frequent. LUCKÉ and SCHLUMBERGER (1957) illustrated several cases in which both adenoma and adenocarcinoma were present in the same kidney and considered that malignant tumours of mature renal tubules nearly all arose in adenomata.

Whether the benign nodules are true tumours or developmental anomalies is open to question.

C. Incidence

I. Benign tumours

NEWCOMB (1937) found that 15.2% of 1172 autopsy cases had benign tumours, either adenomas, mesenchymal tumours, or adrenal rests. APITZ (1943) reported an incidence of 19.5% in 4309 autopsies. Both authors found a low incidence in early life.

II. Malignant tumours

1. Incidence of individual types

Renal cancer is relatively rare; in a review of over 200,000 autopsies collected from the literature LUCKÉ and SCHLUMBERGER (1957) found it present in 0.3 per cent of all cases and in 2.2 per cent of nearly 30,000 cancer cases. GILBERT (1938) gave it a figure of 2 per cent of all tumours in adults and 20 per cent of those in children. GRIFFITHS and THACKRAY (1949) found nearly 1 per cent of renal parenchymal carcinomata in over 8000 cases of cancer admitted to the Middlesex Hospital in a thirteen-year period.

The incidence of the different principal types of tumour in the British Association of Urological Surgeons series of over 2000 cases (1951 RICHES et al) was as follows: —

Adenocarcinoma of the parenchyma . . . . . 75 per cent.
Pelvic tumours . . . . 12.5 per cent.
Wilms' tumour (Nephroblastoma) . . . . . . 8 per cent.
Sarcoma . . . . 2 per cent.

In the Mayo Clinic series of 633 cases reported by Priestley (1939) 79 per cent were adenocarcinomata, 10 per cent pelvic tumours, 6 per cent Wilms' tumours and 5 per cent sarcomata.

2. Age

Except in the Wilms' tumours of childhood the majority of malignant renal tumours both of the parenchyma and the pelvis occur between the ages of 50 and 70. They are very rare below the age of 20 years and uncommon before 40, but the numbers increase in the fifth decade. In the B.A.U.S. series (1951) of 1735 cases of adenocarcinoma 80 per cent occurred in the fifth, sixth and seventh decades and of 241 malignant pelvic tumours 80 per cent were in the same period. There was no significant difference in the age incidence in the two sexes.

3. Sex

On the evidence of surgical statistics there is a considerable difference in the incidence of renal cancer in the sexes. In adenocarcinoma males outnumber females by two to one; in transitional cell pelvic tumours the proportion is between three and four to one, but there is less difference in the squamous cell pelvic tumours. In the total of all tumours of the kidney and ureter in the B.A.U.S. series (1951) there were 1513 males and 801 females, a proportion of nearly two to one.

4. Side affected

The left and right kidneys are affected with equal frequency. In the B.A.U.S. series there were 1124 on the right and 1143 on the left with nine bilateral cases. Lubarsch (1925) collected figures from several sources and found 442 tumours on the right and 441 on the left.

5. Race

Renal tumours occur in all races and in every part of the globe. Uys (1956) found an incidence of 0.84 per cent in 3,707 autopsies in Bantu subjects. Females predominated over males in a ratio of two to one, but there were only fifteen cases of primary tumour of which ten were benign.

There are however considerable differences in the incidence of carcinoma of the kidney in different countries as reflected by the mortality rates. The rate is high in Denmark but low in Japan, Spain and Italy. It is higher in Scotland than in Southern Ireland; in the United States it is higher in the white population than in the non-white (Case, 1964). It is greater in the higher social classes.

D. Classification

Many different methods of classification of renal tumours have been used. Melicow (1944) adopted the site of origin as the basis, naming parenchymal, pelvic and capsular growths, each with subdivisions. A somewhat similar classification was used by Abeshouse and Weinberg (1945), and by Howes (1944). Harvey (1947) simplified the list into cortical adenocarcinoma, pelvic carcinoma, Wilms' tumour and miscellaneous unclassified, whilst Bell's (1939) list com-

prised renal tumours in children and those in adults. He subsequently (1950)
adopted a classification based on the tissue of origin and on the site. FITE (1945)
also used a histogenetic basis and this has been amplified by LUCKÉ and SCHLUM-
BERGER (1957). This method of classification applied to the parenchyma, the
pelvis and ureter, and the capsule is the most rational and will be used here.
The following list includes the main types (which are in italics) and some of
the rare ones. It indicates that for most of the benign tumours there is a malignant
counterpart.

I. Renal parenchyma

Tissue of Origin.	*Benign.*	*Malignant.*
Epithelium	*Adenoma*	*Adenocarcinoma*
Connective Tissue	Fibroma	Fibrosarcoma
	Lipoma	Liposarcoma
	Myoma	Myosarcoma
	Angioma	Angioendothelioma
Mixed		*Nephroblastoma* (WILMS)
	Angiomyolipoma	
		Lymphoblastoma

II. Renal pelvis and ureter

Epithelium	*Papilloma.*	*Carcinoma.*
		1. Transitional Cell
		a) *Papillary*
		b) Solid
		2. Squamous Cell
		3. Adenocarcinoma
Connective Tissue	Fibroma	
	Myxoma	Myxosarcoma
	Haemangioma	

III. Renal capsule

Connective Tissue	Fibroma	Fibrosarcoma
	Myoma	Myosarcoma
	Lipoma	Liposarcoma

IV. Metastatic tumours

V. Cysts

Cystic degeneration of tumours	*Simple ("Solitary") Cysts*
Cysts of chronic nephritis	Multilocular Cysts (Cystadenoma)
Pyelogenic cysts	Dermoid cysts
Paracalyceal and parapelvic cysts	Hydatid cysts
Lymphatic cysts	*Polycystic disease*

E. Clinical features

I. Symptoms

1. The classical triad of symptoms

In conformity with the wide variations in the pathological appearances there
are differences in the symptoms produced by the various types of renal tumour.
In all, however, the cardinal symptoms are haematuria, pain and tumour, and

although they are sometimes late in appearing and do not all occur in every case they give the first evidence of the condition in the majority of patients. Although each may occur as an isolated or presenting symptom it is commoner to find two or sometimes all three in combination. As might be expected haematuria is more frequent in papillary transitional cell growths whilst a palpable tumour is more often present in the solid adenocarcinomata and in Wilms' tumours. Pain in the loin is an important symptom and is present in about half the cases.

a) Haematuria

In the B.A.U.S. series (1951) bleeding occurred alone or in combination in 62 per cent of 1746 cases of adenocarcinoma, in 88 per cent of 246 cases of papillary transitional cell tumour of the pelvis whether benign or malignant, in 18 per cent of 189 Wilms' tumours and in 46 per cent of 43 sarcomata and miscellaneous tumours.

The bleeding may start spontaneously or may occur after exercise whether mild or strenuous. It can be profuse but is usually intermittent until the disease is advanced, when it becomes continuous. The act of micturition is painless unless there are large clots, but there is often pain in the loin on the affected side and sometimes severe colic from the passage of clots down the ureter. The finding of long thin ureteric clots in the voided urine is uncommon but when it does occur it is diagnostic of bleeding from the upper urinary tract. At times the bleeding is so profuse and rapid as to cause clot retention in the bladder.

Haematuria starts early in pelvic tumours, but in adenocarcinomata it does not occur until a calyx or the pelvis is eroded by the growth; its onset depends more on the proximity of the initial tumour to a calyx than on the size of the growth.

Other urinary symptoms are usually absent but the presence of blood sometimes causes increased frequency of micturition.

b) Pain

Apart from the occasional ureteric colic produced by the passage of clots the pain of a renal tumour is of a dull aching character, constant in position in the loin and continuous. It is due to the stretching of the renal capsule by the enlarging tumour or to distension of the pelvis by blood. Tension on the pedicle by the displaced kidney has also been cited as a cause of pain (Herbut 1952). The incidence of pain differs little in the different types of tumour except that in Wilms' tumour it is recorded in only about 20 per cent; this may be because the majority of patients are infants. It is present in about 50 per cent of other cases, and rather more in squamous cell pelvic tumours, probably owing to the frequent presence of infection or stones in this condition. When there are vertebral metastases involving spinal nerves the pain may be very severe.

c) Tumour

The finding of a mass in the loin together with haematuria or pain is strong evidence of a renal growth. The swelling is produced by the renal mass and is most often found when the growth arises in the lower pole of the kidney. As a single symptom it is present in only about 6 per cent of adenocarcinomata, but in combination with pain or bleeding in more than 30 per cent. It is the outstanding feature in Wilms' tumour, being the sole presenting sign in about 55 per cent of patients and in combination in about 70 per cent. In papillary tumours of the pelvis a palpable swelling is present in only from 2 to 5 per cent and it is then often the fluid swelling of a consequent hydronephrosis. In the solid squamous tumours of the pelvis a palpable mass is present in about 10 per cent. In sarcoma and capsular growths there is frequently a large mass to be felt.

2. Other symptoms
a) Pyrexia

Intermittent fever occurs in some cases of renal neoplasm and has been seen in both parenchymal and pelvic growths; the evening temperature may reach

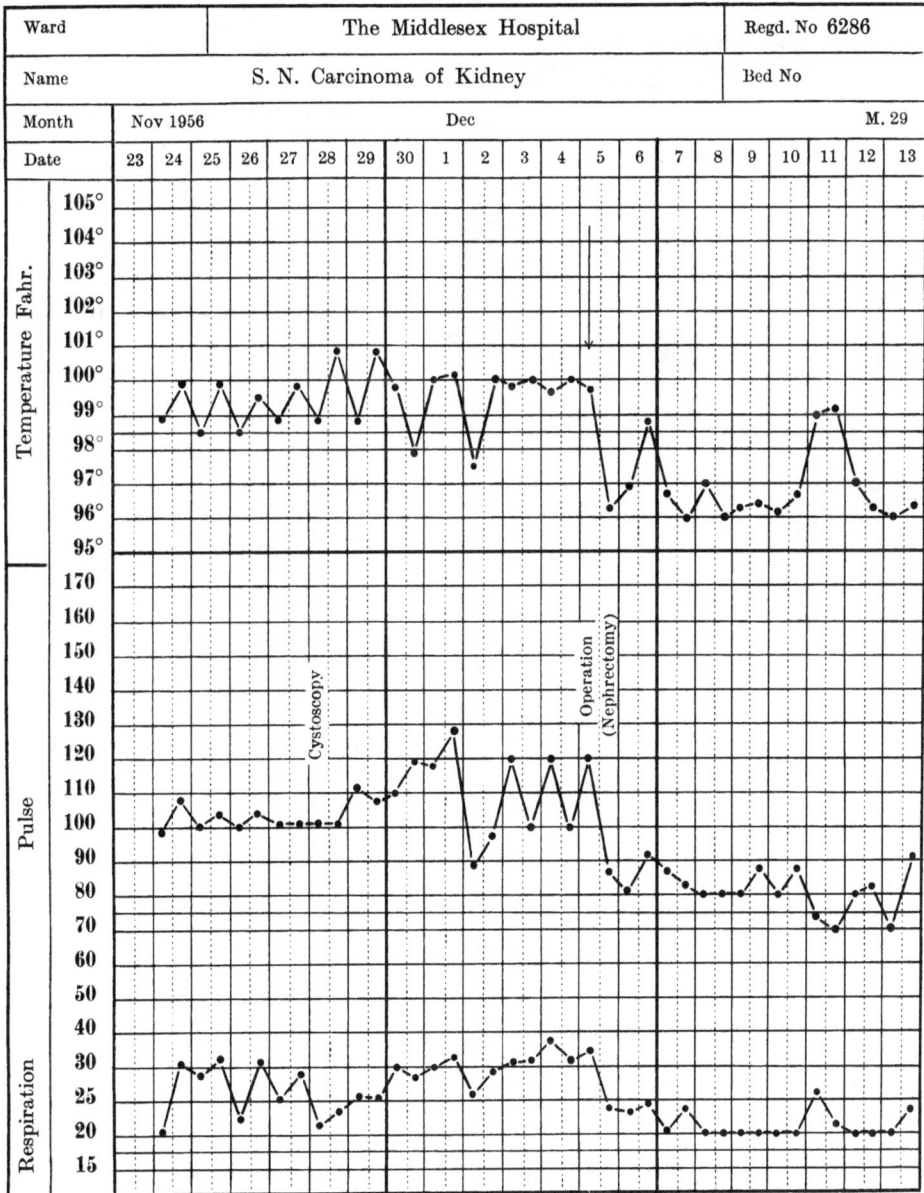

Fig. 3. Temperature chart of a man of 63 with a grade 3 carcinoma of the left kidney and metastases in the liver. The fever abated immediately on nephrectomy

101 to 103° F without any infection of the urinary tract. Hyman (1925) found it in 13 per cent of 60 cases, Griffiths and Thackray (1949) in 12 per cent of 103 cases and Abeshouse and Weinberg (1945) in 21 per cent of their series. It was the sole presenting symptom in the case reported by Rowlands (1951). Whilst the tumour is necrotic in some of the cases with pyrexia, in others there is no necrosis. In the former absorption of pyrogenic substances has been suggested as the cause of the fever, but in the early cases with no

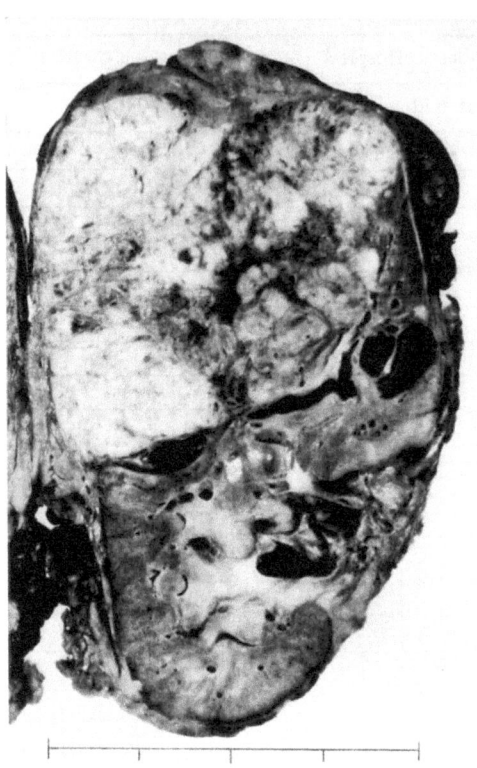

necrosis it may be an anaphylactic effect due to foreign protein from the growth entering the blood stream (Wright 1922). It is said to be commoner in amyloid disease which occurs in about 2 per cent of cases of renal carcinoma (Kimball 1961). In most cases the temperature subsides at once when the kidney is removed and its persistence after nephrectomy has sometimes indicated the presence of metastases. However in the patient whose chart is shown in Fig. 3 there were already metastases in the liver, yet the fever abated as soon as the kidney was removed; the kidney growth was necrotic (Fig. 4) whilst the hepatic secondary deposit was not. In the patient whose pyelogram is shown in Fig. 5 extensive medical investigations had failed to explain a persistent fever lasting for two months. Removal of the kidney containing an undifferentiated spheroidal cell pelvic tumour was followed by an immediate subsidence of temperature, but it rose again later and remained irregular until her death

Fig. 4. Necrotic adenocarcinoma of the kidney from the same patient as Fig. 3 (scale in inches)

six months later from cerebral metastases. As a symptom unexplained pyrexia is sufficiently important to suggest the need for renal investigation; it often indicates a serious prognosis.

b) Loss of weight

Sufficient wasting to be remarked upon by the patient is a late manifestation and is a bad prognostic sign. It may indicate the presence of metastases. Of the 16 patients who complained of it reported by Griffiths and Thackray (1949) eleven died within one year.

c) Varicocele

Stress was formerly laid on the occurrence of varicocele on the same side as an indication of a renal tumour; the enlarged perinephric veins anastomose with the spermatic vein and may cause the varicosity. In practice however varicocele on the left side is so common without any obvious cause as to have no diagnostic importance. It is less common on the right side but in the large B.A.U.S. series (1951) varicocele on the right side which disappeared after the removal of a right renal tumour was only found in one case (0.04per cent). It is a sign of little importance but when present is not a contraindication to operation.

d) Renal rupture

Spontaneous rupture of the kidney is a rare mode of presentation; it occurred in two cases in the B.A.U.S. series (1951).

Levi and Ferrara (1959) found records of less than fifty cases in the literature. Miller and Kaufman (1963) added five cases with symptoms of pain and intra-abdominal bleeding suggesting an acute abdominal catastrophe.

e) Polycythaemia

The occurence of polycythaemia in association with carcinoma of the kidney has been recorded in about sixty cases. In 203 cases of "primary" polycythaemiaLawrence (1955) found only two of renal new growth whilst Damon et al. (1958) found nine in 205 cases (4.4 per

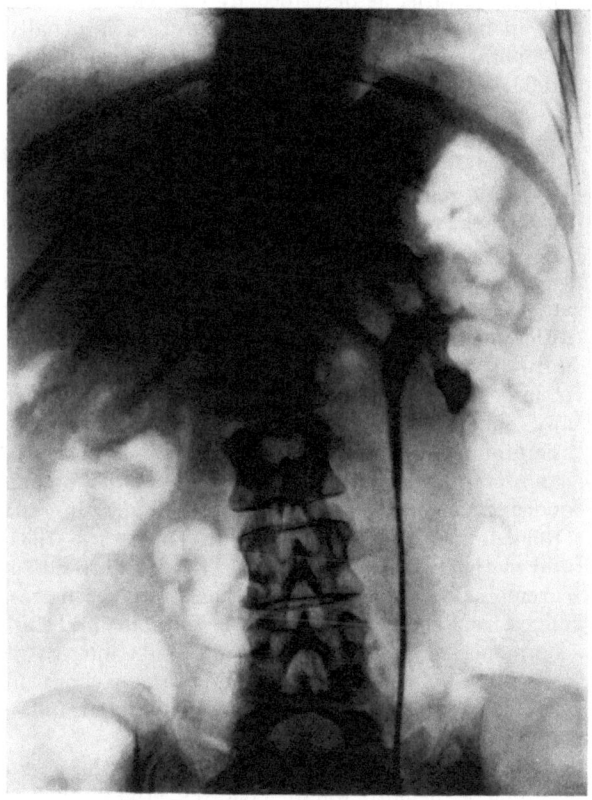

Fig. 5. Pyelogram in a case of carcinoma of the renal pelvis in a woman of 63. The temperature was about 100° F for two months, but subsided on nephrectomy

cent); these nine occured in 350 cases of carcinoma of the kidney (2.6 per cent). Hillas Smith and Riches (1963) found three cases of polycythaemia in 131 of renal carcinoma (2.3 per cent). The incidence thus appears to be between 2 and 3 per cent of cases of carcinoma of the kidney. It is not accompanied by splenomegaly or by a raised leucocyte or platelet count. In a second group of patients, described by Berger and Sinkoff (1957) there is a raised haemoglobin and red-cell count not amounting to polycythaemia; in the series of Hillas Smith and Riches (1963) 16.8 per cent (22/131) had haemoglobin values of at least 15 g. per 100 ml. The criteria for acceptance as polycythaemic defined by Damon et al. (1958) are a haemoglobin value of at least 18 g. per 100 ml., a red-cell count exceeding 6.3 million per c. mm., and a haematocrit exceeding 55 per cent. Polycythaemia has also been found in other renal abnormalities and in a few lesions of other viscera.

The findings of Conley et al. (1957) and of Hillas Smith and Riches (1963) indicate that patients with renal carcinoma without true polycythaemia have a higher proportion of raised haemoglobin values than the normal population. In the true polycythaemics there is often a raised level of erythropoietic hormone in the plasma and in extracts of the tumour or its metastases (Gurney 1960, Hewlett et al. 1960, Hillas Smith and Riches 1963); after nephrectomy the level falls to normal (Barnard 1961). Any subsequent rise may be an indication of a metastasis, as described by Omland (1959). In the second group an initial fall in haemoglobin following operation is usually followed by a return to the preoperative value. In the three cases with polycythaemia described by Riches (1963) survival was unexpectedly long even in the presence of metastases.

3. Symptoms due to metastases

Many renal tumours, particularly adenocarcinoma and the less common sarcoma have a marked tendency to invade the venous radicles in the kidney and spread thence to the renal vein and inferior vena cava. Having entered the systemic blood stream the tumour emboli can be carried to any part of the body, and it is therefore not surprising that metastatic deposits are found in many different situations. As the primary tumour may not give rise to haematuria until it has eroded a calyx or the pelvis, or to pain or swelling until it is large, it happens not infrequently that metastases are found before there is any clinical evidence of the kidney growth. It follows that the presenting symptoms are extremely varied, and as Creevy (1935) has written, renal tumours can reproduce the clinical appearances of an amazing variety of disorders. The incidence of such presentation varies widely according to the literature, for whilst Israel (1896) found it to be 50 per cent, Deuticke (1931) gave a figure of 9 per cent, and in the B.A.U.S. series (1951) there were 56 such cases in 1746 renal adenocarcinomata (3.2 per cent). Of these 56 eighteen presented with paraplegia or hemiplegia from secondary deposits in the spine and this is now recognised as one of the commoner modes of metastatic presentation. Simpson (1933) reported four such cases. In any obscure case of pain in the back a urinary investigation is necessary.

Griffiths and Thackray (1949) found four patients who presented with sciatica, and one with backache.

Other bones are also affected, and in the case of the long bones a pulsating swelling is usually found. Riches (1953) illustrated one such tumour which appeared in the humerus six months before the onset of haematuria.

The lungs are affected by metastases even more frequently than bones; they may be single or multiple and radiologically show as discrete rounded shadows like the "cannonball" secondaries from the testis, or as a diffuse carcinomatous infiltration. The patient may complain of haemoptysis and the first diagnosis may be pulmonary tuberculosis or carcinoma of the bronchus.

Lymph nodes are also the site of metastases; if the cervical nodes are involved it occasionally happens that this is the first manifestion of the disease.

Metastases are found in the skin, although rarely as the first presenting feature. Scorer (1951) reviewed the incidence of cutaneous metastases and reported one case of a vulval deposit from a renal sarcoma, and Hyman (1925) recorded a case in which a cutaneous nodule over the tibia was the first indication of an adenocarcinoma. The tumours are usually mobile at first and can be excised for biopsy; those on the scalp are soft and resemble sebaceous cysts (Scorer 1051).

Secondary deposits appear in the brain, and the first symptoms may be those of a cerebral tumour. Smyth (1939) reported a patient whose first complaints were of headaches, nausea and difficulty in walking and balancing. A cerebellar metastasis was removed six weeks after the removal of a renal adenocarcinoma.

The thyroid gland is sometimes the site of metastases from renal neoplasms, and BENNETT (1952) has reported a case in which the patient presented with a swelling in the neck; it was explored and found to be an adenocarcinoma. The patient subsequently developed a mass in the loin and at autopsy had a necrotic renal adenocarcinoma. He recorded that the first such case was reported by PEMBERTON and BENNETT in 1934 and the second by MOORE and WALKER in 1950.

A secondary tumour in the testis was the first sign of a renal tumour in the case reported by BANDLER and ROEN (1946).

Secondary deposits in the corpora cavernosa of the penis produced priapism as the presenting symptom in the cases reported by BEGG (1928), BURRELL (1948) and HENDERSON (1950); in the first two the primary renal growth was an adenocarcinoma and in the third an intrarenal transitional cell carcinoma. In all three the left kidney was affected.

4. Silent cases

In addition to those patients who present with metastatic signs or symptoms there are some in whom a renal tumour is found quite accidentally during operation for some other condition such as stone, or at autopsy. The incidental finding of adenocarcinoma at post-mortem was recorded in eleven cases of the B.A.U.S. series.

These are the true "silent" cases, although the term is often used to describe those in whom metastatic symptoms are found first. Their prognosis is generally good, but even in those discovered very early the outlook depends more upon the grade of malignancy than on the size of the tumour.

II. Physical signs
1. Local

Unless there is a palpable tumour in the loin there may be a complete absence of abnormal physical signs on examination.

A swelling is most likely to be present in Wilms' tumour and adenocarcinoma. In a child it is often visible as a prominence in one side of the abdomen. On palpation it has the usual characters of a renal mass. It can be palpated in the loin bimanually, it moves on respiration until it becomes fixed by perirenal extension, it is dull on percussion except where it is crossed by the colon. The distinction between a solid and a fluid swelling can usually only be made if the patient is thin.

Other abdominal signs which may be found are enlargement of the liver in the case of metastases, and distended superficial veins if there is thrombosis of the inferior vena cava. The spermatic cord should be examined for the presence of a varicocele.

Palpation of a suspected renal tumour should be limited to a minimum as it can cause venous dissemination of the growth; this applies especially in the case of Wilms' tumour where it should be forbidden when the diagnosis has been made or exploration has been decided upon.

2. General

A general examination may disclose abnormalities in the respiratory, cardio-vascular or nervous systems. Any abnormal swellings must be examined; a bony metastasis is usually painful and produces expansion of the bone with pulsation. Enlarged lymph nodes in accessible situations are often mobile for a considerable time and are not tender. The same applies to cutaneous nodules.

3. Urine

In all cases the urine must be examined for albumin and for red blood cells, both of which are usually present.

4. Blood

Some degree of anaemia may be present even in the absence of gross haematuria, turia, but some patients have polycythaemia or a raised haemoglobin level. The erythrocyte sedimentation rate may also be raised. A complete blood count should be done in all cases.

F. Diagnosis

I. History

If there is a history of haematuria or of pain in the loin the diagnosis of a renal tumour will be suggested and routine investigations will confirm or refute it. In the absence of leading symptoms a history of pyrexia, general malaise or loss of weight should lead to the correct investigations. Renal new growths are notorious in their ability to mimic many other diseases.

II. Physical examination

A palpable tumour in the loin may be expected in about 70 per cent of Wilms' tumours, 30 per cent of adenocarcinomata, 25 per cent or more of sarcomata but only 2 to 10 per cent of pelvic and ureteric tumours. The absence of physical signs on general examination does not exclude the need for special investigations in a case where there is the slightest suspicion of a renal tumour. A search for metastases should not be neglected and radiological examination of the chest should always be included; microscopical examination of the urine for red blood cells may give the first clue to the diagnosis.

III. Special diagnostic methods

The aim of diagnosis is not limited to the discovery of the presence of a neoplasm. It should give information about its site, particularly whether parenchymal or pelvic, its operability, the presence of complications and some idea of the prognosis. The condition is rarely a surgical emergency and although there should be no undue delay in treatment there should be no omission of an investigation which will throw light on any of these factors.

The order in which investigations are done will depend upon the presenting symptoms and the facilities available. Not all the tests are necessary in every case and the experience of the urologist must be the guide in deciding which to do first.

1. Cystoscopy

The value of cystoscopy is in discovering from which side bleeding occurs. As the bleeding is usually intermittent no opportunity should be lost of performing cystoscopy whilst it is actually in progress. The examination also serves to exclude a vesical cause of haemorrhage. If active bleeding is taking place from one kidney the efflux from the ureter on that side will consist of a series of bright red spurts. As the bleeding lessens the efflux tends to become darker and sometimes thicker, and worm-like clots may be seen being slowly extruded from the ureteric orifice. If bleeding is seen from one ureter the other must also be watched as there are some conditions, for example polycystic disease, in which it may be bilateral. In order to be quite certain that the blood comes from the ureteric orifice the medium must be kept clear and a modern cystoscope providing for constant irrigation should be used. A source of error is bleeding from the prostate on contact with the instrument which ADAMS (1950) has termed the prostatic decoy; out of his ten cases the diagnosis of renal tumour had been delayed in two, one for 18 months and one for 15 months from this cause. It rarely happens

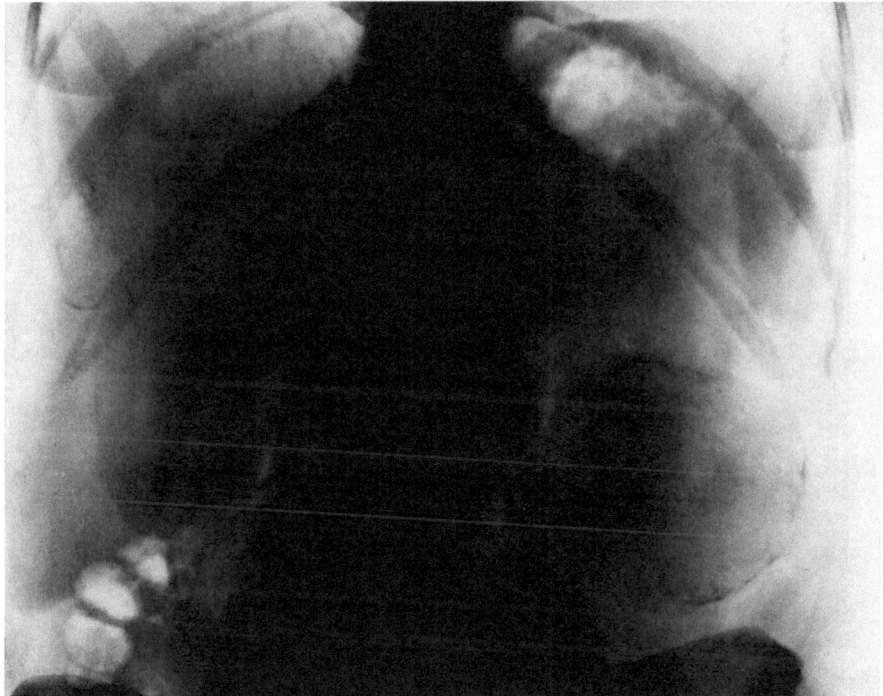

Fig. 6. Plain radiograph showing calcification in a renal adenocarcinoma in a woman of 48. She is alive and well 25 years after nephrectomy (see Fig. 26)

that the bleeding is so copious as to prevent a proper diagnostic view, but if cystoscopy under a local anaesthetic fails to give a diagnosis it should be repeated as soon as possible under general anaesthesia.

At cystoscopy it is sometimes possible to see a tumour of the lower end of the ureter. The growth may appear and recede with each wave of ureteric contraction, or it may develop a pedicle and remain trapped in the bladder as in the case described by GALBRAITH (1950).

2. Radiography

a) Plain X-ray

It occasionally happens that an abdominal X-ray will show the enlargement of a kidney containing a tumour. If there is calcification (as in Fig. 6) the picture may be more suggestive, but it is not usually diagnostic, and the plain X-ray is better kept as a prelude to excretion pyelography after proper preparation.

b) Excretion urography

Intravenous or excretion urography can be a most informative investigation. If there is no haematuria when the patient is first seen it is an advantage to use it as the first investigation; any doubt which may exist about either kidney can be cleared up by retrograde pyelography at the time of cystoscopy.

Excretion urography will show the function of both kidneys. Absence of excretion on the affected side may indicate extensive growth, or blockage of the ureter, or possibly invasion of the renal vein; it is a bad prognostic sign. In the

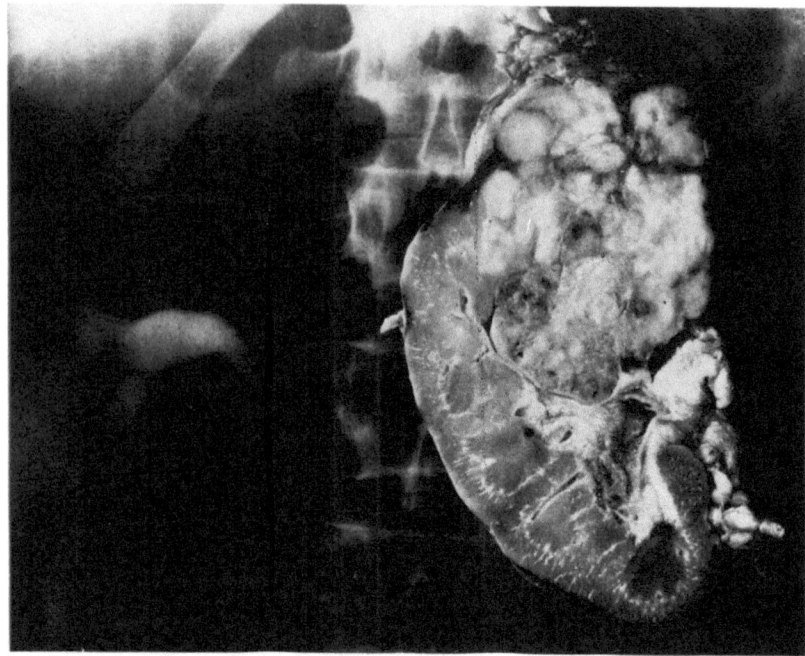

Fig. 7. Compression, elongation and displacement of the upper calyx by an adenocarcinoma in the upper pole. From a man of 62 with a grade 1 tumour but lymphatic metastases who survived for 4 years

B.A.U.S. series (1951) in 212 cases where the vein was involved 43 per cent had no excretion on the affected side whilst in 722 cases where the vein was not involved only 31 per cent had shown no excretion. The survival rates were lower in the cases with no excretion, bearing out WADE's (1933) contention that the extent of malignant activity varies inversely with the renal functional activity.

Excretion urography should never be omitted; it will sometimes show the site and extent of the growth, and it will show whether the other kidney is normal.

c) Instrumental urography

A retrograde pyelogram is usually necessary to show the detail of the filling defect caused by a new growth and to define its extent. It is essential for pre-operative diagnosis if there is no excretion on intravenous pyelography or if that radiograph is doubtful. It fails if the ureter cannot be catheterised and it will only indicate that there is a space occupying lesion which may in practice be a cyst or a new growth. Profuse haematuria or other clinical manifestations may be sufficient to differentiate the two, or other radiographic methods may be necessary.

It is important to distinguish before operation between a parenchymal and a pelvic growth because the treatment of the two conditions differs; this can usually be done by pyelography. In general the parenchymal tumours produce their main effects on the calyces, whilst pelvic tumours, whether papillary or

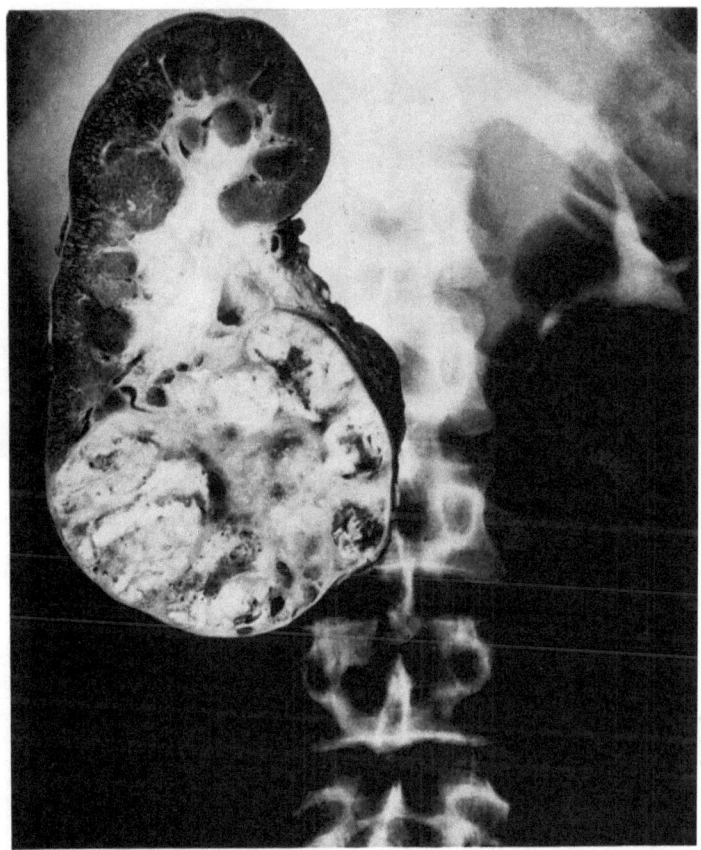

Fig. 8. Obliteration of the lowest calyx by an adenocarcinoma in the lower pole. From a man of 38 with a grade 1 tumour who survives for 6 years

solid cause alterations in the pelvis. The principal changes in the calyces produced by a cortical tumour are as follows: —

Compression, often producing a crescentic deformity (Fig. 7).

Obliteration of an entire calyx (Fig. 8).

Displacement (Fig. 7).

Elongation. This may affect a single calyx or all of them; in the latter case it produces a spider-leg deformity (Fig. 9).

Increase in the angle between two calyces (Fig. 10).

Secondary dilatation of the calyces and pelvis from obstruction. The ureter may be displaced medially by a large growth in the lower pole (Fig. 11).

The pyelographic changes cannot always be reduced to their lowest terms in this way and some unusual types will be encountered which may justify the older use of the terms "dragon", "spider", or "carnation". Fig. 12 shows a good radiological dragon whilst Fig. 13 could be called a "puffin".

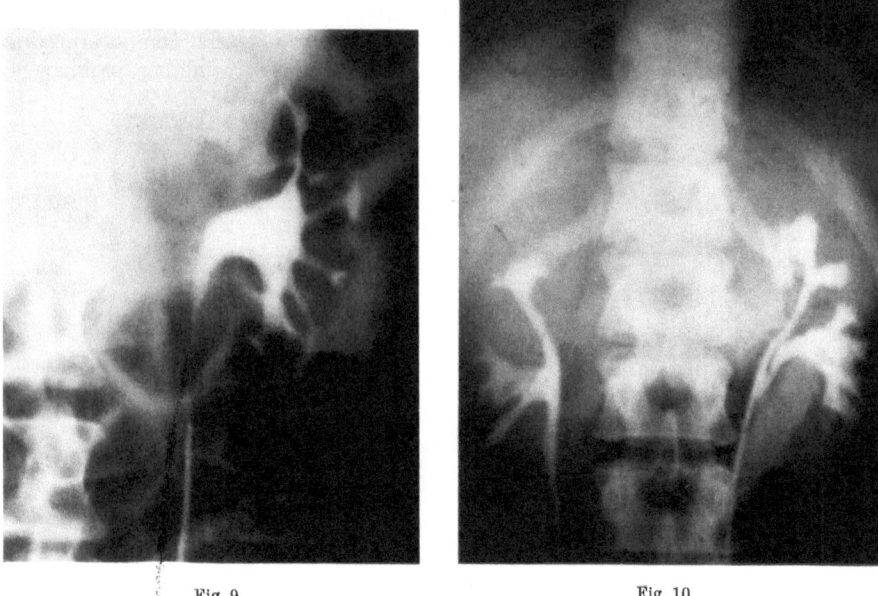

Fig. 9   Fig. 10

Fig. 9. Spider leg deformity produced by a tumour in the upper pole. From a man of 55 with a grade 2 tumour
who survived for one year

Fig. 10. Increase in the angle between the upper and middle calyces produced by a grade 1 tumour in the right
kidney. From a man of 28 who was killed in an accident 4 years after nephrectomy

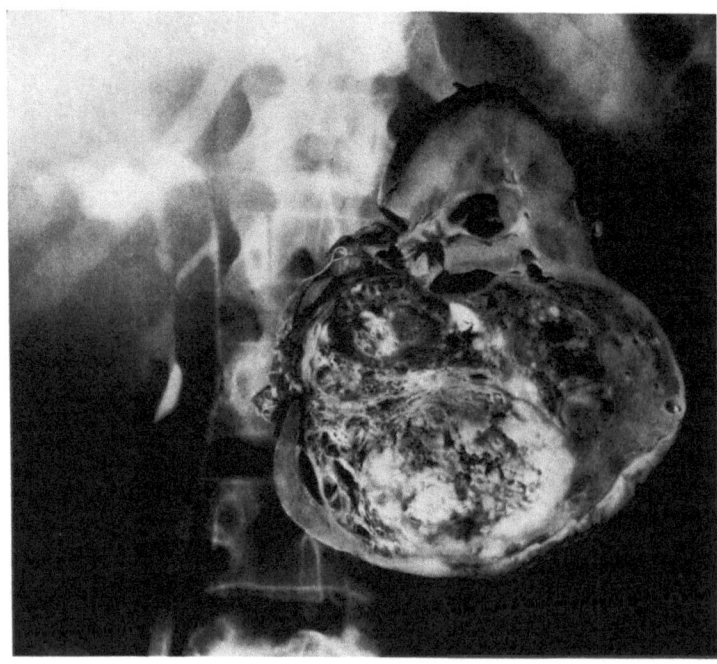

Fig. 11. Medial displacement of the ureter and secondary dilatation of the calyces from a large growth in the
lower pole

Pelvic tumours produce a filling defect in part of the renal pelvis. If the growth is papillary there may be an uneven distribution of the contrast medium between the fronds of the tumour (Fig. 14). It if is solid the pelvic defect is clear cut (Fig. 15). In either case there may be hydronephrosis or hydrocalicosis from obstruction. A cupped lower border to the filling defect is characteristic of a pelvic tumour (Fig. 16).

A squamous cell tumour of the pelvis cannot be distinguished with certainty from a solid transitional cell tumour but if stones are also present the former is more likely.

Benign and malignant tumours of the pelvis cannot as a rule be distinguished from each other by pyelography.

Tumours of the ureter may also be diagnosed by instrumental urography. TRESIDDER (1954) has pointed out the superiority of ureterograms over pyelograms in the diagnosis of lesions of the ureter; he advocates the use of the Braasch bulb catheter rather than the plain ureteric catheter. TRESIDDER and WARREN (1954) showed that even with tumours in the more difficult lower third of the ureter the diagnosis was still possible in this way. In any part of the ureter a complete block will prevent the contrast medium from passing it and the diagnosis of tumour is only conjectural. For tumours in the upper third a ureteric catheter can be passed high enough to show the site of the block (Fig. 16). In the middle third the irregularity of the filling defect can be shown if the medium will traverse the narrow site of the growth; the defect has an irregular outline in most malignant cases (Fig. 17). It is in obstruction of the extreme lower end

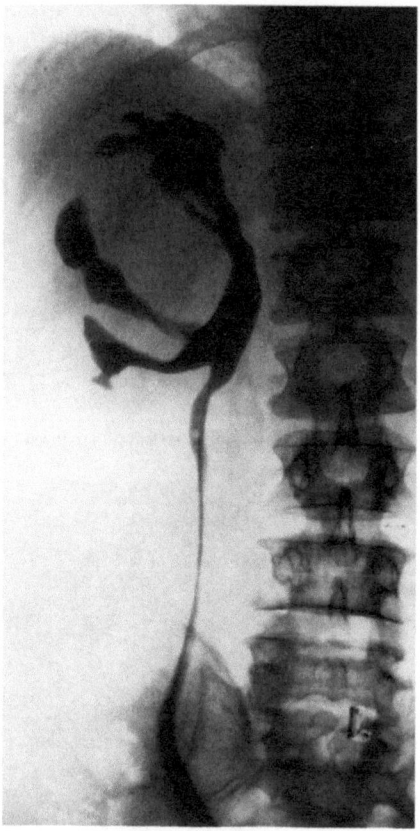

Fig. 12. "Dragon" type of pyelogram given by a grade 2 carcinoma between the upper and middle calyces. From a man of 52 who died from pulmonary metastases after three years

of the ureter that most difficulty arises and even with the Braasch bulb catheter a ureterogram cannot always be obtained. I may however be possible to demonstrate a hydronephrosis and hydroureter without being able to indicate its cause.

Multiple papillary tumours are shown as irregular filling defects in a dilated and tortuous ureter; they are often associated with papillary carcinomata of the renal pelvis (Fig. 18).

A benign tumour can give rise to a regular rounded filling defect which must be distinguished from a non-opaque calculus or an air bubble (Fig. 19).

Any ureteric growth may bleed when a catheter is passed; this is a suggestive sign but the bleeding must be brisk to be of diagnostic value.

d) Aortography

It has been pointed out that whilst retrograde pyelography will demonstrate a space-occupying lesion of the kidney it does not indicate its nature. In

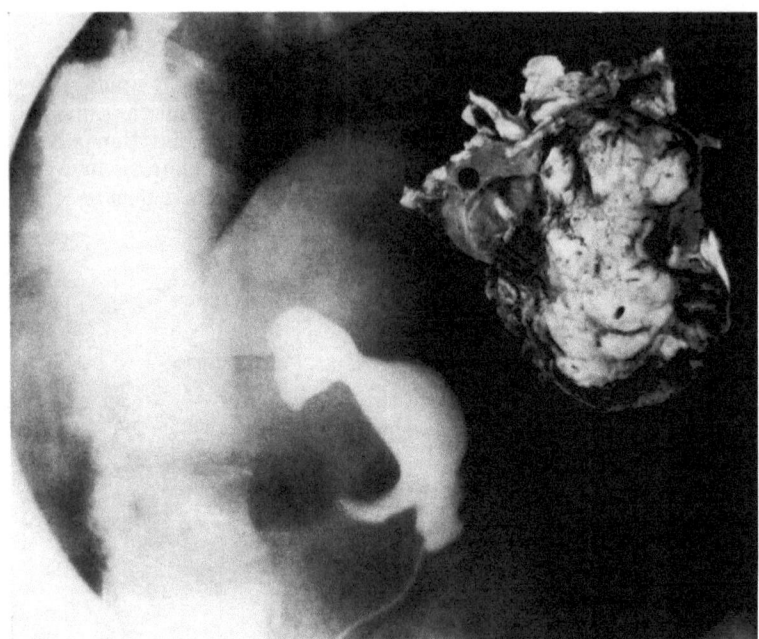

Fig. 13. "Puffin" type of pyelogram produced by a central grade 2 tumour. From a woman of 68 who survived 2 years

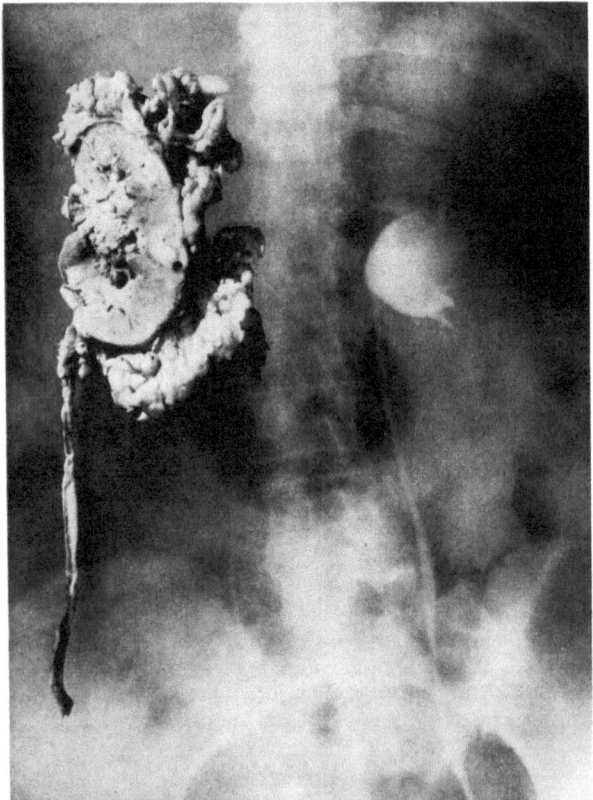

Fig. 14. Papillary carcinoma of the renal pelvis producing a mottled filling defect in the pelvis. From a woman of 68 who is alive and well twelve years after nephroureterectomy

practice the diagnosis rests between a tumour and a cyst and although the probability of either may be inferred by other clinical features the distinction between a vascular growth and an avascular cyst can usually be made by aortography.

The procedure can be carried out by translumbar puncture of the aorta, the method devised by DOS SANTOS (1929) and subsequently followed by NELSON

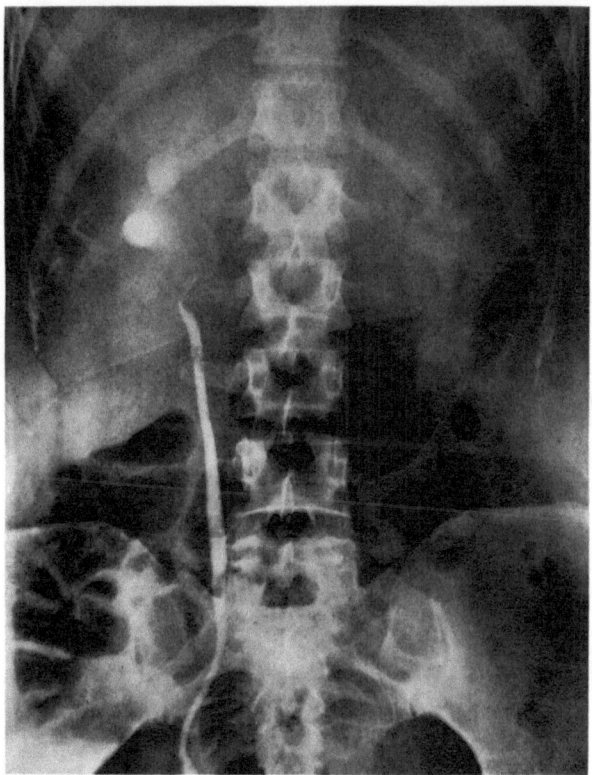

Fig. 15. Complete obliteration of the renal pelvis produced by a solid transitional cell carcinoma. From the same case as Fig. 51

(1942), Doss et al. (1942), GRIFFITHS (1950), WHITESIDE (1953), or by retrograde femoral catheterization as described by FARINAS (1941), PIERCE (1951), LINDGREN (1953) and SELDINGER (1953).

In a typical case of adenocarcinoma the vessels near the growth are displaced, and the tumour shows a deposit of contrast medium which gives it a mottled appearance known as "pooling" (Fig. 20). This appearance persists in the nephrogram. Occasionally a cortical tumour is avascular, as when it has a thickened or calcified pseudo-capsule or when it has invaded the pelvis extensively (RICHES and WHITESIDE 1955). Metastatic tumours are also avascular and will not show pooling. A metastasis in a vertebra from a renal adenocarcinoma however will sometimes take up the contrast medium (RICHES 1955), and we have also demonstrated an unexpected metastasis in the liver.

MURRAY and TRESIDDER (1957) have illustrated a case of angioma of the kidney which showed as a collection of large vessels and enabled a preoperative diagnosis to be made.

2*

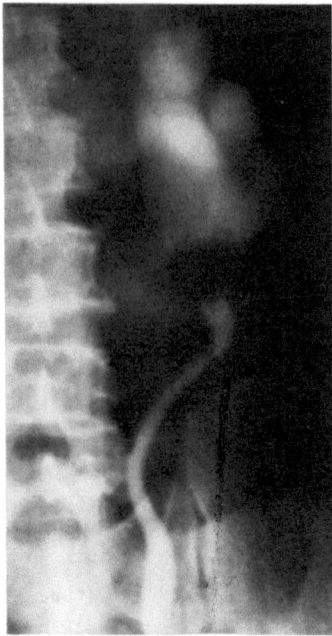
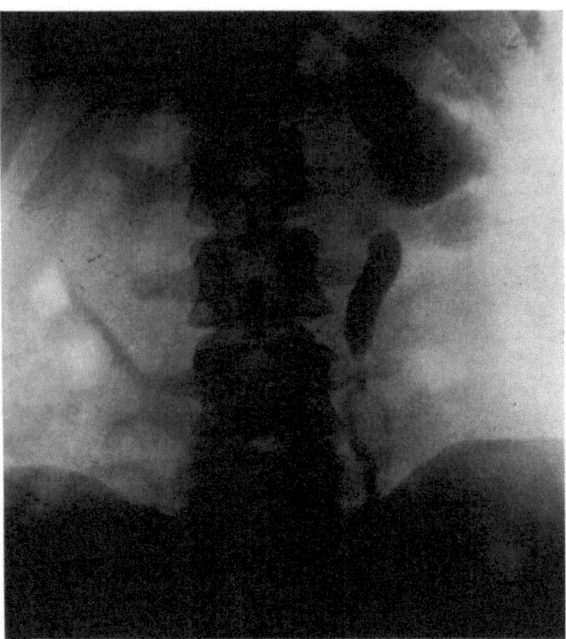

Fig. 16   Fig. 17

Fig. 16. Uretero-pyelogram in a case of papilloma of the pelvi-ureteric junction. Note the cupped lower border
of the growth. From a man of 69 (see Fig. 49)

Fig. 17. Carcinoma of the middle third of the ureter. The ureterogram shows an irregular filling defect and there
is dilatation above it. From a woman of 46 who survived for five years

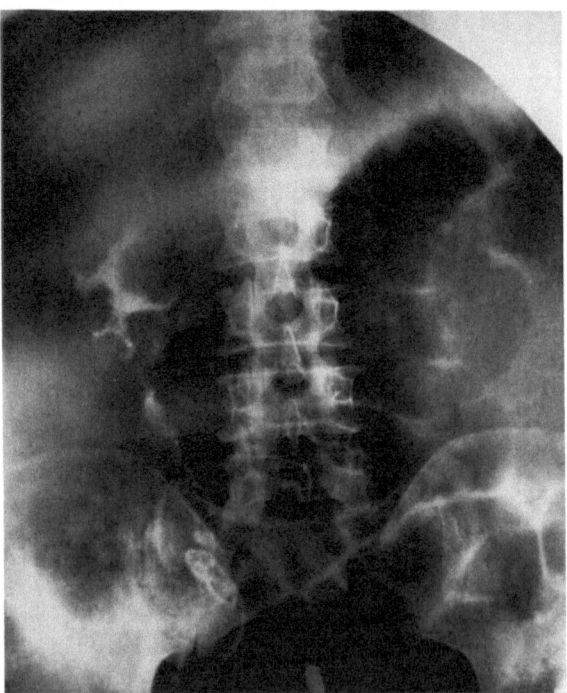

Fig. 18. Irregular filling of the renal pelvis and ureter caused by multiple papillary tumours. From the same
case as Fig. 59

Pelvic papillary growths present no specific signs on aortography unless they are large when the kidney will be displaced and the renal artery elongated. Solid tumours of the renal pelvis are avascular, and a pelvic tumour which

invades the parenchyma may show an avascular defect in the nephrogram (Fig. 21).

Apart from the possibility of distinguishing a tumour from a cyst the main value of aortography is in recognising an *early* parenchymal new growth (Figs. 22 and 23).

If the aortogram would show with certainty whether there was tumour invasion of the renal vein its value would be enhanced, but this demonstration is rare. LIND-BLOM and SELDINGER (1955) have illustrated it in one case. The condition of renal vein thrombosis can sometimes be shown by venography, a subject which has been reviewed by STEINER (1957).

e) Nephrotomography

The distinction between a tumour and a cyst can be made in more than 90 per cent of cases by nephrotomography. Whilst a nephrographic phase is sometimes seen after excretion urography it can be shown with more certainty after the ra-

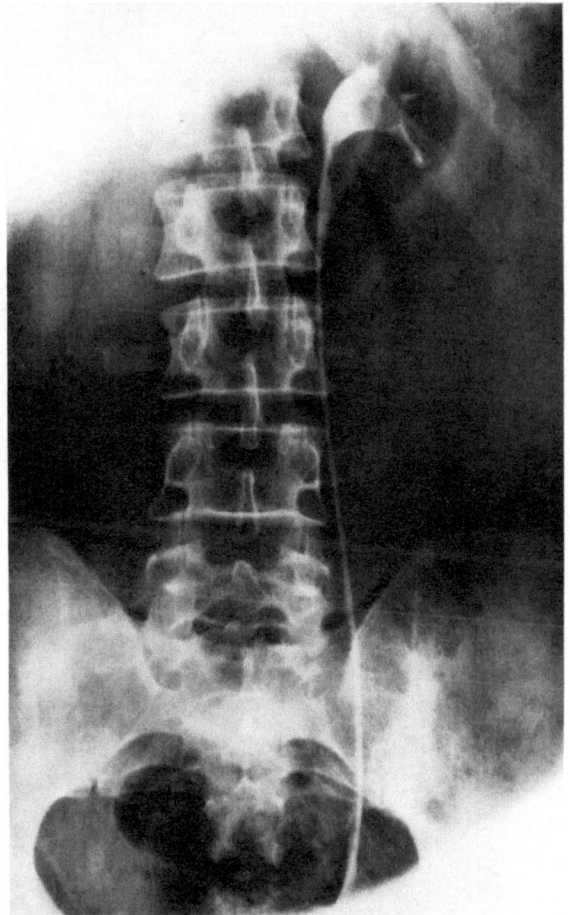

Fig. 19. Rounded filling defect produced by a papilloma of the renal pelvis in a man of 72

pid intravenous injection of 50 ml. of 90 per cent hypaque into the antecubital vein. Tomographic films are taken at one cm. cuts after an interval predetermined by an estimation of the arm to tongue circulation time; this is found by measuring the interval between the injection of 2.5 ml. of 20 percent sodium dehydrocholate (Decholin) and the appearance of a bitter taste in the tongue. To intensify the nephrogram a loading dose of 30 ml. of 50 per cent hypaque is injected just before the injection of the warmed 90 per cent hypaque.

The method was first described by EVANS and his associates (1954, 1955); more recent accounts expressing satisfaction with it have been given by CHYNN and EVANS (1960) who use sodium 3,5, diacetamido — 2, 4, 6, triiodobenzoate (hypaque) and describe the technique in detail, and by WITTEN, GREENE and EMMETT (1963) who use sodium diatrizoate with diatrizoate methylglucamine (hypaque

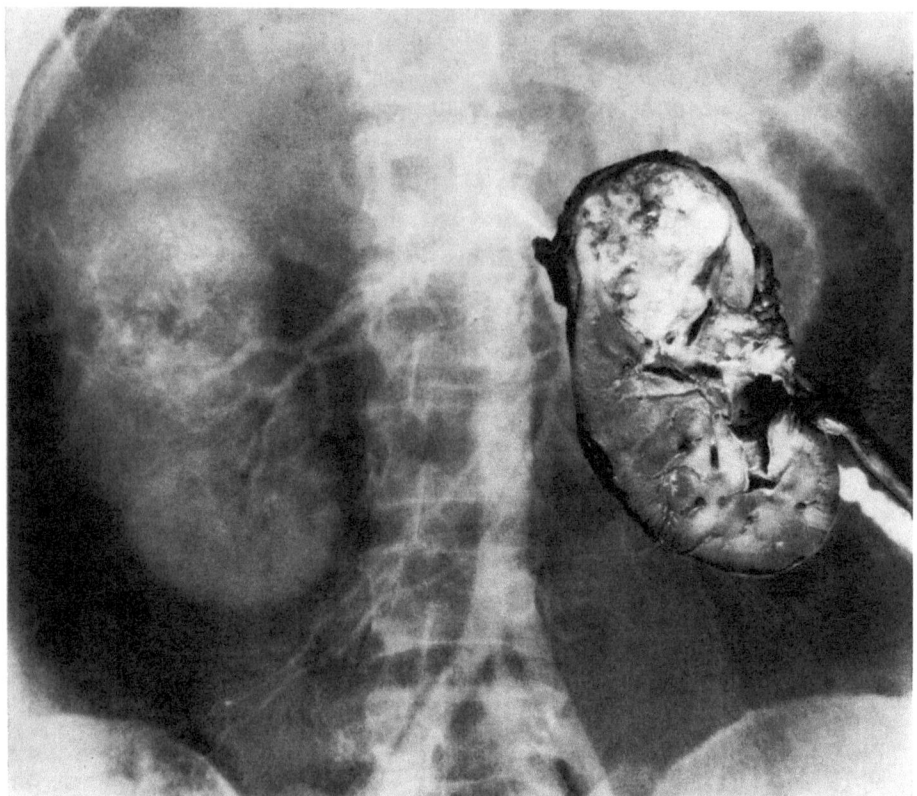

Fig. 20. Pooling of contrast medium in a grade 2 adenocarcinoma of the upper pole of the right kidney.
From a man of 40

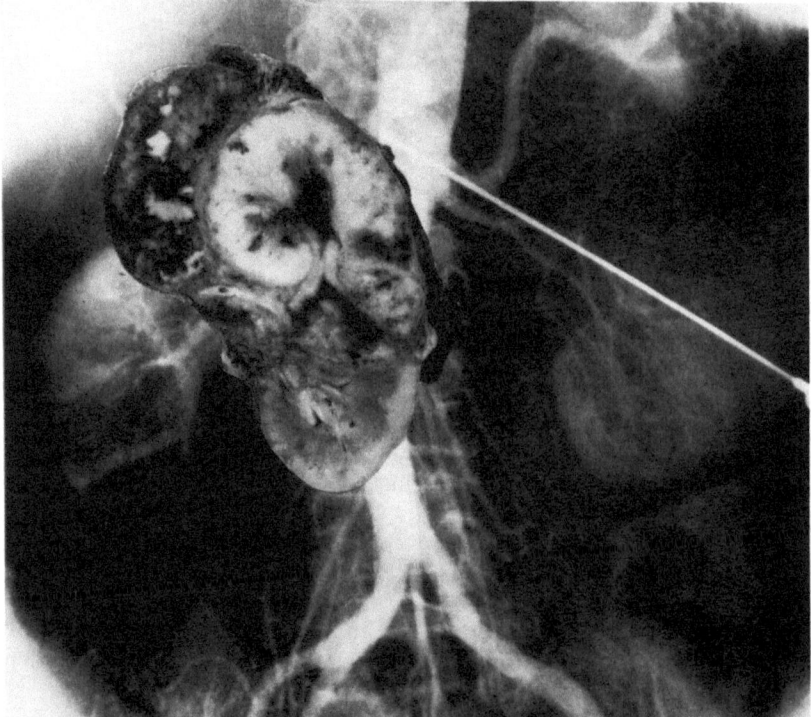

Fig. 21. Avascular defect in the upper half of the left nephrogram produced by a solid transitional cell pelvic
tumour invading the parenchyma and the adrenal. From a woman of 63 who survived only nine months

M) because of its low toxicity and high radiodensity.

The arterial phase is usually seen but the nephrographic phase gives most information. Whilst a cyst remains translucent with a sharply defined margin a tumour is opacified and gives a dense shadow often with an irregular margin. The density is uniform unless there is necrosis in the tumour when it is irregular and blotchy.

SOUTHWOOD and MARSHALL (1958) had only 5 misdiagnoses in 115 cases interpreted as cysts and only 3 in 44 cases interpreted as tumours. CHYNNE and EVANS (1960) had only one diagnostic error in their last 100 cases. All writters stress the importance of careful technique and good films. There have been no serious reactions.

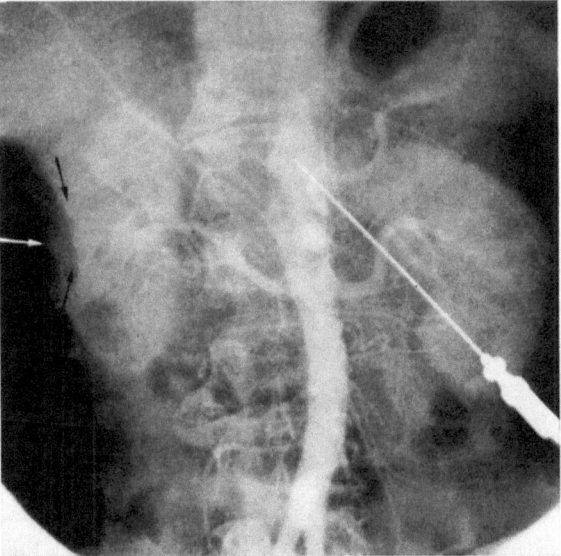

Fig. 22. Aortogram showing a small area of pooling near the centre of the right kidney

The value of a positive differential diagnosis between tumour and cyst lies in the possibility of avoiding operation for a cyst in a poor-risk patient with few symptoms; if however there is any serious suspicion of a tumour from the history, symptoms or other findings exploration is mandatory provided that the other kidney is satisfactory and the patient fit for it. Renal scintillation scanning shows absence of function in either case.

f) Percutaneous kidney puncture

The limitations of renal angiography in distinguishing a tumour from a cyst led some of the Scandinavian workers (LINDBLOM 1946, LINDBLOM and SELDINGER 1955) to puncture the swelling with a needle and inject the contrast medium directly into it. They claim that the differentiation of an expanding focus dis-

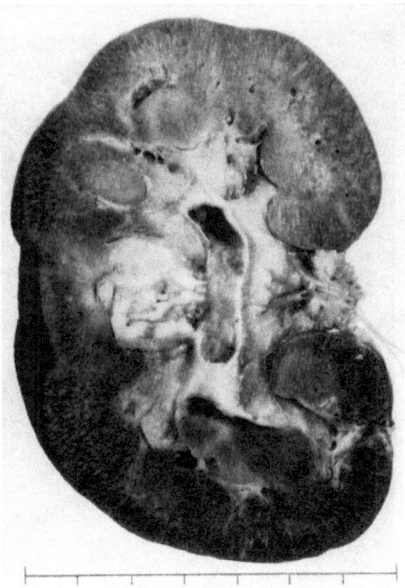

Fig. 23. The excised kidney of Fig. 22 with an early grade 1 adenocarcinoma. From a man of 55, alive ten years after nephrectomy (scale in centimetres)

covered at urography is better made by this method. This claim is substantiated in the demonstration of a cyst where the contrast medium spreads evenly into the rounded cavity. In a tumour the distribution is irregular, but we have hesitated to use the method where a tumour is suspected owing to the risk of disseminating malignant cells.

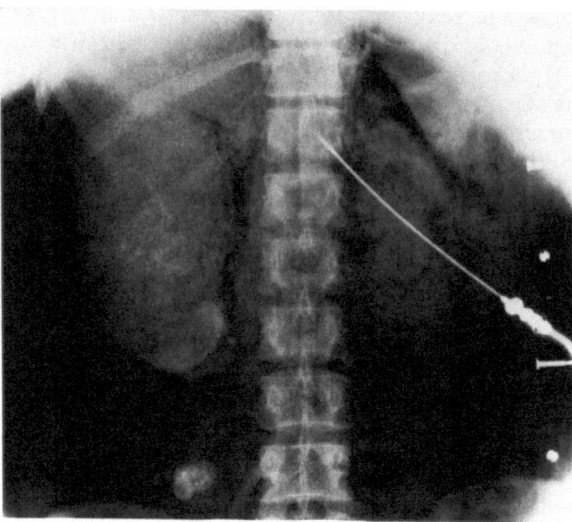

Fig. 24. Perirenal oxygen insufflation with aortography. The large tumour on the right kidney is adherent to the diaphragm. From a woman of 49, alive fourteen years after nephrectomy for a grade 2 tumour

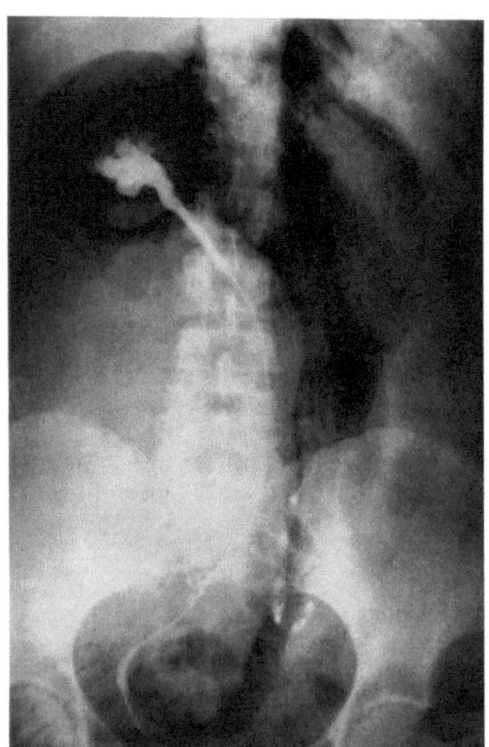

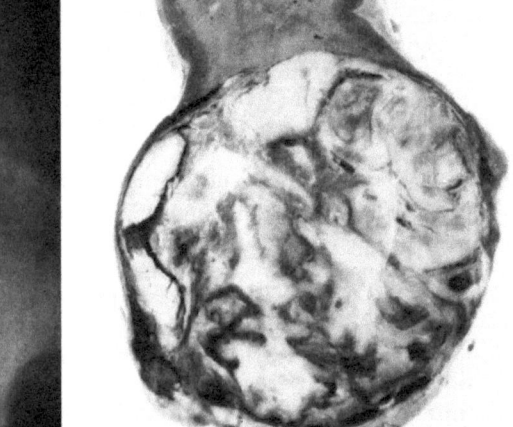

Fig. 25   Fig. 26

Fig. 25. Perirenal oxygen outlining a large retroperitoneal lipoma which displaced the right kidney and ureter. From a woman of 46

Fig. 26. Specimen of grade 1 adenocarcinoma of the lower pole of the kidney in the same case as Fig. 6; it was mistaken for a mesenteric cyst

g) Perirenal oxygen

The radiological investigations already described serve to show the functional capacity, the pelvic and calyceal outline and the arterial arrangement of the kidney. A good radiograph will show the renal outline, but in order to show this more clearly oxygen or air can be injected around it. The gas is introduced by means of a needle passed in front of the sacrum into the retroperitoneal tissue after the method of RIVAS (1948) which has replaced the older method of puncture through the loin. After a successful injection the outline of both kidneys can be clearly seen. If a tumour or cyst is present it will distort the normal outline producing a projection which can be seen on the antero-posterior or oblique films, or on tomographs. If the kidney is fixed the gas will not pass completely around it and this has been used as a preoperative test of infiltration (Fig. 24). After irradiation therapy in the case illustrated the oxygen passed completely round the upper pole of the kidney which was removed subsequently without difficulty.

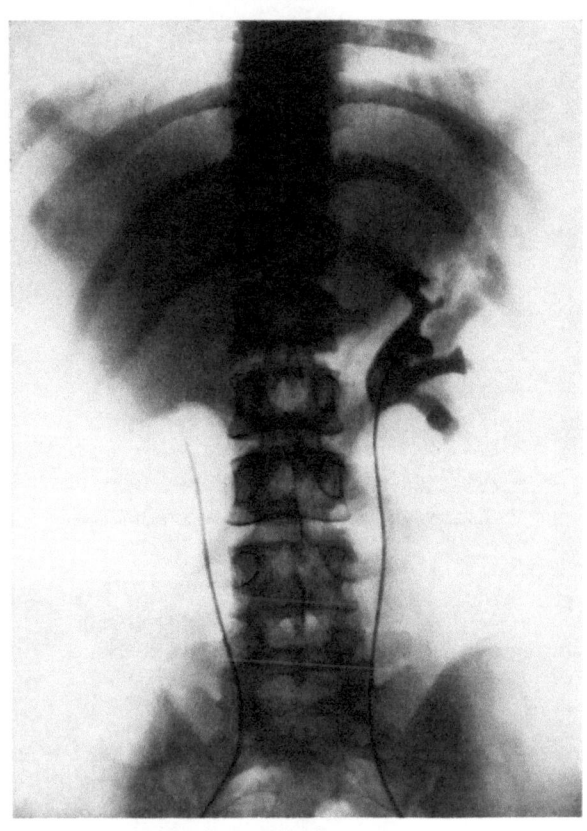

Fig. 27. Uric acid calculus in the left renal pelvis. Compare with Fig. 19. From a girl of 15

Perirenal insufflation will also serve to distinguish a renal from a retroperitoneal tumour (Fig. 25) or from a suprarenal tumour.

3. Exfoliative cytology

The recognition of malignant cells in the centrifuged urine has been used as an aid to diagnosis of tumours of the urinary tract (PAPANICOLAOU 1947, HARRISON et al. 1951). In the case of the kidney the examination is done on the urine withdrawn by ureteric catheter. It has a limited value and is most likely to succeed in papillary tumours of the pelvis.

4. Renal biopsy

It has been suggested that renal biopsy is justified as a method of distinguishing a benign from a malignant tumour (WEYRAUCH et al 1952). The danger of dissemination is an argument against it but if facilities for an immediate section are available it may prove useful in certain cases.

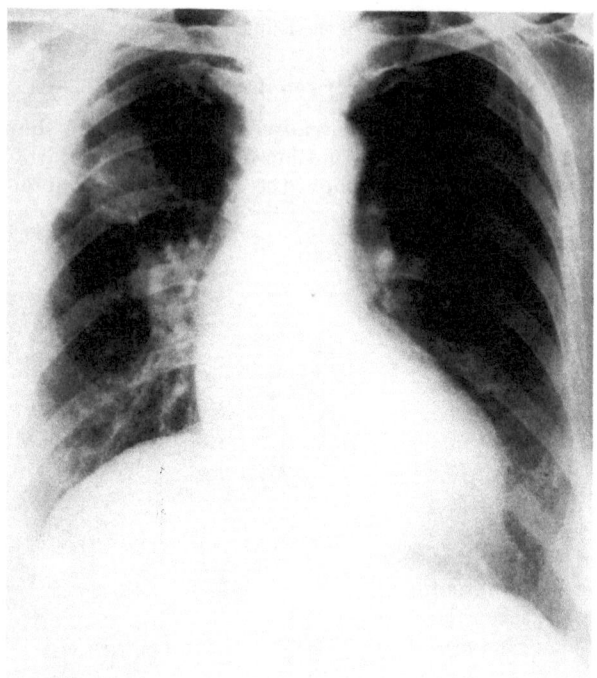

Fig. 28. Radiograph suggestive of a renal metastasis in a woman of 72 with haematuria

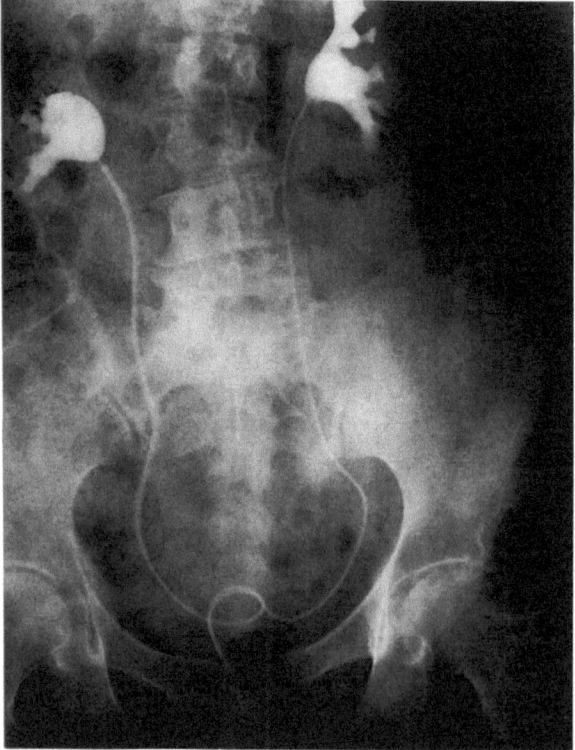

Fig. 29. Pyelogram in the same case as Fig. 28. The right upper calyx is deformed

IV. Differential diagnosis

1. Palpable tumour

A renal tumour which is palpable has the usual characters of an enlarged kidney which have been described under "Physical Signs" (E, II, 1.). In particular it can be grasped bimanually whilst most of the other swellings cannot.

It may be mistaken for an enlarged gall-bladder, enlargements of the liver or spleen, a pelvic tumour, ovarian or uterine, with a long pedicle, a neoplasm of the colon or impacted faeces, a tumour of the omentum or mesentery (Fig. 26), a retrocaecal appendix, a suprarenal tumour or a retroperitoneal tumour, cyst or abscess.

Any of these conditions may cause pain in the abdomen and its exact site and type may indicate its probable origin.

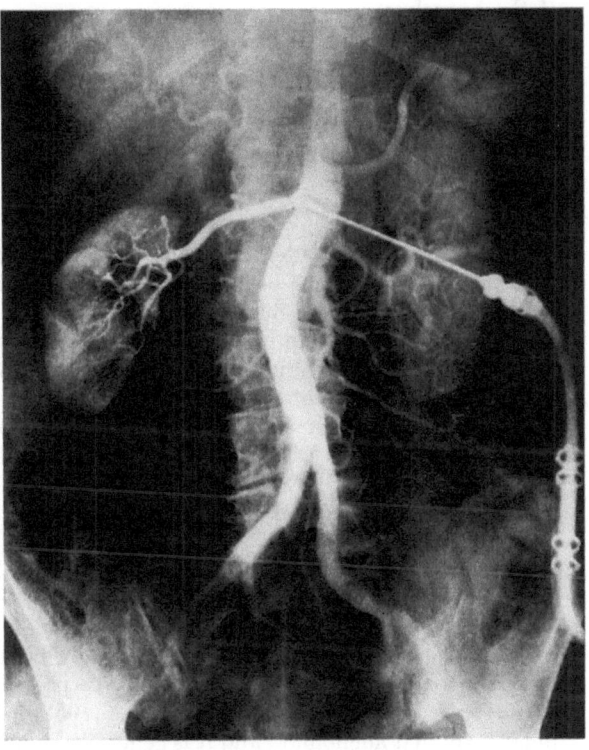

Fig. 30. Aortogram demonstrating a normal right kidney in the same patient as Figs 28 and 29. She died from carcinoma of the bronchus

2. Haematuria

Blood in the urine coming from the kidney is generally dark and may contain clots; it stains the urine evenly and is often copious although intermittent. There may be wormlike ureteric clots and occasionally triangular clots formed in the renal pelvis. A renal origin is suggested by the occurrence of colic but the act of micturition is usually painless.

If the investigations detailed have been systematically performed it should be possible to arrive at a correct diagnosis with more accuracy than in most other systems of the body.

3. Pyelographic appearances

The chief source of error is in connection with polycystic kidney which shows compression or dilatation of calyces, an elongation of the kidney and a generally bizarre appearance; the changes are generally bilateral, but more advanced on one side (see Fig. 64).

A non-opaque calculus may mimic a growth of the renal pelvis. Either may give a negative shadow on pyelography (Fig. 27), but the stone will change its position in the pelvis with that of the patient.

4. Metastases

A suspected metastasis may be subjected to biopsy; the pathologist must have in mind the possibility of a renal primary.

A radiographic shadow in the lung may be a primary tumour or a metastasis from the kidney or elsewhere. The patient whose X-rays are shown in Fig. 28 had haematuria. The pyelogram (Fig. 29) suggested a renal tumour but the aortogram (Fig. 30) showed a normal low kidney. She died later from a carcinoma of the bronchus.

G. Consideration of individual tumours

I. Renal parenchyma

1. Epithelial

a) Benign

Adenoma. Renal adenomata are small and multiple or large and single. The small nodules occur beneath the capsule or in the substance of the kidney. Newcomb (1936) found them in 7.2 per cent of 1,172 post mortems. They are papillary in structure and often occur within a small cyst; the kidney usually shows evidence of arterio-sclerosis or chronic pyelonephritis. Their numbers increase with the age of the patient.

These small benign tumours rarely give rise to symptoms but Riches and Thackray (1956) described one which caused haematuria; it was dissected by Dukes (1948).

The larger tumours which are usually single, are firm greyish masses situated just beneath the capsule and producing a bulge on the surface. They may reach a very large size.

Microscopically they are tubular, papillary or alveolar in structure, and different types can be found in the same kidney. The alveolar type resembles the clear cell renal adenocarcinoma and may be difficult to distinguish from it. Both adenoma and adenocarcinoma may be present in the same kidney (Cristol et al. 1950). Buckley (1944) and Hicks (1954) have recorded malignant change in large tubular adenomata and it is generally held that adenocarcinoma developes from adenoma.

Large adenomata give rise to a tumour in the loin, pain and haematuria. Carver (1935) reported one in which the pain was severe, and Childs and Waterfall (1953) found thirty-three cases in the literature which needed surgical removal and added two more. The distinction between benign and malignant tumours can only be made after their removal but on a few occasions it has been possible to remove the adenoma by partial nephrectomy and conserve the remainder of the kidney. This was done by Nitch (1927), by Kretschmer and Doehring (1929) and by Watts (1955) who pointed out that it should only be done when the diagnosis of a benign tumour was reasonably certain.

The condition known as cystadenoma appears to be identical with multilocular cystic disease and will be considered with other cysts.

b) Malignant

Adenocarcinoma. The name adenocarcinoma is here used to include all the malignant epithelial growths which arise in the renal parenchyma, known formerly as Grawitz tumour (1883), hypernephroma, clear cell carcinoma, granular cell carcinoma, alveolar carcinoma and renal cell carcinoma.

a) Incidence. It forms about 75 per cent of all renal tumours, affects males twice as often as females, and the right and left sides equally. 80 per cent of the cases occur in the fifth, sixth and seventh decades. It is extremely rare in childhood. CLINTON-THOMAS and ROBINSON (1956) reported one case in a girl of ten with metastases in the skull, sternum, vertebrae and humerus. They reviewed the literature and noted that CAMPBELL (1951) had found only four cases in children recorded in the American Tumour Registry.

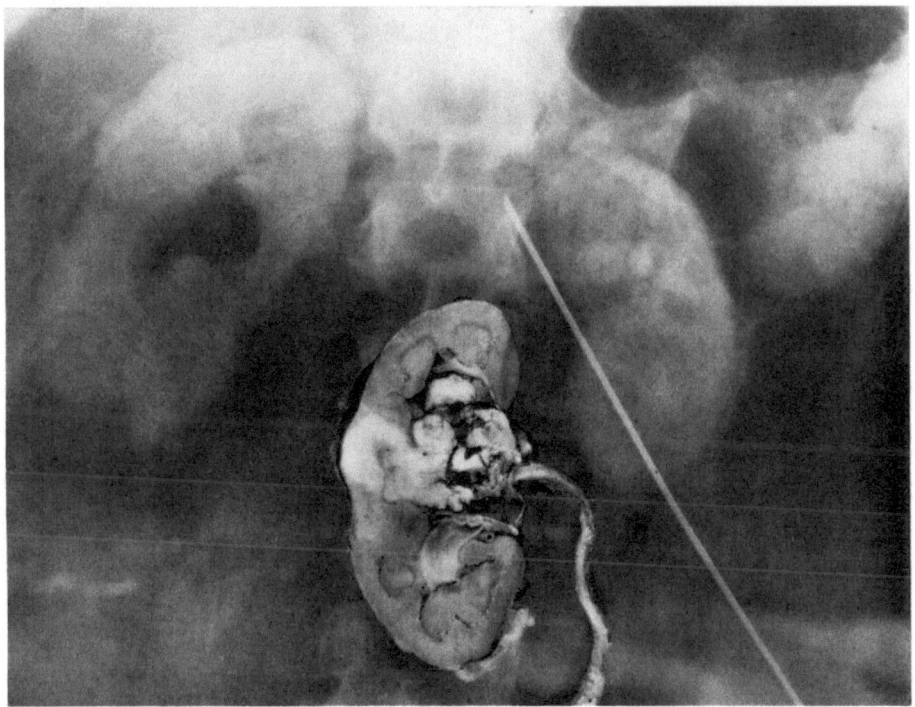

Fig. 31. Adenocarcinoma of the right kidney (Grade 2) invading the renal pelvis. It was avascular on the nephrogram. From a woman of 58 who died after eight months from metastases in the liver, lungs and spine

β) Pathological anatomy. **Macroscopic.** The growth is usually single and unilateral; the occasional finding of multiple tumours in one kidney or of bilateral tumours may be explained on the grounds of metastases or of origin from multicentric foci.

The growth starts in either pole or in the centre of the kidney. It forms a rounded swelling which sooner or later causes a prominent projection on the surface of the kidney. It is firm and solid but with increasing size the centre may undergo necrosis and softening. There is often an apparent capsule but histologically it is composed only of compressed renal tissue and it offers no barrier to spread of the growth. The cut surface displays a variety of colours, red or brown from haemorrhage, orange or yellow from lipoid material and blue or grey from mucoid degeneration. The different areas are separated by fibrous trabeculae which fuse with the pseudo-capsule at the periphery. In some cases this capsule is absent and the growth appears to be continuous with the kidney substance, which it may replace completely. In many cases the tumour extends into the renal pelvis (Fig. 31) and often there is visible evidence of spread through the capsule of the kidney or into the renal vein (Fig. 32). Quite

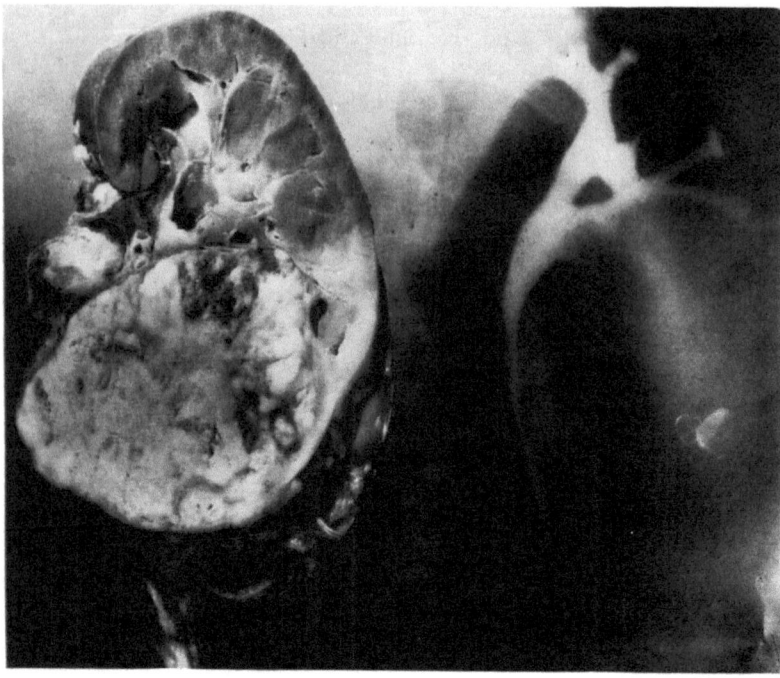

Fig. 32. Grade 1 adenocarcinoma of left kidney with gross invasion of the renal vein. From a man of 56 who
died eight months after nephrectomy with metastases in the spine

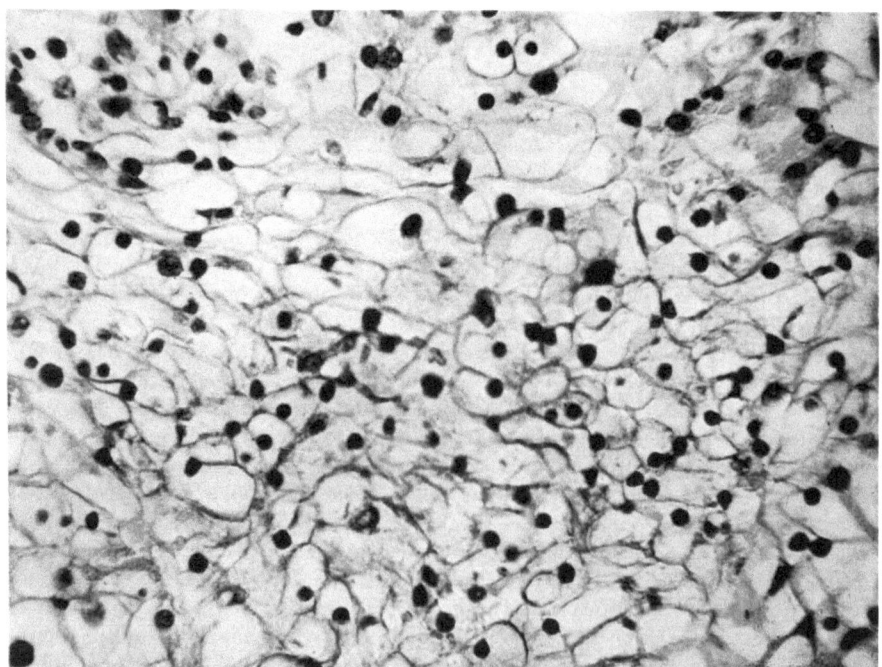

Fig. 33. Clear-cell adenocarcinoma, Grade 2 (× 265)

often the cut surface of the tumour shows little or no evidence of this degree of differentiation but appears a uniform greyish colour. Occasionally there are calcified areas or calcification in the pseudo-capsule (see Fig. 6).

Microscopic. Any description of the variable histological picture of a renal adenocarcinoma must include an account of the cells, their arrangement in the tumour and of the stroma.

The *cells* are of three principal types, clear, granular and anaplastic; tumour giant cells are also encountered. The *clear cells* are large and polyhedral and have well defined margins. The small deeply staining nucleus is central or eccentric, and the cytoplasm is clear (Fig. 33). In life it is filled with crystalline cholesterol ester which LEARY (1950) considers provides the stimulus for progressive growth and for malignant change in a benign tumour. Similar cells are found in the adrenal cortex and it was from their appearance and arrangement that GRAWITZ (1883) suggested an adrenal origin for the growth and the name hypernephroma.

The *granular cells* are smaller, with only a moderate amount of cytoplasm which is finely granular

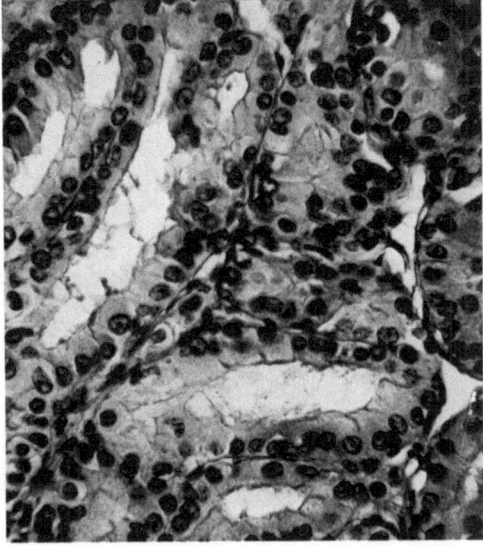

Fig. 34. Granular-cell adenocarcinoma, Grade 1 (× 265)

(Fig. 34). Differences in malignancy and behaviour have been attributed to the clarity or granularity of the cell (MUIR and GOLDSMITH 1935), but both clear and granular cells are often found in the same tumour.

Anaplastic cells vary in size, shape and regularity. The cell boundaries may be indistinct and the cytoplasm vacuolated or granular. The nuclei are round or oval and hyperchromatic, and mitotic figures are numerous (Fig. 35).

At times the spindle shape of the cells is suggestive of a sarcoma.

Giant cells with large nuclei are often present in the more anaplastic tumours.

The *arrangement* of the cells may be tubular or papillary or they may form sheets or cords or clusters; different arrangements may be found in various parts of the same tumour. In the tubular type the cells are usually clear and the tubules resemble the convoluted tubules of the kidney; this is the true adeno-carcinoma (Fig. 36). In the papillary type there are cystic spaces filled with branching papillary processes; the cells are often granular (Fig. 37). To this type the name papillary cyst-adenocarcinoma has been applied but the origin of both is in renal epithelial cells and either clear or granular cells may compose either type.

It is in the tumours with anaplastic cells that an arrangement in sheets or clusters is more likely to be found.

The *stroma* consists of loose vascular connective tissue in thin strands. The vascular spaces may be large and cavernous. There is sometimes calcification in the stroma.

Grading. Opinions on the value of histological grading and its effect on prognosis vary. HAND and BRODERS (1932) recognised four grades of malignancy; GRIFFITHS and THACKRAY (1949) arbitrarily suggested three. FOOT et al (1951)

recognised the better outlook in the cases with clear regular cells than in those
with pleomorphic granular cells. Thackray (1953) has described the method
of grading. Three factors are taken into consideration; (i) the degree of tubule
or papillary cystic formation; (ii) regularity in size, shape and staining of the
nuclei; (iii) the number of hyperchromatic nuclei and mitoses. Each factor is
noted as being present in slight, moderate or marked degree and points are
allocated according to which the tumour is placed in grade 1 (low), grade 2

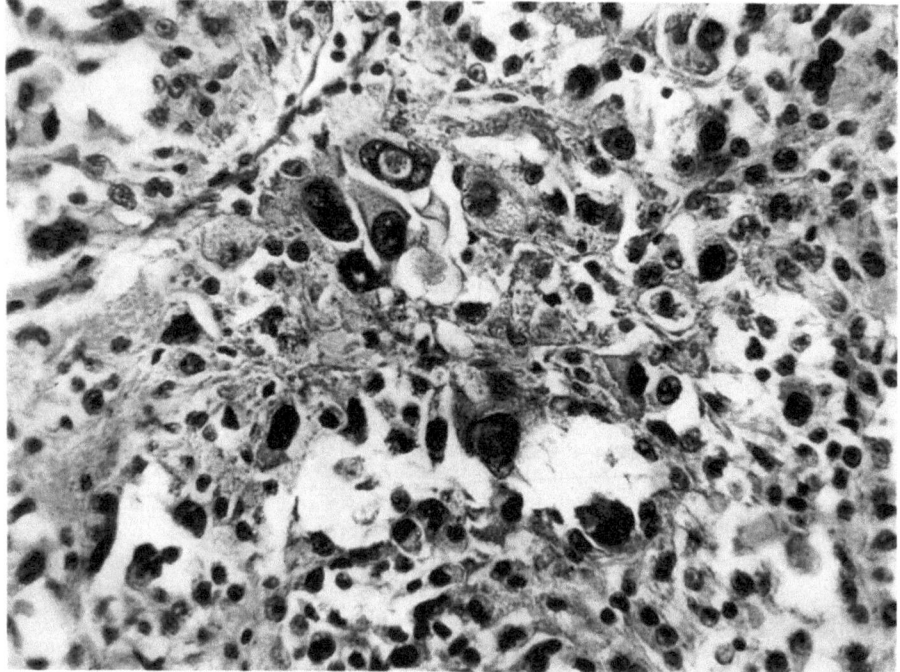

Fig. 35. Anaplastic adenocarcinoma, Grade 3 (×265)

(average) or grade 3 (high) malignancy. It is obvious that grading must be done
by a competent pathologist and that consideration must be given to the most
malignant part of the individual tumour. When this is done there is a definite
relationship between grade and prognosis; in 100 cases in the B.A.U.S. series
(1951) the survival rates for the three grades were as follows: —

Table 1. *Percentage surviving*

Years	1	3	5	10
Grade 1.. .	89	82	77	43
Grade 2.. .	72	43	31	0
Grade 3.. .	48	26	8	0

A similar relationship held good in a later ten-year follow up of cases from the
Middlesex Hospital (Thackray 1957).

γ) *Mode of spread.* Adenocarcinoma of the kidney spreads by way of the
blood stream, by lymphatics, and by direct extension. Spread by the urinary
stream down the ureter is very rare. Macalpine (1948) reported one case and
collected from the literature twelve others which he considered authentic. Since
then Howell (1951) has added one more of his own and one from the literature.

Spread by the *blood-stream* is the factor which does most to account for the lethal nature of this tumour. The growth invades the radicles of the renal vein by erosion at an early stage and spreads thence to the main renal vein, the inferior vena cava and the right side of the heart. From there tumour emboli can be carried to the lungs or through them to any part of the body. Retrograde spread down the genital vein may produce metastases in the ovary, vagina, testis or epididymis. Macroscopic invasion of the renal vein or its main tributaries occurred

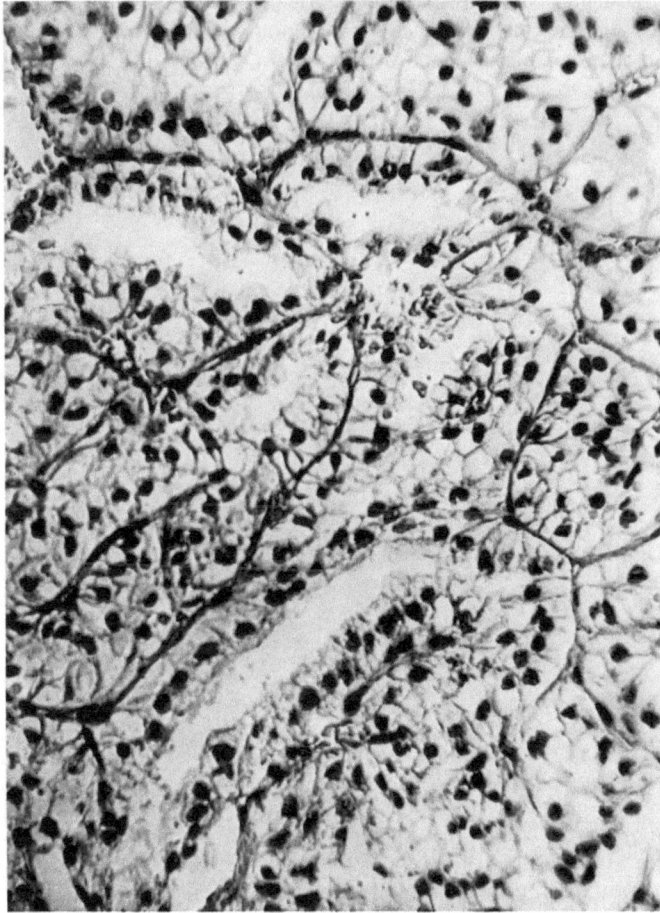

Fig. 36. Clear-cell tubular adenocarcinoma of the kidney; Grade 1 (× 265)

in 23 per cent of the B.A.U.S. series (1951). In a smaller series of 80 cases it was present in 29 per cent of which 19 per cent were in grade 1, 28 per cent in grade 2, and 45 per cent in grade 3.

In a personal series of 110 operable cases it occurred in 37 per cent of which 18 per cent were in grade 1, and 53 per cent in grades 2 and 3 (RICHES 1963). It is less common in tumours of low grade malignancy and there is a correlation between the two factors of high malignancy and venous involvement. MACDONALD and PRIESTLEY (1943) reported an incidence of venous involvement as high as 54 per cent.

The kidney contains many *lymphatic* vessels which accompany its arteries. As these are eroded by the growth extension can take place into main lymphatic trunks and lymph nodes are invaded. The hilar glands and the para-aortic glands may be enlarged and spread by the thoracic duct may lead to implication of cervical glands. Invaded lymph nodes were found in 151 out of 562 cases in the B.A.U.S. series (1951). There appears to be some connection between the histological grade of the tumour and lymphatic invasion.

By *direct extension* the growth invades the renal capsule, spreads through it and involves the perinephric tissues causing fixity of the kidney. Evidence of such local extension was present in 24 per cent of 76 cases operated upon and was more evident in the higher grades of tumour. There may be direct invasion of the pancreas, the intestine or other viscera.

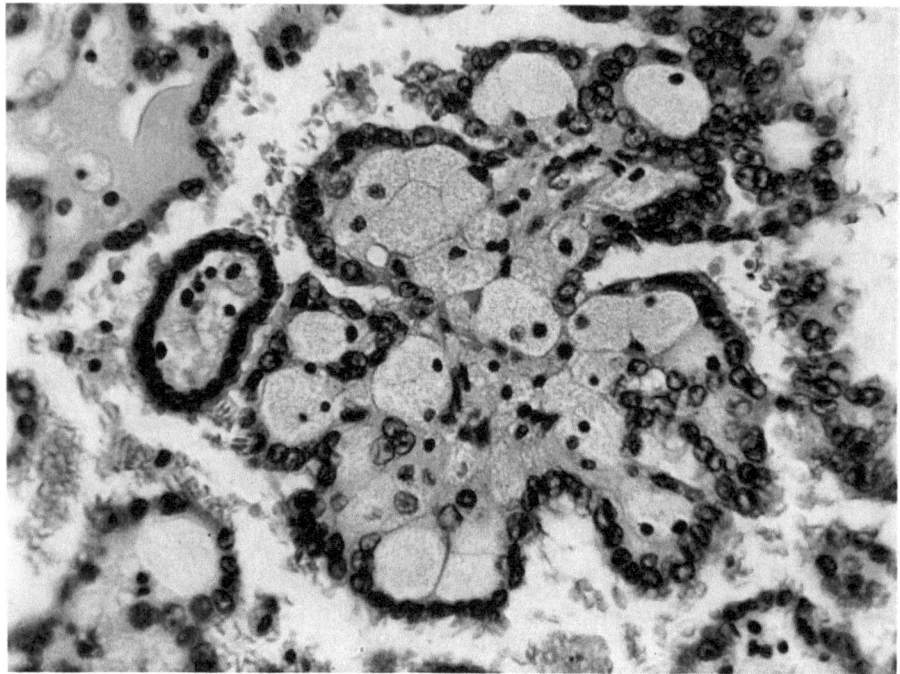

Fig. 37. Papillary adenocarcinoma of the kidney, Grade 1. The papilla is filled with foam cells. (× 265)

d) Site of metastases. From the foregoing it will be evident that metastases may be found in any part of the body, and such is the case. They may be late in appearance, and Snelling (1949) reported a local recurrence 15 years after nephrectomy.

The lungs are most frequently affected. Pulmonary metastases, noted clinically or found at post mortem were present in 314 out of 562 cases in the B.A.U.S. series (1951). Lucké and Schlumberger (1957) collected these and two other large series and found them present in 56 per cent of 320 autopsies for renal adenocarcinoma. They can be detected in life by their physical signs or when small by the rounded shadows seen in a radiograph (Fig. 38). In the same collected series of 320 autopsy cases there were lymph node metastases in 38 per cent, hepatic in 35 per cent and osseous in 33 per cent. The adrenals were involved in 19 per cent, the opposite kidney in 7 per cent, the brain in 7 per cent, the spleen in 5 per cent, the heart in 5 per cent, the skin in 3 per cent and the thyroid in 1 per cent. Other sites accounting for less than 1 per cent each were the intestine, muscle, spinal cord, testis, pancreas, ovary and penis. In addition there were in the B.A.U.S. series (1951) clinical metastases in the vagina in two cases and the bladder in one.

Of the bones invaded those containing red marrow were most often involved, as is general in secondary carcinoma; in order of frequency they were vertebrae, pelvis, ribs, femur, humerus, skull, clavicle and tibia. The bony metastases are usually osteolytic and may be responsible for pathological fractures in long bones (Fig. 39).

ε). **Staging.** In carcinoma of the kidney grade is all important as an index of malignancy (see p. 31) while in the case of solid tumours of the bladder it is the stage of the growth which gives the best idea of prognosis. The methods of staging renal growths which have been suggested all contain some anomalies. FLOCKS and KADESKY (1958) described four stages according to the extent of the primary lesion as follows: —

1. Limitation by the renal capsule,

2. Invasion of the pedicle and/or the perinephric fat,

3. Involvement of regional lymph nodes,

4. Distant metastases demonstrable.

This takes no account of spread within the kidney or of invasion of the renal vein.

PETKOVIC (1959) recognised the importance of extension through the

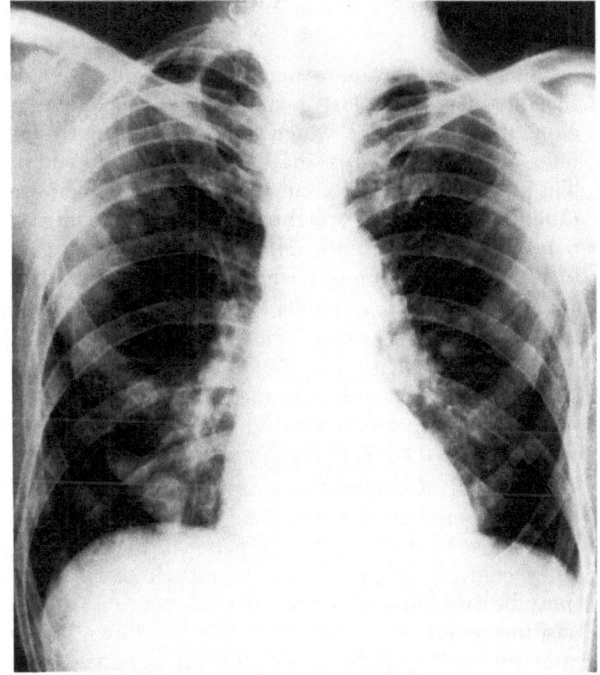

Fig. 38. Multiple rounded pulmonary metastases in a case of adenocarcinoma of the kidney

pseudo-capsule of the tumor within the kidney as a stage in spread. His four stages are: —

a) Intracapsular; the pseudo-capsule of the tumour is intact.

b) Intrarenal, extracapsular; the pseudo-capsule is broken or non-existent.

In the stages a and b neither veins nor lymphatics are invaded.

c) Periorganic: there is no pseudo-capsule or only a trace of one, the tumour having metastasized within the kidney and spread into the perinephric tissue. Veins and lymphatics are invaded.

d) Systemic involvement: distant metastases are present.

This scheme is better but still contains some anomalies; while the concept of metastasis within the kidney is important multiple tumours may have been present from the outset. A tumour may be both intracapsular and intrarenal (stage a) but still have gross invasion of the renal vein (Fig. 32) or it may be largely extrarenal whilst still within an intact pseudo-capsule.

If staging is to be of any value the scheme should include a separate stage for invasion of the renal vein or lymph nodes. The following scheme of pathological staging is suggested as one which includes most of the possibilities:

I. Encapsulated: the tumor is enclosed in a pseudo-capsule.

II. Intrarenal, extracapsular or no apparent capsule.

III. Perinephric extension or invasion of the pedicle.

IV. Invasion of the renal vein or regional lymph nodes.

V. Distant metastases present.

2. Connective tissue tumours

a) Benign

Fibroma. Fibromata occur in the subcapsular region of the cortex as small firm nodules or as larger tumours in the medulla. They consist of bundles of elongated fibroblasts which compress the surrounding renal tissue. The deeper ones may undergo myxomatous degeneration and may include some renal tubules. They rarely produce clinical evidence of their presence, but Gordon-Taylor (1930) removed one weighing 22 pounds from a patient aged 20. It contained a few epithelial tubules. Nightingale and Lytle (1937) reported the removal of another large tumour which was a pure fibroma.

Lipoma. This is met as an incidental post-mortem finding in the cortex of the kidney. Occasionally it may reach such a size as to produce symptoms. Beadles and Urich (1952) recorded one measuring 8.5 cm. in diameter into which haemorrhage had occurred. It was radiotranslucent, a point which might suggest a correct preoperative diagnosis. They added two to the fourteen previously reported by Robertson and Hand (1941) as being of surgical significance.

Myoma. *Leiomyomata* occur as multiple small subcapsular tumours which are discovered at autopsy, or as large single tumours which give rise to symptoms. They arise in the capsule, the pelvis or calyces or the blood vessels. Histologically they contain elongated smooth muscle cells, but the distinction from fibromata may be difficult and require special staining (Patch 1937). Owen (1953) reported a fibromyoma of 3 inches diameter which was undergoing malignant change, and this is the tendency of all such tumours. Clinton-Thomas (1956) removed a leiomyoma weighing 82 pounds (37,195 g) from a Malayan male aged 23. He reviewed the literature and stated that probably less than 50 of these rare renal tumours have been reported.

Rhabdomyoma is excessively rare and Herbut (1952) states that the only recorded case is the one described by Constance (1947). It was large and contained a high proportion of striated muscle fibres.

Angioma. Haemangioma within the kidney occurs around the base of a pyramid (Swan and Balme 1935) but it is often found at the apex of a papilla when it affects the pelvis. It may occur anywhere within the kidney however and must therefore be considered in both situations. The tumours may be single or multiple; they are generally small but Lucké and Schlumberger (1957) have depicted a kidney containing multiple tumours of several centimetres in diameter.

The condition is not common but it is important as being one of the chief causes of "Essential Haematuria". In 1951 Weyrauch and Berger had found records of 76 cases including one of their own which was bilateral. They placed 48.7 per cent in the mucosal and submucosal areas, 42.1 per cent in the medulla and 9.2 per cent in the cortex.

Histologically angiomata are either capillary, with small thin-walled blood vessels, or cavernous, with wider vascular spaces (Fig. 40). DUKES 1949 suggests that they are not neoplastic but are due to vascular reparative tissue secondary to trauma or infection. The case described by WATERFALL (1950) does not support this view.

Clinically they give rise to haematuria which is profuse and intermittent. Pyelography and aortography are usually not diagnostic, although MURRAY and TRESIDDER (1957) have demonstrated a cavernous angioma by aortography. A clue to the cause of the bleeding may be given by the presence of other angiomata on the body surface.

If the bleeding is severe and is known to be unilateral with a normal kidney on the other side the treatment is nephrectomy. The possible multiplicity of the lesions generally precludes partial nephrectomy, but this might be necessary in the case of a solitary kidney.

Lymphangioma is very rare. In 1950 WYNN-WILLIAMS and MORGAN described one in a pregnant woman of 33 years. They failed to find a similar case in the literature. HIGGINS, WILLIAMS and NASH (1951) depicted a specimen of renal lymphangioma removed by Barrington-Ward in a girl of 4 months and mention another removed by Browne.

The specimens showed cystic dilatation of a large part of the kidney and macroscopically resembled multilocular cysts. Microscopically the cysts were lined by flattened endothelial cells. The presenting symptom was haematuria in the first case and a palpable tumour in the second. The preoperative diagnosis was renal neoplasm.

b) Malignant

All these apparently benign tumours of connective tissue are liable to become malignant, and each has its malignant counterpart.

Fibrosarcoma. Of the several types of renal sarcoma fibrosarcoma is the commonest. MINTZ (1937) reported it in 53 cases out of 93 renal sarcomata recorded in the literature. PRIESTLEY (1941) found 5 per cent of sarcomata in 568 renal

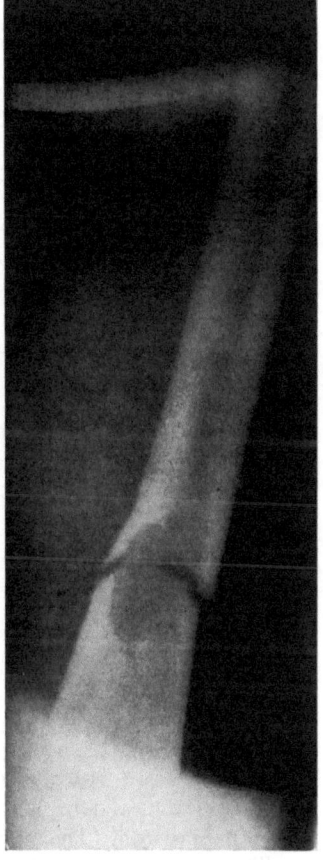

Fig. 39. Pathological fracture of the femur through a metastasis from a grade 3 adenocarcinoma of the kidney in a man of 64

tumours. The age of maximum incidence is 40 to 60 years, and males and females are affected equally.

The tumour is composed of pleomorphic cells with rounded or oval nuclei and fibroblasts. It grows rapidly and spreads by direct extension, by the renal vein and by lymphatics. Metastases occur in the lungs.

Rarely the fibroblasts undergo metaplasia to osteoblasts and lay down bone, producing an *osteogenic sarcoma* (HAMER and WISHARD 1948).

Liposarcoma is less common. NEWMAN and REED (1949) could find only twelve cases to which they added one. FISH and McLAUGHLIN (1946) found eleven cases, including one of their own, of whom seven had tuberous sclerosis.

Mogg (1957) added a further case and noted that it was radiolucent. The tumours may be unilateral or bilateral and are composed of adult fat cells, lipoblasts and large multinucleated cells. A single tumour is likely to be large, but multiple small ones also occur. It spreads by direct extension and metastases have not been recorded.

Leiomyosarcoma is rare and only 16 reported cases were found by Lazarus and Friedman (1954). Subsequent cases have been reported by Higbee and Atkins (1954) and by Bruce and McNaught (1954). This tumour was large and occured in a post-menopausal woman, as most do. It was removed transperitoneally through an incision medial to the ascending colon, in the interval

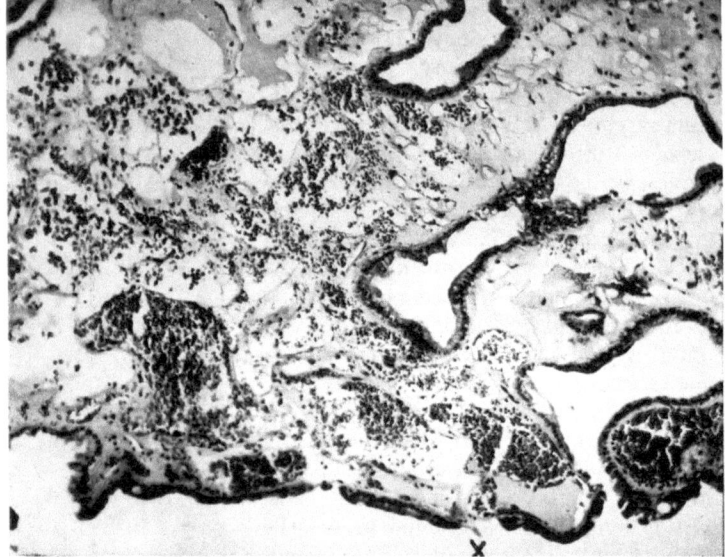

Fig. 40. Angioma of the kidney. The site of rupture of a vessel into the pelvis is marked X

between the ileocolic and right colic vessels. There was recurrence locally and in the inguinal and axillary lymph nodes within nine months, which proved to be radio-resistant. Kerr (1954) removed a gastric leiomyosarcoma in March 1951 which was followed by a similar growth in the right kidney in February 1952. After nephrectomy the man was alive and well in December 1952.

This tumour occurs at any age; it reaches a large size and contains interlacing smooth muscle cells with giant cells. It gives rise to widespread blood-borne metastases.

A case of *rhabdomyosarcoma* containing striated muscle fibres was described by Messinger and Jarman (1937); Weisel and his colleagues (1943) also reported a case. Doubts have been expressed on the existence of rhabdomyosarcoma as an entity separate from a Wilms' tumour (nephroblastoma) with a predominance of striated muscle.

3. Mixed tumours

a) Benign

Amongst the tumours arising in mesenchyme are some which are composed of blood vessels, muscle and fat in varying proportions. The name *Angio-*

myolipoma describes them. They form one variety of renal hamartoma, the term applied by ALBRECHT (1904) to swellings having a quantitative increase in tissues normally present without qualitative change.

The tumours may be small and multiple or large and single. In gross appearance they resemble haemorrhagic fat. Microscopically the blood vessels are surrounded by smooth muscle fibres which radiate from them; the amount of fat is variable.

In many cases, possibly 50 per cent, they are associated with the condition known as tuberous sclerosis, a congenital disorder of which the chief clinical features are mental deficiency, epilepsy and adenoma sebaceum of the face; the last is said by FISH and McLAUGHLIN (1946) to be present in all cases of tuberous sclerosis but sometimes it is only mild as in one case reported by LE BRUN, KELLETT and MACALISTER (1955). Sometimes there are cutaneous or subungual fibrous nodules, and retinal phacomata may occur. The cerebral lesions consist of nodular gliosis of potato-like appearance, hence the term "tuberous".

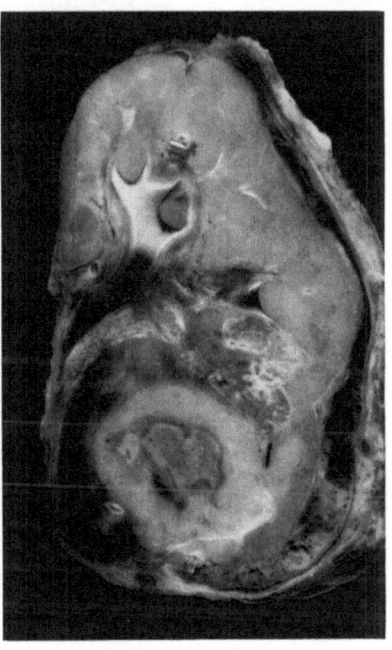

Fig. 41. Wilms' tumour (Nephroblastoma) in the lower pole of the kidney of a boy a $4^1/_2$. He died from pulmonary metastases

The renal lesions may be small and unrecognised by the usual methods, but sometimes they are large and the presence of a renal tumour is shown by pyelography, having been suggested by the clinical evidence of pain in the loin and and swelling.

Whether or not associated with tuberous sclerosis these tumours when large show a great tendency to cause massive retroperitoneal haemorrhage, as in MACALISTER'S (1955) third case, and in those reported by RUSCHE (1952) and MORGAN et al (1951). MOOLTEN (1942) has discussed the connection between renal hamartoma and tuberous sclerosis.

b) Malignant

Nephroblastoma (Wilms' tumour). The many names which have been applied to this tumour give some idea of the conflicting views regarding its nature; CULP and HARTMANN (1948) gave a list of 53 different appellations. It is of embryonic origin and appears to arise in immature renal parenchyma, but as it contains mesenchymal and epithelial elements it is included here in the group of mixed tumours. Current views (LUCKÉ and SCHLUMBERGER 1957) are that it has its origin in the metanephrogenic blastema which can develop into all the different tissues found in the tumour. The older views (EBERTH 1872, WILMS 1899) attributed to it an origin in the mesonephros.

α) Incidence. Wilms' tumour accounts for from 6 to 8 per cent of all malignant renal growths. It is predominantly a tumour of infancy and childhood and in 189 cases in the B.A.U.S. series (1951) nearly 50 per cent occurred between the ages of 2 and 4 years. It has been found at birth in the full term foetus (NICHOLSON 1931) and even in a $6^1/_2$ months foetus (BARRETT and McCAGUE 1954). It occurs rarely in adults; HILL (1946) collected 35 cases from the literature and in 1947 Esersky et al brought the number in adults up to 57. Clay (1930) reported a case in a woman of 80. It affects the right and left sides and the two sexes with equal frequency, and is usually unilateral. Sheach (1953) reported a bilateral case and collected 18 others from the literature. L. S. Scott (1954) reported a 3.6 per cent incidence of bilateral cases in 906 reviewed, and considered that some of them were metastatic and some double primaries, whilst Rickham (1957) described two children with bilateral tumours one of whom was living and well 34 months after nephrectomy on one side and excision of the tumour on the other.

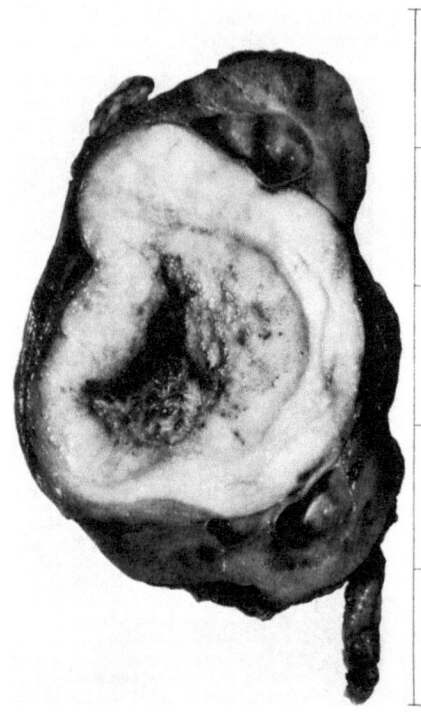

Fig. 42. Wilms' tumour (Nephroblastoma) compressing the remainder of the kidney. From a girl of 3 who developed another tumour in the opposite kidney nearly two years later; this was excised (scale in inches)

β) *Pathological anatomy.* **Macroscopic.** The growth arises in any part of the renal parenchyma and forms an ovoid or spherical swelling which is generally of large size by the time of diagnosis. It is of grey or dirty white colour and usually solid and homogeneous; at a later stage haemorrhage and necrosis occur and cause changes both in the colour and consistence of the mass. It is usually covered by a thin pseudocapsule but sometimes invasion of the renal parenchyma is evident; fibrous traebculae intersect the tumour mass in some cases. The remainder of the kidney may appear normal (Fig. 41) or compressed to a thin shell by the enlarging tumour mass (Fig. 42). There may be visible evidence of invasion of the renal vein or of extension down the ureter.

Microscopic (Fig. 43). The tumour is cellular with a loose connective tissue stroma of variable amount but containing few vascular spaces. The cells are small, rounded or elongated, with scanty cytoplasm and large irregular basophilic nuclei. They are arranged in cords, sheets, masses or tubules and in some cases the tubules are well formed and lined by columnar or cubical epithelium. There is often a general resemblance to the developing foetal kidney, and proglomeruli may be prominent. The stroma may undergo metaplasia to adult connective tissue, fat, myxomatous tissue, muscle, either smooth or striated, or may show areas of calcification, cartilage or bone formation. Thus the histological picture is extremely variable and although it is possible to recognise extreme degrees of differentiation they cannot as yet be correlated with prognosis in the individual case. There is however some evidence that the more highly differentiated tumours are more radio-resistant, and Twistington Higgins and his associates (1951) have suggested that these tumours have a better prognosis than the poorly differentiated type which shrink rapidly under radiotherapy but nearly always recur (see also Vol. XV).

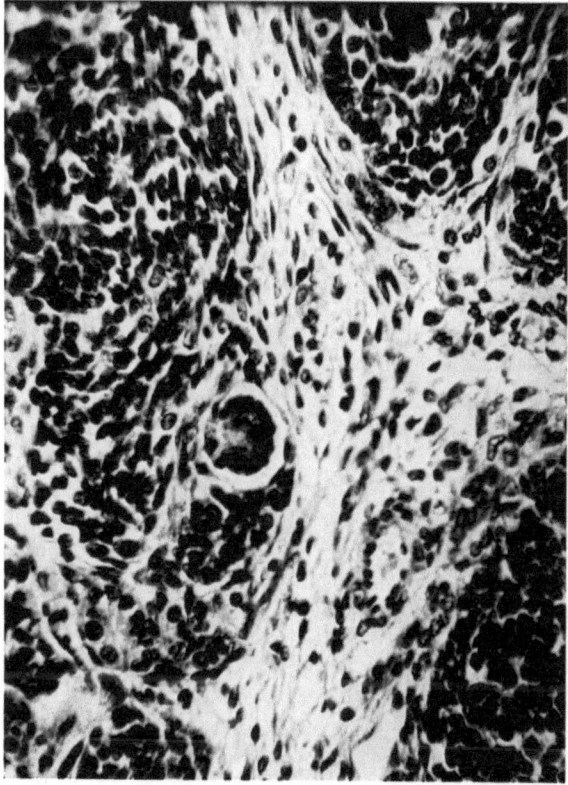

Fig. 43 Microscopic section of Wilms' tumour showing masses of of small cells in a stroma containing few vascular spaces. There is a proglomerulus in the centre (x 390)

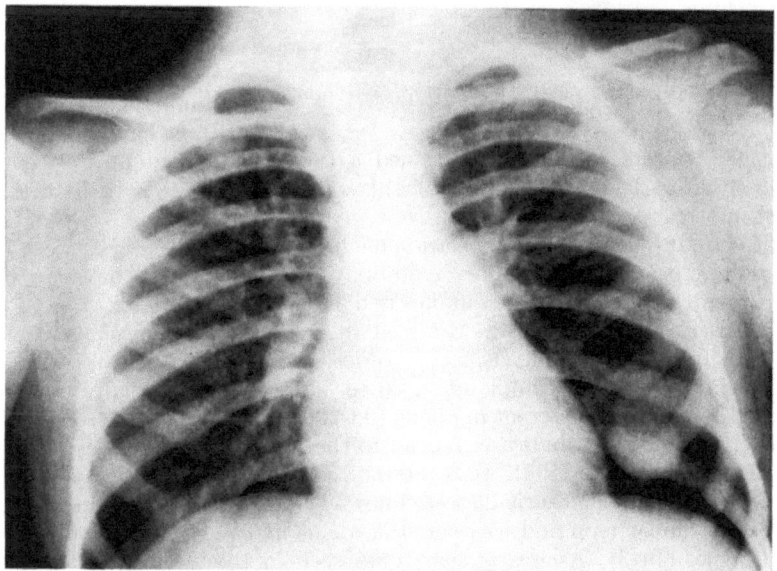

Fig. 44. Solitary rounded metastasis in the left lung in a case of Wilms' tumour

γ) Mode of spread. Although the tumour may remain localised until it has reached a large size it spreads ultimately by direct extension within the kidney and through the capsule, infiltrating the posterior abdominal wall, the peritoneum, the diaphragm and the suprarenal gland. Lymphatic spread is less common, but it does occur, and as in other renal tumours spread by the venous blood stream accounts for most of the lethal effects.

δ) Site of metastases. Because of the invasion of the renal vein, which was present in 12 out of 30 specimens reported by Weisel, Dockerty and Priestley

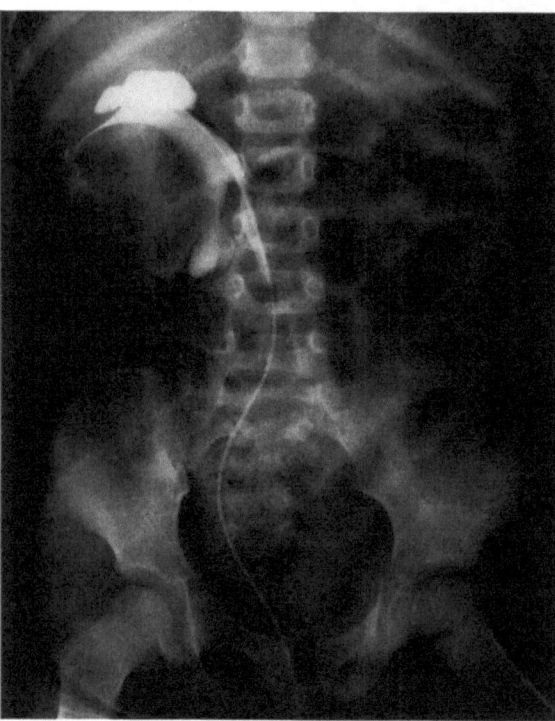

Fig. 45. Retrograde pyelogram of a Wilms' tumour, from the same case as Fig. 42

(1943), metastases are found most often in the lungs (Fig. 44). In their whole series of 44 cases the lungs were involved in 13, lymph nodes in 8 and the liver in 2. In the B.A.U.S. series (1951) renal vein invasion was present in 9 per cent of the 138 cases submitted to operation. Fifty-two had metastases either noted clinically or found at post mortem, and they included 34 in the lung, 17 in the liver, 12 in lymph nodes and 8 in bone. The para-aortic and mediastinal nodes were those most often invaded, and the skull led amongst the bones. Metastases were also found in the abdomen and were present at times in the opposite kidney, the brain, the spleen, the pancreas, and in muscle. Higgins et al (1951) have reported subcutaneous deposits and have mentioned a case with spread up the vena cava into the right auricle. Blomfield (1954) has illustrated three patients with intraorbital metastases causing proptosis.

ε) Clinical features. As there are some important differences between Wilms' tumours and the renal tumours of adults already described, the main clinical features and diagnosis will be summarised here.

The classical symptoms are reversed in frequency. A palpable tumour is present in from 68 per cent (B.A.U.S. series, 1951) to 85 per cent (Abeshouse 1957); pain may be the complaint in 20 to 30 per cent, especially in the older children; haematuria is present in about 10 to 20 per cent. There is some evidence that haematuria, which indicates spread to the renal pelvis, is a particularly bad prognostic sign (White 1951, L. S. Scott 1954). In the B.A.U.S. series (1951) out of 34 who had haematuria 23 were known to be dead within eighteen months. Fever is not uncommon and was noted in about half their cases by Weisel and his colleagues (1943). Ascites is sometimes seen in advanced cases, and listlessness, loss of appetite or of weight, and anaemia are often present. Hypertension has been mentioned as present in a quarter to one half the patients (BRAD-LEY and PINCOFFS 1938) and the blood pressure may return to normal after nephrectomy or irradiation therapy (DANIEL 1939).

The *diagnosis* rests mainly on the finding of a tumour, but is strengthened by the occurrence of any of the other symptoms. Excretion pyelography may show deformity of the calyces or pelvis, but frequently indicates a functionless kidney. It serves a most useful purpose however in demonstrating the presence and function of the other kidney. Retrograde pyelography gives a clear picture of the renal deformity (Fig. 45) and is justifiable as

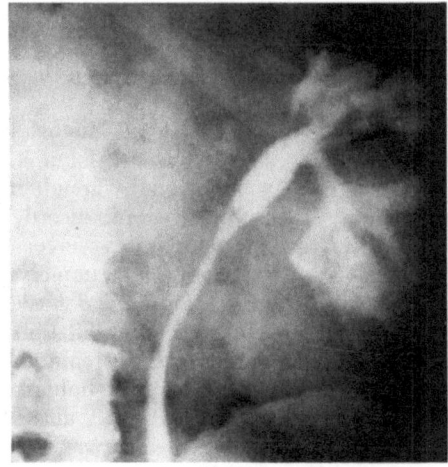

Fig. 46. Pyelogram from a case of reticulum cell sarcoma in a man of 73

an immediate pre-operative step. A plain abdominal X-ray often shows a soft tissue shadow displacing the colon downwards. In cases where the only clinical finding is a tumour in the loin it is wise to use at least one of the radiographic aids to diagnosis. Although the final answer may only be known by operation a Wilms' tumour can usually be distinguished by excretion or retrograde pyelography from the other conditions with which it may be confused. These are given in the following order of frequency by ABESHOUSE (1957) after an enquiry into 856 cases from 81 surgeons: — neuroblastoma, hydronephrosis, serous cyst, polycystic disease, retroperitoneal sarcoma, cyst or tumour of the adrenal, intra-abdominal tumour and ectopic kidney. HIGGINS and his colleagues (1951) point out that in a young child

Fig. 47. Section of kidney in reticulum cell sarcoma

a tense hydronephrosis may be translucent and that a neuroblastoma is more deeply seated and medial in position. They also mention pararenal teratoma as a rare growth which should be considered in the differential diagnosis. We have seen one case of hydatid cyst of the kidney which was thought to be a nephroblastoma until exposed at operation (see Fig. 63).

In all cases a radiograph of the chest must be taken to exclude pulmonary metastases.

Lymphoblastoma. Under this heading are included lymphosarcoma, reticulum cell sarcoma, leukaemia, myeloma and Hodgkin's disease. In any of these the kidney may be involved, although it is likely that the renal lesion is a local manifestation of the general disorder; Gibson (1948) has stated that primary renal involvement probably never occurs. He described four types of renal invasion by lymphosarcoma namely perirenal encompassment, nodular infiltration, diffuse infiltration, and a large solitary tumour. The large tumour is the most frequent in sarcoma, whilst in leukaemia, whether lymphatic or myelogenous, multiple tumours are commoner. Watson, Sauer, and Sadugor (1949) reviewed 1073 cases of lymphoblastoma and found involvement of the kidneys in 66, the renal pelves in 12 and the ureters in 14. The renal involvement is shown clinically by a mass in the loin, pain, and sometimes haematuria, and unless differential blood counts, urine tests, X-rays of bones, marrow biopsy or chromatography are diagnostic, or biopsy of an enlarged gland has been undertaken a diagnosis of carcinoma of the kidney is likely to be made and nephrectomy performed; this has been uniformly unsuccessful in curing the disease. Pyelography may show the calyceal changes expected in a renal neoplasm or may reveal displacement of the kidneys or ureters by a retroperitoneal mass. Fig. 46 shows the pyelogram of a man of 73 who complained of pain and one episode of haematuria. Nephrectomy was undertaken and a large renal mass weighing 4 lbs. 2 ozs. removed. It consisted of a friable tumour infiltrating the muscles behind, the peritoneum in front and the pedicle medially; there were two nodules in the ureter. Removal was incomplete and he died of uraemia six days later. Sections (Fig. 47) showed extensive invasion of the kidney, ureter and surrounding tissues by reticulum cell sarcoma. It was thought to be secondary although preliminary investigations had failed to show any other primary lesion.

If the diagnosis is made treatment by irradiation may afford some palliation.

II. Renal pelvis and ureter

1. Epithelial

a) Benign and malignant

α) Incidence. There were 315 cases of tumour of the renal pelvis in the B.A.U.S. series (1951) or 12.5 per cent of the whole. Of these 23 per cent were simple papillomata, 55 per cent transitional cell carcinomata and 22 per cent squamous cell carcinomata. No cases of adenocarcinoma were found.

The borderline between simple papilloma and papillary carcinoma is ill-defined and some pathologists regard all papillary tumours of the pelvis as papillary carcinomata. Nevertheless such distinction can be made on pathological evidence and the prognosis in the two groups is sufficiently different to justify it. The five year survival in simple papilloma is about 50 per cent, and in papillary carcinoma 35 per cent. Corresponding figures at ten years are 44 per cent and 25 per cent. The clinical and diagnostic features are the same for each, haematuria being present in about 90 per cent, pain in nearly one half, and a palpable tumour in only about 5 per cent. The principal diagnostic method for both pelvic and ureteric tumours is retrograde pyelography and ureterography (see Fig. 14).

β) Pathological anatomy. **Macroscopic.** Most of the tumours are papillary; a small proportion, given as 15 per cent by C. C. HIGGINS (1953), are solid. The papillary tumours are similar to those in the bladder but are more often sessile than pedunculated. They consist of a mass of pink delicate fronds which are extremely friable. Confinement within the pelvis or a calyx may give the whole tumour a rounded shape (Fig. 48). They arise anywhere in the pelvis, in a calyx, or at the pelvi-ureteric junction (Fig. 49) and their site of origin determines the degree of obstructive change produced in the kidney. Although sometimes found singly at an early stage (Fig. 50) they are usually multiple or confluent. The surface remains clean unless infec-

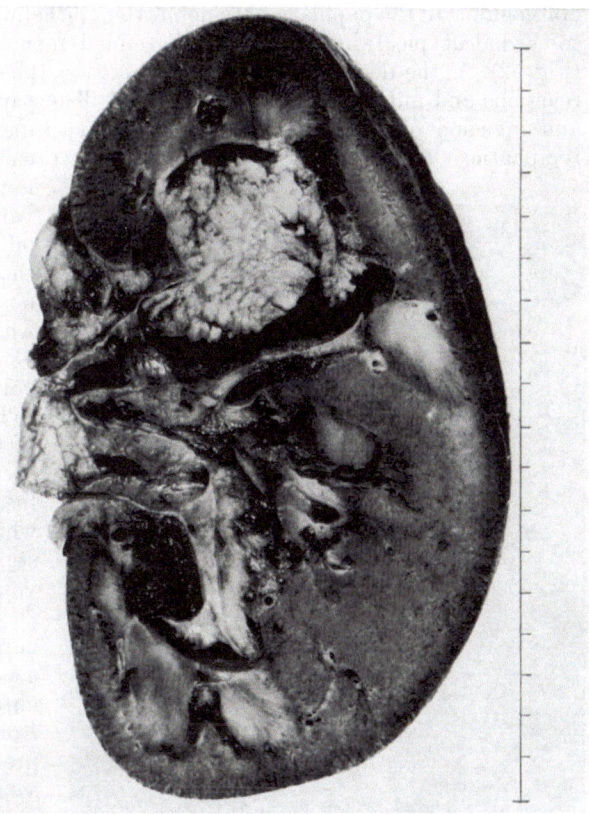

Fig. 48. Papillary carcinoma of the upper calyx in a man of 73 (scale in centimetres)

tion supervenes when it becomes flecked with phosphates or pus. Similar tumours may be found in the ureter on the same side or anywhere in the bladder.

The condition is usually unilateral but bilateral tumours have been reported by MACALPINE (1947) and by COLSTON and ARCADI (1954), who mention two other cases, those of SANFORD (1931) and GIBSON (1953).

The solid tumours are always infiltrating; they start as a diffuse thickening of the pelvic epithelium and grow to form a firm mass which usually causes obstruction (Fig. 51). They may infiltrate the substance of the kidney as well as extending through the wall of the pelvis (see Fig. 21).

Microscopic. In the simple papilloma the slender villi consist of a core of loose connective tissue covered by regular transitional

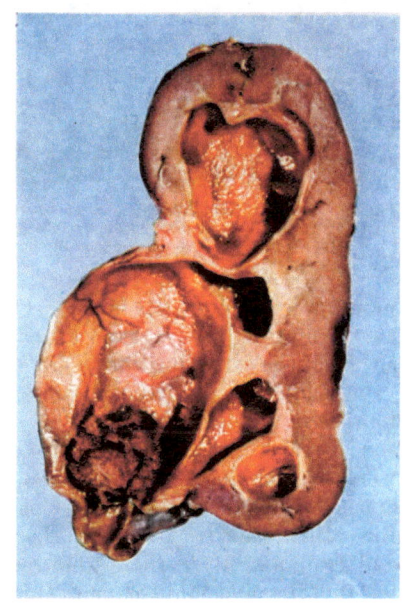

Fig. 49. Papilloma at the pelvi-ureteric junction producing hydronephrosis. From a man of 69. The same case as Fig. 16

epithelium. In the papillary carcinoma the epithelium is thicker and the cells are somewhat atypical; invasion of the basement membrane can generally be found (Fig. 52). As the degree of malignancy increases the cells vary in size and staining reactions and mitoses are present (Fig. 53). The papillae are shorter and thicker and invasion of the pelvic wall is apparent; tumour cells may be seen in the lymphatics. Thackray (1951) graded pelvic transitional tumours into those of low or high malignancy; the value of this grading was brought out by the better survival of those of low grade, but the prognosis also varied according to whether there was ureteric or bladder involvement. This involvement however was found to be commoner in the tumours of high malignancy.

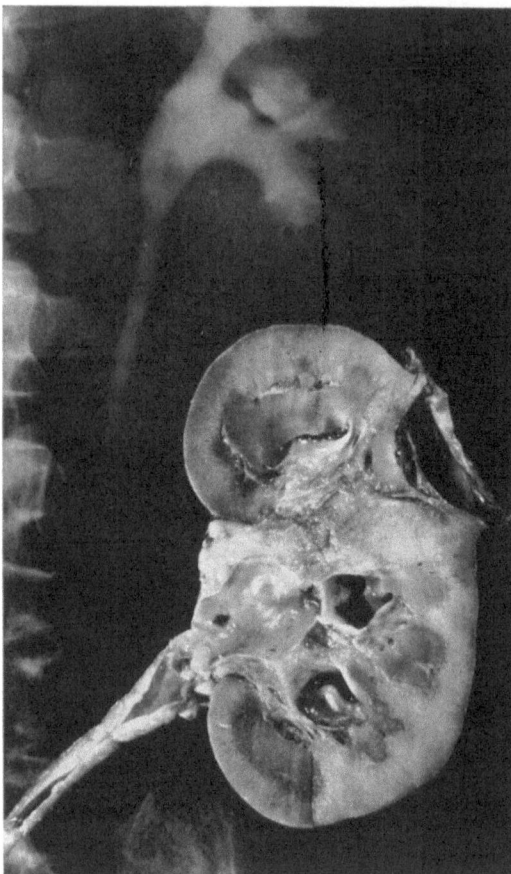

Fig. 50. Constant filling defect just above the pelvi-ureteric junction produced by a small single papilloma. From a man of 46 who had also a papilloma of the bladder

The solid growths show diffuse masses of transitional cells which permeate into the substance of the kidney and beyond it.

Squamous metaplasia occurs in the pelvic epithelium, especially in connection with chronic infection or irritation from calculi. It may be the precursor of the *squamous cell carcinoma* which accounts for a small number of renal pelvic growths. They form low, solid, nodular tumours which infiltrate the renal pelvis and parenchyma and may spread down the ureter by direct continuity. They are often associated with renal calculus which was present in 29 per cent in the B.A.U.S. series (1951) and in 48 per cent of 100 cases collected from the literature by Gahagan and Reed (1949) (Fig. 54). Five of the tumours were in horseshoe kidneys. Clinically they cause pain more often than haematuria and they have a poor prognosis. There were no five year survivals in either of the series recorded above.

Glandular metaplasia is still less common but is the presumed origin of the four cases of *Adenocarcinoma* of the renal pelvis which have been reported in the literature (Lucké and Schlumberger 1957). They were all associated with urinary infection and stone, and histologically showed mucin-secreting epithelial cells. Although one patient lived for six years after nephrectomy the others survived for less than a year. In the case described by Ragins and Rolnick (1950) there were radiological signs of pulmonary metastases after ten months, and multiple metastases were present at death in the others.

γ) *Mode of spread.* Although papillary carcinomata of the renal pelvis do not possess the same inherent degree of malignancy as adenocarcinomata of the parenchyma they have a peculiar property of being associated with similar tumours of the ureter and bladder. The assumption that these are seedling growths, shed from the main tumour, carried in the urinary stream and implanted on the epithelium lower down is supported by the fact that almost all are unilateral. The alternative theory of multicentric foci of origin in a urinary tract which is in a pre-cancerous state gains ground from a study of industrial bladder cancer and from the knowledge of the effects of carcinogenic substances excreted in the urine in aniline dye workers. Experimentally papillary tumours of the bladder can be produced in dogs by feeding them with β-naphthylamine (BONSER 1943, SCOTT and BOYD 1953), and one of the few recorded cases of bilateral papillary pelvic tumours occurred in a dye worker (MACALPINE 1947). It is likely that other carcinogenic substances are excreted and that the whole condition is of urogenous origin, but further work is needed for confirmation of this theory. In addition to this urine-borne spread, if such it is, papillary carcinoma of the pelvis does in time infiltrate the underlying tissues locally and invade the renal vein and the lymphatics although not to the same extent as the solid tumours.

δ) *Site of metastases.* Apart from involvement of the ipsilateral ureter or the bladder, which was present in 65 of 143 cases (46 per cent) in the B.A.U.S. series (1951), distant metastases were found in the lungs (Fig. 55),

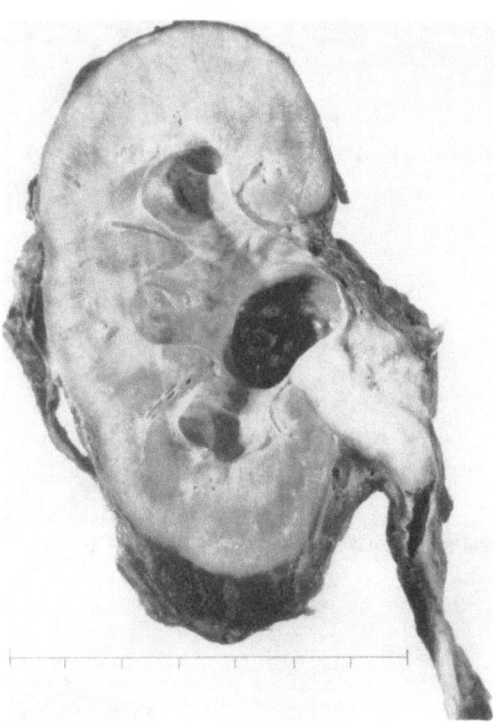

Fig. 51. Solid transitional cell carcinoma of the renal pelvis from a woman of 30 who died within six months with multiple metastases. (The same case as Fig. 15) (scale in centimetres)

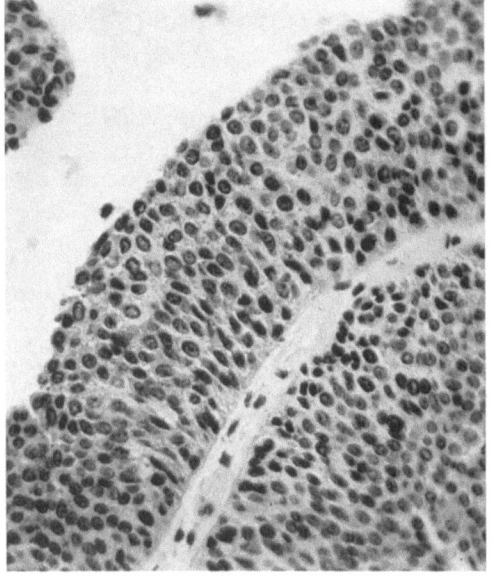

Fig. 52. Papillary transitional-cell carcinoma of the renal pelvis of low malignancy (x 265)

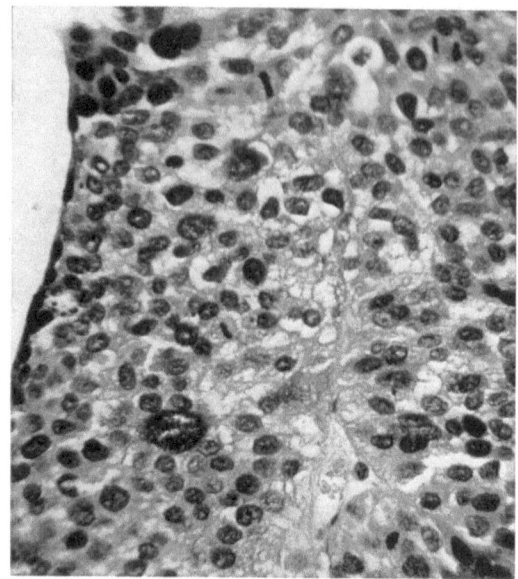

Fig. 53. Papillary transitional-cell carcinoma of the renal pelvis of high malignancy. The cells are of variable size and mitoses are present (x 265)

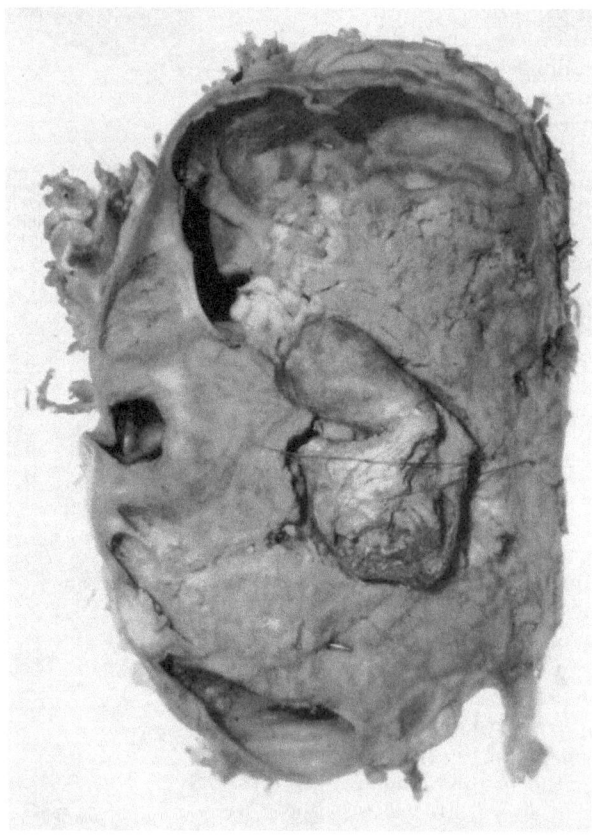

Fig. 54. Squamous cell carcinoma of the renal pelvis associated with a large calculus. From a man of 52 who died from multiple metastases

liver, lymph nodes, bones, adrenal, brain, heart, vagina and opposite kidney. They were more frequent in squamous than in transitional cell carcinoma. One patient with a solid transitional cell growth developed widespread metastases including nodules in the skin.

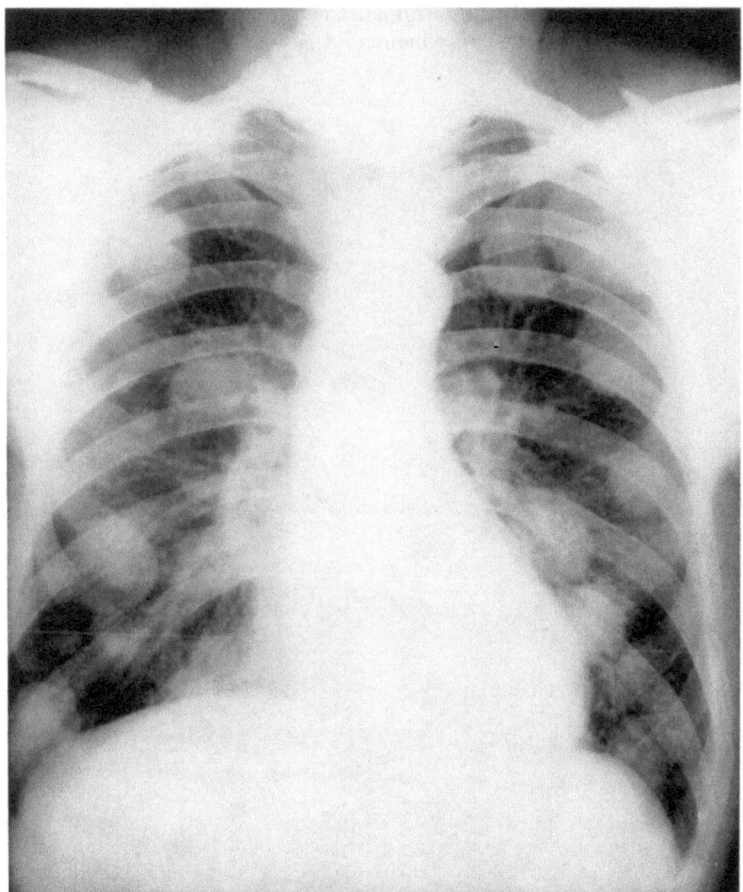

Fig. 55. Multiple pulmonary metastases in a case of carcinoma of the renal pelvis. From a man of 60

2. Connective tissue
a) Benign

Fibroma. This is a very rare tumour of the renal pelvis. IMMERGUT and COTTLER (1951) reported one case of pedunculated fibroma in which haematuria had occurred from torsion of the pedicle. They found a record of only one other case (BOROSS 1929).

Myxoma One case was reported by HÜSCH in 1949.

Haemangioma. Whilst both cavernous and capillary angiomata may occur in any part of the kidney many of them are situated immediately beneath the mucous membrane of the pelvis or a calyx (SWAN and BALME 1935) and must therefore be considered amongst the pelvic lesions. They may give rise to severe haematuria (RAPPOPORT 1945) sometimes with a localising lumbar pain

but frequently with no other symptoms. Cystoscopy during bleeding is mandatory as pyelographic and aortographic studies are usually negative. Haemangioma is probably a congenital lesion and may be a different pathological entity
from the varicosities sometimes found in the renal papillae although there is no
clinical distinction (Fig. 56). Nephrectomy must be performed if the bleeding is
sufficient to cause constitutional effects; it should always be seriously considered in a patient over 40, or whenever there is the slightest suspicion of new
growth.

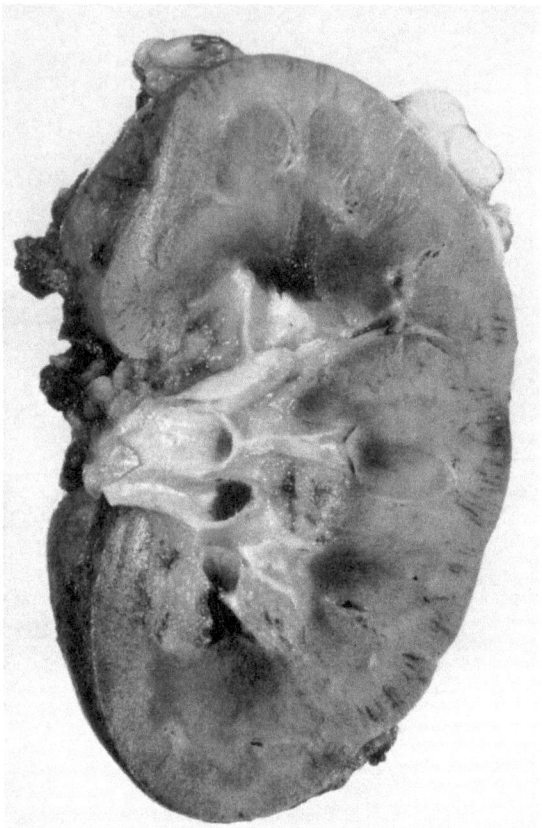

Fig. 56. Haemangioma of the upper calyx which gave rise to profuse haematuria in a man of 64. Pyelography
and aortography were normal

b) Malignant

Myxosarcoma. Whilst it is theoretically possible for malignant connective
tissue tumours such as fibrosarcoma to arise in the pelvis, in practice they do not
occur. We have seen one case of myxosarcoma of the renal pelvis in a woman
of 36. She had been known to have stones in the left kidney for seven years
and presented with left renal pain, occasional haematuria and a large tumour
in the left loin. X-ray (Fig. 57) showed a radiotranslucent mass with faint calcification at its periphery; there was very little excretion on intravenous pyelography but the right side was normal. Retrograde pyelography showed obstructive and compression changes consistent with a renal new growth, and at
operation a solid renal mass weighing 2 lbs. 12 ozs. was removed. On section

it showed a pale yellow gelatinous tumour arising from the pelvis and filling and distending it. The kidney tissue was displaced and thinned (Fig. 58). Microscopically it showed invasion by malignant myxomatous connective tissue and was classed as a myxosarcoma. The patient was given postoperative X-ray treatment and made a good recovery; it is too early to assess the ultimate prognosis, but she is alive and well nine years later.

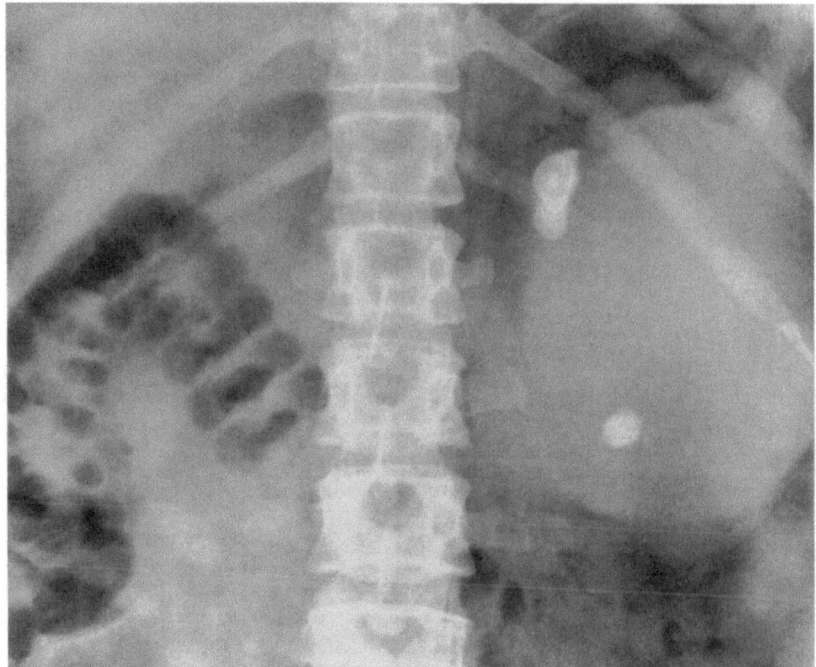

Fig. 57. Myxosarcoma of the renal pelvis in a woman of 36. Plain X-ray showing two stones and a radiolucent renal mass

III. Ureter

Incidence. The same pathological types of primary tumour occur in the ureter as in the renal pelvis. They are not so rare as was formerly supposed, and malignant growths outnumber benign by about three to one. MORTENSEN and MURPHY (1950) collected 77 benign tumours and 245 of carcinoma from the literature, whilst BARON and GREEN (1953) found 262 cases of carcinoma alone. The frequent age of incidence is from 50 to 70; males are affected twice as often as females, and the right side more often than the left.

Pathological anatomy. More than half the reported cases involve the lower third of the ureter.

1. Epithelial

Simple *papilloma* is a transitional cell tumour which is histologically benign.

Transitional cell carcinoma is often multiple and associated with a similar tumour in the renal pelvis. It may be papillary or solid. It infiltrates the wall of the ureter and spreads into the posterior abdominal wall.

Squamous cell carcinoma, like its counterpart in the renal pelvis, is less common and is usually associated with chronic infection and stone.

Adenocarcinoma is extremely rare and is also associated with stone.

2. Connective tissue

Ureteric growths arising in connective tissue are quite rare.

Fibroma usually occurs as a fibrous polyp, often with a long pedicle.

Haemangioma also tends to become pedunculated, as in Galbraith's (1950) case in which a vascular mass was extruded through the ureteric orifice of the bladder at the end of a narrow tapering stalk 6 cm long.

One case of *lymphangioma* was described by Jeppesen (1953).

Sarcoma. There was one case of spindle cell sarcoma in the B.A.U.S. (1951) series. It was in a woman of 65, and most of the few other reported cases have occurred in women in this age group.

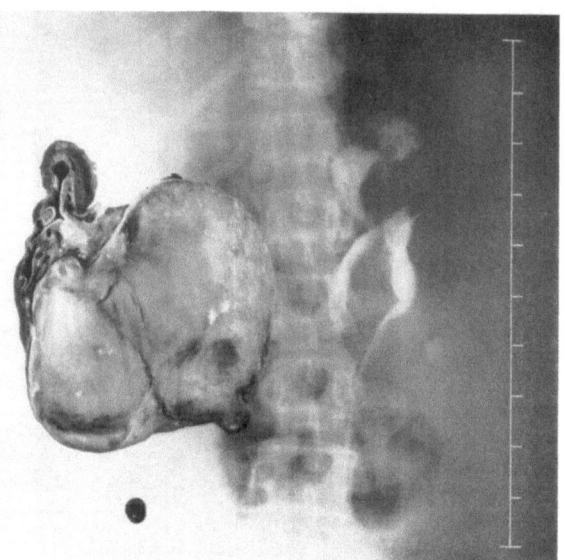

Fig. 58. The kidney from the same patient on its pyelogram. The tumour was gelatinous (scale in inches)

Mode of spread and metastases. All of the malignant ureteric tumours spread by direct extension, but they also invade the regional lymph nodes. Occasionally remote metastases are found in the liver or lungs. Their most important feature however is the frequency with which similar tumours are found in the bladder, whether from multicentric foci or by urogenous seeding. In one patient this necessitated treatment by cystodiathermy on 13 occasions in five years following nephro-ureterectomy (Riches 1941). Downward spread from a papillary tumour in the renal pelvis may fill the ureter with multiple similar tumours (Fig. 59). Retrograde spread by antiperistalsis in an obstructed ureter was postulated by Macalpine (1950) and by Mortensen and Murphy (1950) to account for the occasional appearance of a small papillary carcinoma in the renal pelvis in association with a large ureteric tumour. Tumours of the vesical end of the ureter may arise in the juxtavesical, intramural or submucous part (Fig. 60); the two latter are usually visible on cystoscopy and Masina (1950), who described 40 cases, pointed out that the infiltrating tumours produced radiological evidence of ureteric dilatation whilst the non-infiltrating tumours rarely did so. These tumours show a tendency to early fixity to the wall of the pelvis.

Clinical features. Of the classical symptoms haematuria is the one most frequently observed. W. W. Scott (1943) found it in 70 per cent of 182 cases; pain was present in 64 per cent, mostly of the obstructive type, sometimes due to colic, and sometimes metastatic. Tumour was noted in 40 per cent, but it was the swelling of a distended kidney in the majority of cases.

Diagnosis can be difficult in an early case but should be possible by routine investigations. The excretion pyelogram may show a functionless kidney, or may disclose hydronephrosis or hydroureter. Retrograde ureterography is more effective than pyelography in demonstrating a filling defect in the ureter (Fig. 17).

A ureteric tumour often has a cupped or crescentic margin (Fig. 16). The passage of a ureteric catheter may be obstructed or may cause brisk bleeding. A tumour which appears at the ureteric orifice, even if intermittently, may be subjected to biopsy. A tumour at the vesical end may be palpable on bimanual examination under full anaesthetic relaxation.

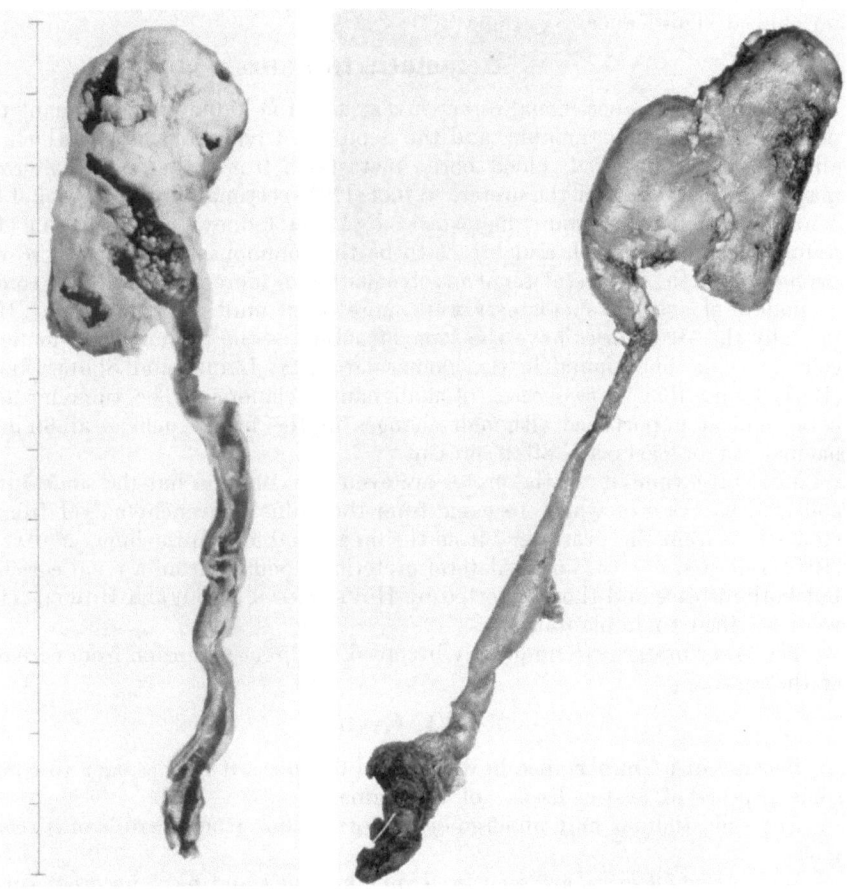

Fig. 59   Fig. 60

Fig. 59. Solid and papillary carcinoma of the renal pelvis with multiple tumours in the ureter. From a man of 72 who had bled for three years. The upper pole of the kidney is thinned almost to the point of rupture. (The same case as Figs. 18 and 66) (scale in inches)

Fig. 60. Infiltrating carcinoma of the intramural ureter in a man of 66. The tumour could be everted into the bladder for biopsy. It has caused hydroureter and hydronephrosis

A source of confusion can be the distinction between a new growth and a non-opaque ureteric calculus. An air bubble can also mimic a benign growth on the ureterogram.

IV. Renal capsule

The renal capsule is subject to any of the connective tissue tumours which occur in the kidney itself. Many of these, especially fibromata and leiomyomata arise in the capsular area of the kidney, whilst many so-called capsular tumours arise in the deep layer of the capsule and penetrate the cortex. There must always be therefore some doubt about the exact site of origin. All these tumours, including lipomata, may undergo malignant change into the corresponding sarcomata but the numbers recorded are small.

Colvin (1942) used the term "capsuloma" to describe the small benign tumours, mostly leiomyomata or fibromata, found at autopsy. In 144 cases 99 per cent occurred in subjects more than 40 years of age. They have little or no clinical significance.

V. Metastatic tumours

Apart from the occasional direct invasion of the kidney by malignant retroperitoneal or adrenal tumours, and the deposits of lymphoblastoma which have already been considered, blood borne metastases from other primary growths may reach the kidney or the ureter. Willis (1948) estimated that a total of 8 per cent of malignant tumours metastasised to the kidneys and Klinger (1951) found the lung, stomach and breast to be the commonest sites of origin of the primary growth. Bilateral renal involvement was more than twice as common as unilateral and the metastases were more often multiple than single. Histologically the secondaries have the same structure as the primary, and malignant cells are sometimes found in the glomerular tufts; Lucké and Schlumberger (1957) found this in two cases of malignant melanoma. The tumours are of little clinical importance although changes in the urine such as albuminuria, haematuria or casts are often present.

Metastatic tumours of the ureter are even less common but the same authors collected 65 cases of which 15 came from the kidney parenchyma, 11 from the stomach, 9 from the prostate, 7 from the breast and 5 from the lung. Macalpine (1948) reported one case of ipsilateral ureteric secondary from a renal carcinoma but both his case and those reported by Hovenanian (1950) and Howell (1951) were ascribed to implantation.

The lower ureters are frequently involved by direct extension from carcinoma of the cervix.

VI. Cysts

Because of its importance in differential diagnosis it is necessary to consider some aspects of cystic disease of the kidney.

Any solid tumour may undergo *cystic degeneration* from necrosis of its central part.

Multiple small cysts are seen in *chronic nephritis* and have no great surgical interest.

The so-called *pyelogenic cysts* are calyceal diverticula which communicate with the renal pelvis. Barrie (1953) has described cases of *paracalyceal cysts* in elderly subjects; they are caused by avascular degenerative changes in the fat around the calyces. Parapelvic cysts produce a rounded impression on the pelvis in a pyelogram; they may contain hamartomatous tissue in their wall (Turner-Warwick and Thomson, 1964).

Lymphatic cysts occur on the course of the main lymph channels in the hilum of the kidney. They are uncommon, and usually small. Henthorne (1938) reported twenty cases, and Scholl (1948) drew attention to the distortion they produce in the renal pelvis.

Simple cysts often attain a large size and may be confused with tumours. They are usually single and unilateral, hence the alternative name of "solitary" cysts, but bilateral and multiple cases occur. 75 per cent of cases occur in the fifth decade (Braasch and Hendrick 1944). They are frequently symptomless and are discovered as an incidental finding during routine investigations, but they may give rise to pain and a swelling and rarely to haematuria.

On pyelography a cyst produces a space occupying lesion (Fig. 61). It is avascular on aortography with vessels spread around it (Fig. 62) but it can be filled and shown by percutaneous puncture as a regular rounded cavity (LIND-BLOM 1946).

Treatment is necessary if symptoms are present or if there is any doubt about the diagnosis. Carcinoma has been known to occur in a simple cyst. Nephrectomy is only necessary in very large or infected cases; usually the cyst can be treated by enucleation, partial nephrectomy or by decapitation (GLASER 1952). The prognosis is good.

Multilocular cysts are similar to simple cysts except for the presence of intracystic septa. It is probable that the condition called cystadenoma is similar although there are some pathological differences. FRAZIER (1951), who collected 31 cases, noted in four a gross resemblance to cystadenoma.

Multilocular cysts may be found in infancy (CARVER 1949) or in middle life (O'FLYNN 1953). They produce a renal mass with

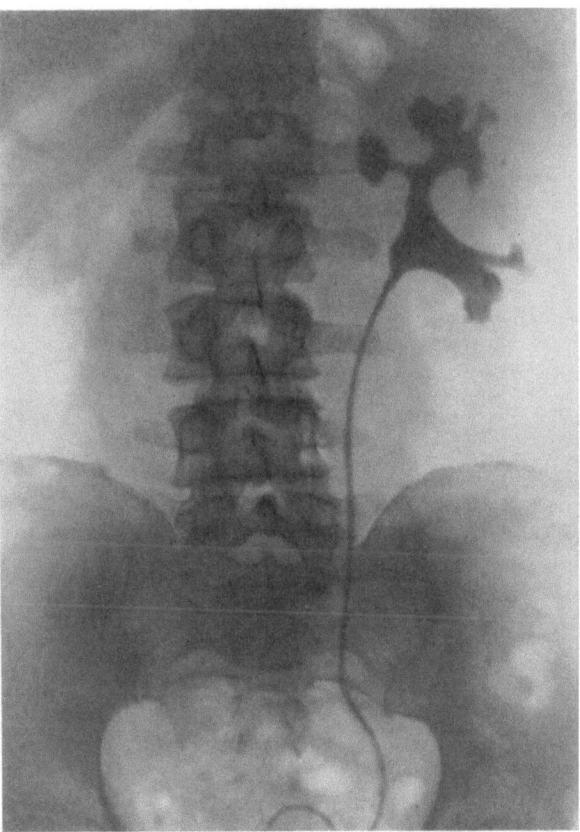

Fig. 61. Pyelogram showing a space-occupying lesion between the upper and lower groups of calyces in a man of 56. It was caused by a simple cyst

pain and sometimes haematuria. Pyelography shows a filling defect which cannot as a rule be distinguished from that of a new growth. POWELL, SHACKMAN and JOHNSON (1951), who collected thirteen cases from the literature and added two of their own, state that these cysts are unilateral, solitary, multilocular, with separate non-communicating loculi and having no communication with the pelvis. They have an epithelial lining but no renal elements within the cyst and the residual kidney tissue is normal. POWER (1955) described a cystadenoma which ROBINSON (1957) identified as a second case of PERLMANN's tumour (PERLMANN 1928). This had been described originally as a lymphangioma. The lymphangioma depicted by WYNN-WILLIAMS and MORGAN (1949) bears a striking resemblance in gross appearance to some of the multilocular cysts. Treatment is by partial or total nephrectomy according to the amount of renal tissue unaffected.

Dermoid cysts (teratomata) of the kidney have been reported (Baldwin 1915) but it is doubtful if true renal dermoids occur. Tumours containing hair or pultaceous material are sometimes found near the kidney; they arise in perirenal tissues as in the case reported by Langley (1950) in which there was a mass of bone resembling half a mandible and containing many teeth.

Hydatid cysts are found rarely, even in countries where hydatid disease is common. Howarth (1950) stated that there were only 44 examples of renal

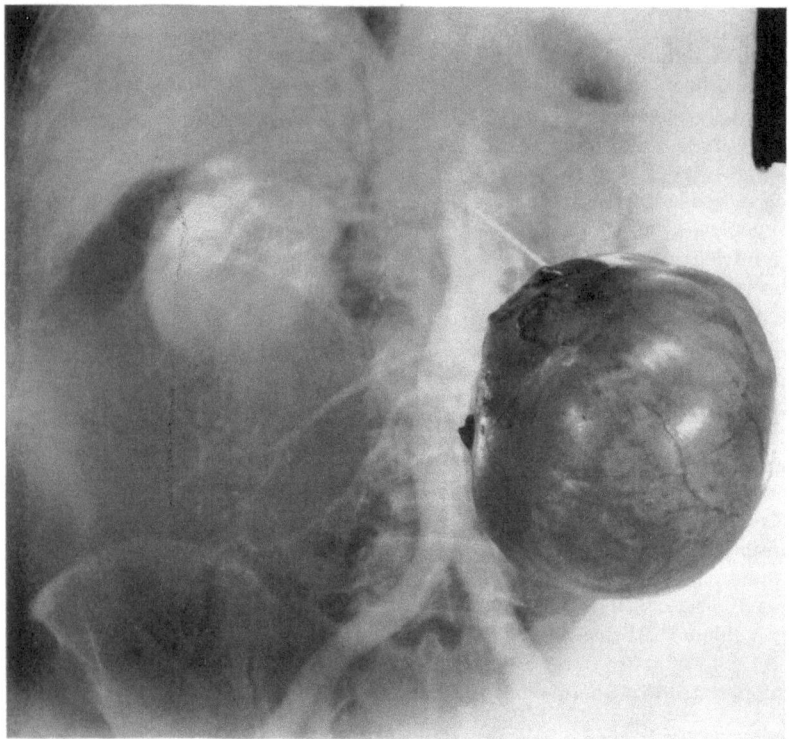

Fig. 62. Aortogram in a case of simple renal cyst in a man of 55. The cyst is avascular, the remaining nephrogram normal. The excised cyst is superimposed

hydatid cyst out of 2002 of hydatid disease in the Hydatid Registry of the Royal Australasian College of Surgeons. The cyst commonly produces a palpable tumour but there may be no other symptoms. Pyelographic distinction from a new growth is often impossible but calcification is more likely to be found in hydatisd. The Casoni test and complement fixation test may be positive. In an early case the cyst may be excised, but in many nephrectomy is necessary (Fig. 63). (See also volume IX/2.)

Polycystic disease (congenital cystic kidneys). In this congenital condition there are many small cysts replacing the renal parenchyma. It is almost invariably bilateral. The familial tendency has been emphasised by Fergusson (1949), who found 307 examples in 84 families, and by Rall and Odel (1949).

It has two peaks of incidence, the neonatal and the adult. In most of the neonatal cases death from uraemia occurs shortly after birth, and still-births are common (Küster 1902). Other congenital abnormalities are frequently found (Campbell 1954).

Symptoms in the adult arise most often between the ages of 30 and 60 years. Pain in the loin leads to the discovery of a mass on one or both sides. Early uraemic manifestations may be found, and hypertension is not uncommon. Haematuria may occur and albuminuria is the rule. Infection and stone formation are the most frequent complications.

The diagnosis is made by retrograde pyelography. The pelvi-calyceal system is elongated, the pelvis compressed and the minor calyces clubbed. Crescentic

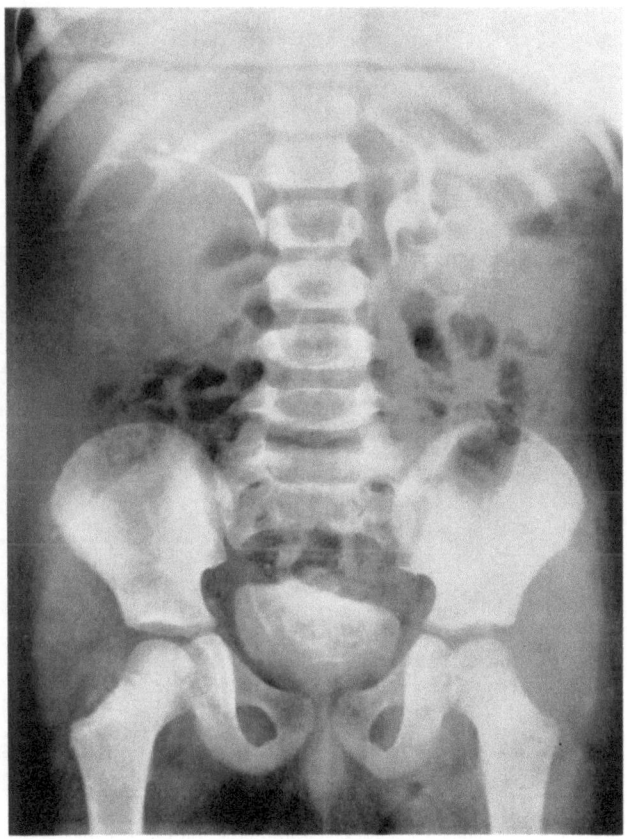

Fig. 63. Pyelogram of hydatid cyst of the right kidney in a boy of 4. It was excised. Note the resemblance to the Wilms' tumour in Fig. 45

impressions of the cysts may be noted (Fig. 64). Renal function tests show impairment of function.

The disease may pursue a slow course and terminates in death from uraemia or intercurrent infection.

The treatment is mainly medical by means of a low protein diet and the prevention or treatment of urinary infection. Surgical treatment is needed if pain on one side is severe either from tension, infection, or stone formation. If the condition is much more advanced on one side nephrectomy may be justifiable. In other cases Rovsing's operation, which consists of surgical decompression by puncture of the cysts, may produce some improvement. Patton and Bricker (1953) found no permanent improvement in renal function after the operation, but YATES-Bell (1956) after treating 18 patients claimed relief of pain, haematuria and hypertension.

H. Treatment

Malignant tumours of the kidney are treated by surgical methods, by radiotherapy, and by a combination of the two. Surgical ablation of the malignant process is the prime method of treatment.

Fig. 64. Pyelograms in a case of polycystic disease of the kidneys. The pelves are elongated and there are some crescentic filling defects. From a man of 36 with hypertension

It is axiomatic that before any operation on the kidney which may lead to nephrectomy the presence of a functioning kidney on the other side must be proved. This will have been done in the course of the investigations already detailed of which the excretion pyelogram is the most important. Other tests of renal function are of less importance in dealing with renal neoplasms, but a test depending on the retention of nitrogenous products in the blood, (e.g. the blood urea) and a urine excretion test (e.g. indigo-carmine or P S.P.) afford valuable confirmation.

I. Surgical

Partial nephrectomy or excision of the tumour alone is applicable in the case of a known benign tumour, or when the presumption of innocence is very strong or can be confirmed by immediate section. It was done for benign adenomata by

NITCH (1927), KRETSCHMER and DOEHRING (1929) and WATTS (1955). It is also indicated in bilateral malignant tumours, or in cancer involving a solitary kidney. It was carried out in one case of the B.A.U.S. series (1951) for bilateral adenocarcinoma by MILLIN, whose patient was alive two years later, and by RICKHAM (1957) in a child with bilateral nephroblastoma who was alive and well nearly three years later.

Total nephrectomy is the method of choice for malignant tumours and for most benign tumours. The various routes of approach — lumbar, abdominal, or thoraco-abdominal — are described in volume XIII. It is important that the exposure should be adequate in order to allow gentle handling, complete excision of the kidney with the perinephric fascia and fat and the para-aortic lymph glands, and ligation of the vascular pedicle at an early stage of the operation so as to avoid venous dissemination of the growth during manipulation. Contrary to opinion the anterior transperitoneal route does not give this early exposure of the pedicle claimed when the tumour is large or is situated in the upper pole. DONOVAN (1955) has claimed that early ligation of the pedicle can be done from a posterior approach through the

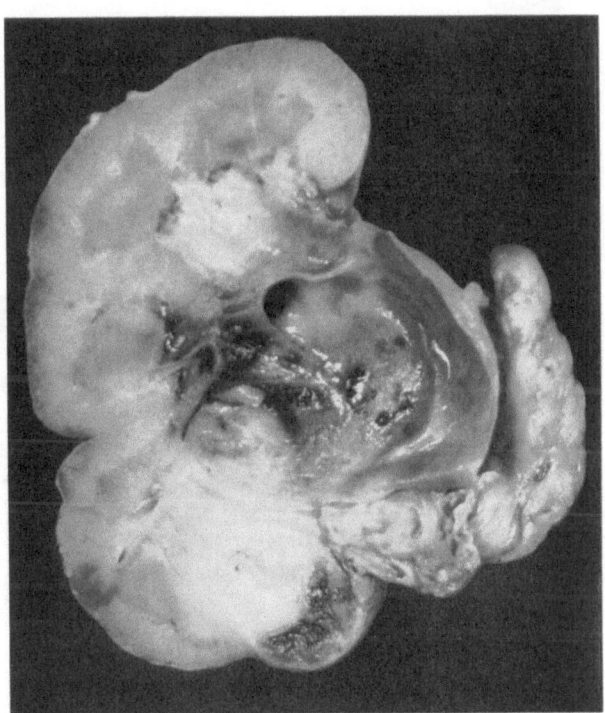

Fig. 65. Adenocarcinoma of the right kidney with extension into the vena cava, removed by incision of the vein. The tumour is in the lower pole. (Mr. I. H. GRIFFITHS case)

bed of the excised twelfth rib if the perinephric fascia is opened far back, and GRAHAM (1953) has extolled the advantages of the thoraco-abdominal approach for the same purpose and for effecting a radical nephrectomy. This approach was used by MORTENSEN (1948) for the removal of a large complicated tumour, and there is no doubt that it gives the best exposure for an extensive tumour on the left side. HIGGINS and his associates (1951) also recommend it for dealing with Wilm's tumours in children. It does however add to the severity of the operation, and our own preference in a tumour of moderate size is for an approach through the bed of the excised twelfth rib which can subsequently be extended if necessary by the addition of a lumbar limb and division of the necks of the eleventh and tenth ribs after the manner of NAGAMATSU (1950), or by an abdominal incision passing downwards and inwards towards the pubic spine. Fey's antero-lateral incision (1956) has the advantage of allowing early ligation of the pedicle before mobilisation of the kidney.

If there is palpable tumour in the renal vein or in the vena cava it is sometimes possible to remove it. One patient in whom growth was removed from the
inferior vena cava lived for four years before dying from recurrence. DONOVAN
(1955) has described a method of procedure. The renal vein is tied; blood transfusion is started in the arm; the patient is tilted head down and the vena cava
is controlled by digital compression below the renal vein; the ligature is cut
from the renal vein and the clot extracted in stages. This can be aided by suction.

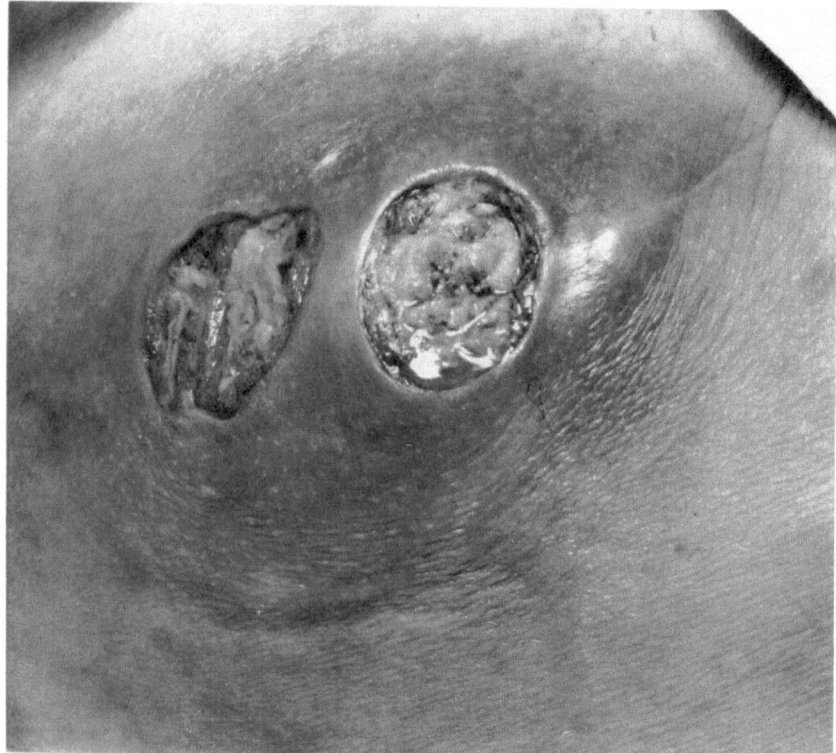

Fig. 66. Ulcerating recurrences in the scar following nephroureterectomy in the patient shown in Fig. 59

There is more danger from air embolism than from haemorrhage. This method
would not allow the removal of continuous tumour embolus in the renal vein
and vena cava in one piece. FEY, DOSSOT and QUÉNU (1956) remove the tumour
by placing a ligature loosely round the renal vein, incising the anterior surface
of the vein, extracting the tumour embolus and tying the vein at once. In some
cases however the vena cava must be opened deliberately if the whole extent
of the tumour embolus is to be removed. This was done by GRIFFITHS (1957)
in the case illustrated in Fig. 65. This patient was alive and well more than 9 years
later.

In papillary tumours of the renal pelvis or ureter nephroureterectomy is the
procedure of choice, the ureter being clamped and divided flush with the bladder
and its intramural part being treated by cystoscopic diathermy. MACALPINE
(1947) advises this in preference to total ureterectomy including a cuff of bladder
owing to the danger of spill of tumour-containing urine if the urinary tract is
opened. There is no doubt that if urine does escape from a papillomatous kidney
recurrence in the wound is almost inevitable and practically untreatable (Fig. 66).

In the cases of bilateral renal pelvic tumour reported by COLSTON and ARCADI (1954) and GIBSON (1953) successful local excision was carried out with preservation of the kidney.

The treatment of a single benign tumour of the ureter by nephroureterectomy sacrifices a kidney which may be healthy, and when the diagnosis is in little doubt it may be justifiable to excise the affected part of the ureter and replace it by an isolated loop of small intestine (PYRAH and RAPER 1954).

Metastases. The known presence of a metastasis in renal adenocarcinoma is not necessarily a bar to nephrectomy. Skeletal metastases are often solitary and if they are accessible to surgical treatment they should be excised. In BLAND-SUTTON's patient (1922) excision of a secondary growth in the humerus was followed by a survival for six years. We have followed nephrectomy for a low grade adenocarcinoma by excision of a metastasis in the seventh rib; the patient and died from cerebrovascular disease with no recurrence (Fig. 67). SMYTH (1939) removed a metastasis from the cerebellum six weeks after nephrectomy for a renal adenocarcinoma and the patient was alive and well five years later. There were seven cases in the B.A.U.S. series (1951) in which a solitary metastasis had been excised with varying degrees of success. Bold-

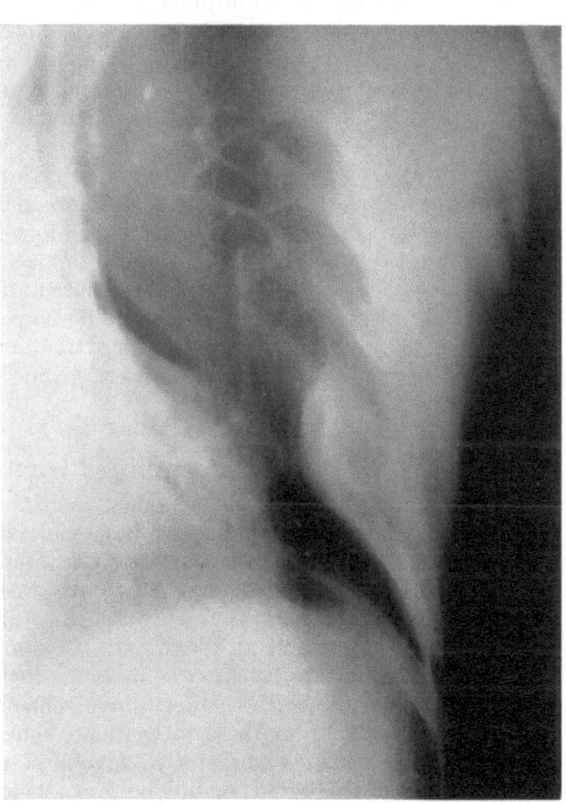

Fig. 67. Tomograph of a metastasis in the seventh left rib from an adenocarcinoma of the right kidney of low grade malignancy. A man of 65; after nephrectomy and excision of the affected rib he lived for seven and a half years

ness in treating recurrences after nephrectomy may also occasionally be well repaid. SCHOFIELD (1952) reported a case in which Wells excised a local recurrence together with one in the left half of the colon two years after nephrectomy. Eighteen months later three further intraperitoneal metastases were excised and the patient was well three months later. As a general rule however the appearance of a local recurrence is the herald of remote metastases.

Even when there are pulmonary metastases nephrectomy should be seriously considered. Spontaneous regression of these metastases following nephrectomy has been recorded in 14 cases and there are others not reported. JENKINS (1959 and 1965) described such a patient who lived for 13 years; he died eventually from local recurrence in the wound and there were viable carcinoma cells encapsulated in the lung.

Regression of pulmonary and osseous metastases has also followed hormonal treatment (BLOOM and WALLACE 1964).

II. Radiotherapeutic

The principles and application of radiotherapy in renal tumours are discussed elsewhere and it is only necessary here to mention some points of special importance to make the summary of methods of treatment complete.

Despite its value as an adjunct to surgery there are no authentic cases of a cure of an adult malignant renal tumour by radiotherapy alone. In the treatment of metastases it may be the only method applicable, and in this connection it may cause temporary disappearance of secondary deposits, useful palliation and prolongation of life. This applies especially to lymphatic metastases (RICHES 1956).

The radiosensitivity of renal tumours varies; in general nephroblastomata are most sensitive, then sarcomata, then adenocarcinomata, then papillary tumours of the pelvis and ureter. Regular, low grade tumours are less sensitive than anaplastic tumours of high grade, but as BLOMFIELD (1954) has pointed out the most radiosensitive are not the most radio-curable as they tend to recur more readily. In the B.A.U.S. series (1951) 25 per cent of tumours were deemed inoperable; 362 of them were given no treatment and the five year survival was 0.5 per cent. 83 similar cases were given X-ray treatment and their five year survival was 6 per cent.

III. Combined treatment

Radiotherapy has its greatest value when used in conjunction with surgery, and there is little doubt that supervoltage methods with their deeper penetration and absence of skin reaction have great advantages over conventional methods. In the B.A.U.S. series (1951), treated before supervoltage therapy was generally available, the added survival given by postoperative X-ray treatment was 6 per cent at 1 year, 9 per cent at 3 years, 19 per cent at 5 years, and 10 per cent at 10 years. Moreover the benefit was most marked in those cases, usually of high grade malignancy, in which there was involvement of the renal vein. Subsequent figures (RICHES 1954) for the same patients showed slightly less good statistical results but the balance was still in favour of radiotherapy.

Whether the treatment should be given before or after operation, or both is a matter for consideration in each individual patient. *Preoperative treatment* offers early control of the activity of cancer cells with a reduced risk of operative dissemination; it will generally cause a diminution in size of a tumour, particularly a nephroblastoma, and may make an adherent or inoperable tumour operable, as has happened in three of our cases. On the other hand it has some disadvantages. In full doses it may lower the general resistance of the patient, reduce the number of white cells and cause erythema and oedema of the skin, interfering with healing. It is a source of delay in the removal of the tumour, and metastases may possibly appear during the period of treatment. If it is to be given, the diagnosis of malignant renal tumour must be certain, and this implies the use of aortography in nearly every case. It should be given when there is evidence of rapid growth or of fixity of the tumour either clinically or on oxygen insufflation, or where the tumour is so large as to warrant an attempt to reduce its size before operation. If it is given we consider it should be limited to about 3000 r given in three weeks and followed by nephrectomy at least three weeks from its termination.

Riches (1966) reported the results of preoperative radiotherapy in 15 cases of renal carcinoma; 4 patients were living at 18, 14, 7 and 5 years and 2 at 1 year. Of the long term survivors 3 had tumours of high grade with vein invasion and one had extensive perinephric spread. In general the effect of preoperative therapy limited to 3,000 r was to reduce the size of the large perinephric veins normally encountered and in some cases to make the whole tumour operable. If a delay

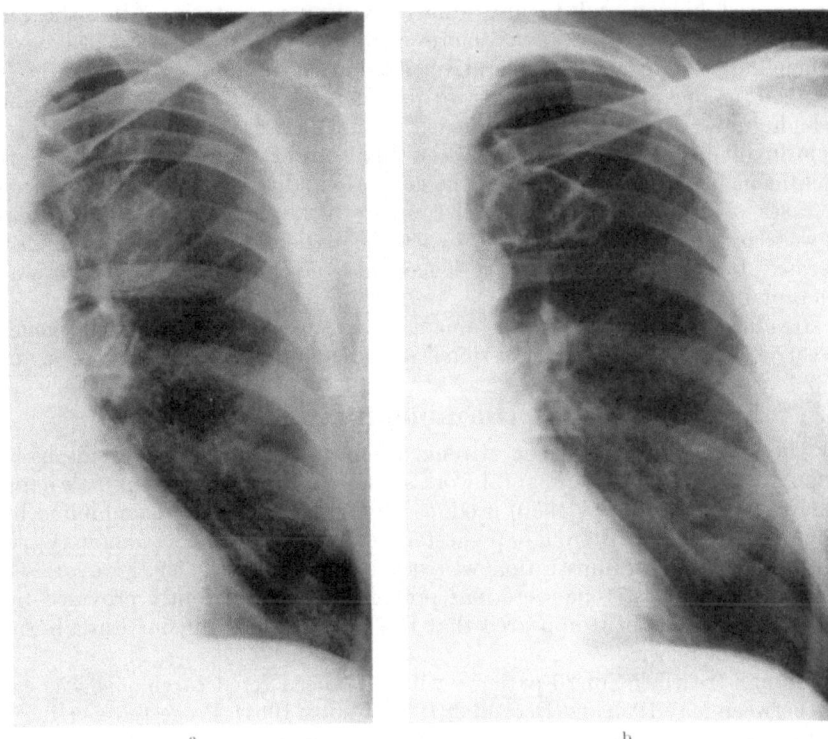

a   b

Fig. 68. From a man of 55 with a grade 1 adenocarcinoma of the right kidney and metastasis in the neck of the 6th left rib. Preoperative X-ray treatment to the kidney and to the rib followed by nephrectomy. (a) Metastasis in ribs before treatment. (b) The ribs nine months after treatment showing regression and increased calcification in the metastasis

of 3 weeks was observed between radiotherapy and surgery the operation was easier than expected; operation within a week was generally difficult owing to oedema and vascularity.

Other workers have advocated preoperative radiotherapy. In the past it was used by Waters et al. (1934), Wharton (1935), Bothe (1935) and Bixler and his colleagues (1944). More recently it has been used by Flocks and Kadesky (1958), Reboul et al. (1962) and Paces and his co-workers. Wells (1954) has also advocated the more frequent use of preoperative radiotherapy.

In view of the successes in a small number of highly malignant cases and of the increased operability rate it is now our practice to give preoperative radiotherapy to all diagnosed cases of renal adenocarcinoma. Anaemia is first corrected by blood transfusion; supervoltage therapy to a dose of about 3000 r is given in 2 to 3 weeks usually by 2 ports of entry; an interval of 3 weeks is allowed to elapse during which the patient may be sent home if convenient; radical nephrectomy is then carried out through an adequate antero-lateral incision.

If it were possible to know the histological grade of the tumour and the state of the renal vein before operation there would be further indications for preoperative treatment. If there has been no preoperative treatment postoperative treatment is more widely used. It is manifestly impossible to irradiate every site where metastasis could occur, and treatment is therefore given to the bed of the excised kidney and to any known existing metastasis. After the kidney has been removed there is a more complete picture of its pathology and although we advocate postoperative irradiation for most patients who have not had preoperative treatment we feel it is especially necessary for tumours of high grade, when the vein is invaded, when there is local extension or lymph node involvement, or if there has been any spill of tumour tissue at operation. Treatment has been witheld when the general condition of the patient was poor, in cases of gross obesity, in small tumours of low grade malignancy or where it was considered important not to let the patient suspect he had malignant disease. It has been curtailed if leucopenia developed or the patient was intolerant of it.

Irradiation of metastases not amenable to surgical removal should generally be carried out; satisfactory regression is seen in a number of cases (Fig. 68a and b).

IV. Hormone treatment

The rationale of treatment of renal tumours by hormones was discussed by Bloom et al. (1963). Experimental work showed that the tumours in male hamsters of Kirkman and Bacon (1950) produced by oestrogens could be inhibited by the administration of testosterone or progesterone. In female hamsters tumours appeared only if oestrogen administration was started after removal of the ovaries or at times of low progesterone secretion; protection was apparently provided by the natural progesterone. It appeared that the hamster renal tumour was a hormone-dependent neoplasm.

In man treatment of 20 patients with advanced renal carcinoma was carried out between May 1959 and December 1963 (Bloom 1964). Progesterone (Provera) was given orally (300—500 mg daily) for eight to 12 weeks; if the response was not good testosterone propionate was substituted. It was given by intramuscular injection (100 mg daily) followed by methyl testosterone orally (50—100 mg daily). Prednisone (10 mg thrice daily) was added for its general inhibitory action.

10 of the 20 patients showed some subjective amelioration but in four there was objective improvement evidenced by regression or disappearance of pulmonary or osseous metastases. The method is being continued and after 40 patients had been treated objective improvement occurred in eight (20 per cent). (Bloom, personal communication 1966). In four of my patients regression of pulmonary metastases has occurred in only one case and the treatment is still under trial.

V. Chemotherapy

The results of chemotherapy in adenocarcinoma of the kidney are disappointing. Many substances have been tried including alkylating agents (e.g. Chlorambucil) antimetabolites (e.g. 5-fluouracil), purine antagonists (e.g. 6 mercapto-purine, Methotrexate) and alkaloids (e.g. vinblastine). Whether given by intravenous injection or by intra-renal perfusion through a catheter passed into the renal artery no worthwhile improvement has been obtained. Intracavitary injection of Thiotepa in a case of metastasis in the lung with pleural metastases offers some hope of temporary amelioration.

There are many adverse effects from chemotherapy; leucopenia and thrombocytopenia are amongst the more serious. In a patient given intravenous Thiotepa peoperatively the leucocytes fell to 2500 per ml and the platelets to less than 90,000; operation was thereby delayed and metastases in the lung developed during treatment.

The antibiotic Actinomycin D which appears to be effective in treating nephroblastoma is ineffective in adenocarcinoma.

There is more promise in the other directions. NAIRN and his associates (1963) have produced a precipitin against renal cancer.

Treatment of nephroblastoma. The controversy regarding the use of X-ray therapy reaches its height over the treatment of WILMS' tumour. It is now generally agreed that irradiation alone does not cure the condition, viable tumour cells being always found in kidneys so treated. Preoperative irradiation is also somewhat out of favour except as a means of reduction in size of very large tumours. LADD and WHITE (1941) considered that it even encouraged metastasis not only because it delayed nephrectomy but also because of its softening effect on the tumour. Moreover the intensive irradiation formerly used can produce radiation nephritis in the other kidney and asymmetrical growth disturbance of the lumbar vertebrae leading to permanent scoliosis (WHITEHOUSE and LAMPE 1953). GROSS (1953) for a time treated alternate cases with preoperative irradiation but found that the mortality rate "skyrocketed" in those so treated and he gave it up within a year.

There is general agreement that nephrectomy should be performed at the earliest possible moment after the diagnosis has been made. If retrograde pyelography is required to confirm the diagnosis it should be done immediately before operation and under the same anaesthetic. LADD and WHITE (1941) and GROSS and NEUHAUSER (1950) are strongly in favour of an anterior transperitoneal approach and attribute the improvement in their results largely to its use. INNES WILLIAMS (1964) favours a long transverse incision above the umbilicus through which both kidneys can be inspected. The incision may be vertical or transverse and we have found a transverse limb added to the vertical incision an advantage in removing a very large tumour. The transthoracic route is in increasing favour (O'CONOR and HEAD 1951, HIGGINS et al 1951), but most tumours of average size can be removed through the bed of the excised twelfth rib or by the NAGAMUTSU incision.

Postoperative irradiation is advised by most urologists, and it should start immediately after operation; GROSS and NEUHAUSER found their 2 year survival rate improved from 32 per cent to 47 per cent during the period of its use. BROWN (1951), supported by McWhirter, also favours postoperative treatment to the whole abdomen as giving a lessened operative risk than preoperative treatment and WHITE (1951) also advised its use. It is noteworthy however that his two longest survivors (23 and 14 years) and my own longest, now 32 years, had no X-ray therapy.

Both preoperative and postoperative irradiation have been given extensively and ABESHOUSE (1957), as the result of a questionnaire answered by 81 surgeons, found this to be the most widely used plan in the last 20 years. HARVEY (1950) suggested that it gave the best survival rate. It possesses all the disadvantages described for preoperative irradiation but the advantage of tumour shrinkage on increased operability which may be of benefit to the surgeon who treats few cases. L. S. SCOTT (1956) reviewing 1,141 treated cases culled from the literature found that it gave the highest 2 year survival rate of any method. If preoperative

therapy is used it should be limited to about 1500 r given over a period of 7 to
10 days and followed by immediate nephrectomy. Close watch should be made
on the leucocyte and platelet count during treatment.

Our own view is against preoperative irradiation in WILMS' tumour owing
to its debilitating effect. We believe that early nephrectomy followed by controlled postoperative treatment is likely to give the best results. Metastases are
treated by more radiotherapy but although they may disappear for a time they
become radio-resistant and always recur. We have seen no permanent recovery
when once there has been a recurrence. Tumours which are very advanced when
first seen are treated by radiotherapy and life may be prolonged for some years.

In contrast with its effects in adenocarcinoma in adults Actinomycin D is considered to serve a useful purpose in nephroblastoma (FARBER et al. 1960) by poten-
tiating the effects of X-rays.

I. Prognosis

I. Adenocarcinoma

The average survival rate from the time of treatment has been estimated
in the various series reported (B.A.U.S. 1951. GRIFFITHS. and THACKRAY 1949,
PRIESTLEY 1941) and the figures show an approximate survival in all cases of
80 per cent at one year, 50 per cent at 3 years, 40 per cent at 5 years and 20 per

Table 2. *Survival by Grade.*

	Total			Low grade			High grade		
	Possible	Survived	%	Possible	Survived	%	Possible	Survived	%
1 Year	110	81	74	50	45	90	60	36	60
3 Years	102	63	62	46	38	83	56	25	45
5 Years	86	42	49	42	30	71	42	12	29
10 Years	64	18	28	30	12	40	34	6	18

cent at 10 years. DEMING (1946) recorded a survival of 9 per cent for more than
10 years. Whilst each case is a law unto itself it is possible from a study of all
the known factors after operation to gain an approximate idea of the probable
outlook for any individual patient. The factors in prognosis have been analysed
(RICHES 1958). The *histological grade* of the tumour, which is based solely on
the microscopic appearance (see p. 30), appears to be of the greatest importance.
The survival of patients with tumours of grades 2 and 3, the anaplastic growths,
shows relatively little difference and they can be considered together as tumours
of high grade; those of grade 1, the well differentiated tumours, are of low grade.
In 110 patients treated by nephrectomy with or without radiotherapy the total
5 year survivals were 49 per cent; those of low grade reached 71 per cent and
those high grade only 29 per cent, the rising scale of malignancy being reflected
in a decreasing survival rate (Table 2, RICHES 1963). *Invasion of the renal vein*
also affected the outcome adversely; it was present in 31 per cent of cases
and only one quarter of them lived for 5 years, whilst 76 per cent died ultimately
from metastases. In the B.A.U.S. series (1951), of the patients treated by
nephrectomy alone 35 per cent with no venous involvement lived for 5 years
but only 16 per cent when the vein was invaded. The figures were improved in
the patients who had X-ray treatment in addition to nephrectomy. Taking
the two factors of grade and invasion of the vein together THACKRAY (1957) gave
the following survival figures to show the two extremes of prognosis (Table 3).

Involvement of lymph nodes has not the same immediate lethal effect as there is a temporary response to irradiation therapy; two patients lived for $7^1/_2$ and 6 years respectively despite massive lymphatic metastases at the time of operation. *Local extension*, which occurs more often in the higher grades of tumour, is also best combated by X-ray treatment, preferably given before operation and 5 out of 9 patients with such extension lived for 5 years. When *metastases* were present at the time of operation there was one 5 year survivor, who lived for seven and a half years after excision of a metastasis in the ribs; he had polycythaemia. The beneficial effect of *X-ray treatment* in conjunction with surgery has been mentioned (p. 63).

Early diagnosis and treatment is undoubtedly an advantage, and 5 out of 7 patients in whom the growth was an incidental finding during nephrectomy for some other condition lived for 5 years or more. It cannot always overcrome the influence of a high grade of malignancy even if the tumour is very small. Pyrexia and loss of weight are adverse factors whilst peripheral calcification may confine the tumour and prevent early spread.

Table 3. *Percentage surviving*

	Years since nephrectomy			
	One	Three	Five	Ten
Grade 1, Vein free . . .	100	87	75	50
Grade 3, Vein invaded .	17	17	0	0

Metastases not suitable for excision are treated by X-ray therapy; some palliation may be expected but further recurrence is the rule. Out of 83 inoperable cases treated by X-ray therapy 3 survived for 5 years; of 362 similar cases untreated there was one 5 year survivor (B.A.U.S. series 1951).

In its natural history carcinoma of the kidney follows two distinct courses. Patients with low grade tumours and no adverse factors may live for many years; those with high grade growths are likely to die early. This duplicity is characteristic of the disease and explains its so-called capricious behavious. The prognosis is more nearly predictable only when all the pathological features are known after nephrectomy but even then there is no certainty.

II. Pelvic and ureteric tumours

Papillary tumours in general have less intrinsic malignancy than solid growths, which are always infiltrating. Transitional cell carcinoma of the renal pelvis shows variations in the histological picture which justify grading; McDONALD and PRIESTLEY (1944), adopting BRODERS (1940) grading, found a five year survival of 63 per cent in grades 1 and 2 and only 13 per cent in grades 3 and 4. THACKRAY (B.A.U.S. series 1951) found a higher proportion of survivors with low grade tumours than with high. In tumours of the renal pelvis however there are other factors which affect the outcome, namely involvement of the ureter and bladder, and the effects of obstruction. Ureter and bladder involvement was present in 43 per cent of 147 cases of transitional cell carcinoma (B.A.U.S series 1951). The five year survival with involvement was 33 per cent, and without involvement 38 per cent, and this feature is of less importance than a high grade of tumour in its influence on prognosis. McDONALD and PRIESTLEY (1944) found such involvement more often in the low grade papillary tumours than in the high grade infiltrating type. They also found hydronephrosis in 71 per cent of their cases and invasion of the renal vein or the perineural lymphatics or both in 45 per cent. The latter had the effect of reducing the 5 year survival rate from 47 per cent to 4 per cent.

Simple papilloma of the renal pelvis gave a 50 per cent five year survival in the B.A.U.S. series (1951).

Solid transitional cell carcinoma has a much worse prognosis than the papillary type and only four of 15 cases we have seen have lived for one year. Swift-Joly (1933) collected records of 29 such tumours and stated that a cure had only rarely been obtained.

Squamous cell carcinoma of the pelvis is also extremely lethal; there were only 6 out of 32 three year survivors in the B.A.U.S. series (1951) and none lived for 5 years after nephroureterectomy. X-ray therapy for inoperable cases gave no three year survivors.

Carcinoma of the ureter has a relatively poor prognosis largely owing to delay in diagnosis. Even in the papillary type survival beyond 5 years is unusual but Baron and Green (1954) report three living for 7, 11, and 17 years respectively. Awareness of the condition and the more extensive use of full investigation should improve the outlook.

III. Nephroblastoma

Although this is recognised universally as a most serious and lethal condition in children there are grounds for hope that the situation is improving with earlier recognition and combined treatment. In assessing prognosis from published statistics there are some difficulties in correlation of the standard of cure. Ladd and White (1941) stated that if no recurrence appeared within four months the outlook was hopeful; the longest period before recurrence in their series was twenty months. Gross and Neuhauser (1950) from the same school considered survival for eighteen months as a probable cure but adopted the period of two years freedom from recurrence as a standard in their published figures, and this standard was also used by L. S. Scott (1956). Despite this Gross and Neuhauser (1950) reported metastases as late as five years after operation and Falkinburg, Kay and Sayer (1954) have described a retroperitoneal recurrence eight years after nephrectomy. It is also relevant that in the figures collected by Abeshouse (1957) from six different sources the 3, 5 and 10 year survivals nearly all show a progressive diminution whatever the method of treatment; this might be expected in adults but not to the same extent in children. Nevertheless the two year standard has been generally adopted as being approximately correct.

Age. It was pointed out by the Boston workers that the prognosis was better in children under one year old than in those over this age. The figures of Gross and Neuhauser (1950) gave the first indication that the prognosis was not so hopeless as had been thought. They are quoted in full (Table 4).

In period A nephrectomy alone was used. In period B transperitoneal nephrectomy was the rule and anaesthesia and pre- and postoperative care had improved. In period C postoperative X-ray treatment was added. The influence of age at the time of treatment on prognosis is confirmed by the figures collected by Harvey (1950) and by Scott (1956).

Delay in treatment, and repeated examinations are an adverse factor, as is a long duration of symptoms before advice is sought.

Haematuria usually indicates invasion of the pelvis and makes the prognosis worse. Williams however (1964) believes that it can be due to rupture of obstructed vessels around the calyces. It can thus draw attention to the condition and lead to earlier treatment and better prognosis.

The presence of *metastases* or extrarenal extension of the growth at the time of operation makes the possibility of cure remote.

Neither *sex* not the *side* of the tumour appear to affect prognosis; bilateral tumours are likely to be uniformly fatal although treatment should not be witheld and RICKHAM'S (1957) case is notable.

The different tissue types described by BODIAN and WHITE (1951, 1955) appear to have a bearing on the prognosis.

The operative mortality has decreased with increasing care, better anaesthesia and improved technique. GROSS and NEUHAUSER (1950) had no operative deaths in 69 cases between 1931 and 1947, and ABESHOUSE (1957) reported a mortality of only 2.7 per cent in 777 operations.

The overall 3 year survival rate in the B.A.U.S. (1951) series was 26 per cent (21/81); HARVEY (1950) in 472 cases reported an eighteen month survival of 22 per cent (102/472); of the patients under one year 41.5 per cent survived (27/65) whilst of those over one year the figure was only 18.4 per cent (75/407). The best figures reported are those in the Boston series (LADD, GROSS and NEUHAUSER).

Table 4

GROSS and NEUHAUSER

	Below 1 yr.	Above 1 yr.	All ages
A. 1914—1930 27 cases, 4 cures	42.8%	5%	14.9%
B. 1931—1939 31 cases, 10 cures	71.4%	20.8%	32.2%
C. 1940—1947 38 cases, 18 cures	80%	43.3%	47.3%

The histological grading of these tumours and the preparation of the microphotographs have been done by Dr. A. C. THACKRAY and I owe him much for his help. The photographs are the work of Mr. M. TURNEY, senior Photographer to the Middlesex Hospital.

Figs. 5, 11, 16, 24, 33—39, 46, 50, 52, 53, 60 and 67 have appeared in the British Journal of Urology (E. S. Livingstone Ltd.); Figs. 12, 20—23, 25, 31, 51 in Modern Trends in Urology or British Surgical Practice (Butterworth & Co., Ltd.); Figs. 17, 28—30 in the British Journal of Surgery (John Wright & Sons, Ltd.); Figs. 40 and 56 in the Proceedings of the Royal Society of Medicine. I am grateful to the Editors and Publishers for permission to reproduce them.

References

ABESHOUSE, B. S.: The management of Wilms' tumour as determined by national survey and review of the literature. J. Urol. (Baltimore) 77, 792 (1957). — ABESHOUSE, B. S., and T. WEINBERG: Malignant renal neoplasms. Arch. Surg. (Chicago) 50, 46 (1945). — ADAMS, A. W.: A review of nephroma and ten nephrectomies in 1949. Brit. J. Surg. 38, 210 (1950). — ALBRECHT, E.: Über Hamartome. Verh. dtsch. path. Ges. 7, 153 (1904). — APITZ, K.: Die Geschwülste und Gewebsmißbildungen der Nierenrinde. Virchows Arch. path. Anat. 311, 285 (1943).

BALDWIN, J. F.: Dermoids of the kidney. Surg. Gynec. Obstet. 20, 219 (1915). — BANDLER, C. G., and P. R. ROEN: Solitary testicular metastasis simulating primary tumour and antedating clinical hypernephroma of the kidney; report of a case. J. Urol. (Baltimore) 55, 663 (1946). — BARNARD, H. F.: Polycythaemia and renal carcinoma. Brit. med. J. 1, 1214 (1961). — BARON, A., and J. A. S. GREEN: Primary carcinoma of the ureter. Brit. J. Surg. 41, 576 (1953). — BARRETT, W. A., and E. J. McCAGUE: Histopathology in prognosis of kidney tumors. J. Urol. (Baltimore) 71, 684 (1954). — BARRIE, H. J.: Paracalyceal cysts of the renal sinus. Amer. J. Path. 29, 985 (1953). — BEADLES, R. O., and R. W. URICH: Intrarenal lipoma; report of a case. J. Urol. (Baltimore) 67, 460 (1952). — BELL, E. T.: Classification of renal tumours. J. Urol. (Baltimore) 39, 238 (1938). — Renal diseases, 2nd ed., p. 424. London: Henry Kimpton 1950. — BENNETT, W. E. J.: A further case of hypernephroma metastasising to the thyroid. Brit. J. Urol. 24, 116 (1952). — BERGER, L., and M. W. SINKOFF: Systemic manifestations of hypernephroma. Amer. J. Med. 22, 791

(1957). — Bland-Sutton, J.: Tumours innocent and malignant; their clinical characters and appropriate treatment, 7th ed. p. 428. London: Cassell 1922. — Blomfield, G. W.: Irradiation therapy in urology. Brit. J. Urol. **26**, 301 (1954). — Bonser, G. M.: Epithelial tumours of the bladder in dogs induced by pure β-naphthyl-amine. J. Path. Bact. **55**, 1 (1943). — Boross, E.: Fibroma of renal pelvis. Orv. Hetil. 1055, (1929). — Braasch, W. F., and J. A. Hendrick: Renal cysts, simple and otherwise. J. Urol. ,Baltimore) **51**, 1 (1944). — Bradley, J. E., and M. C. Pincoffs: Association of adeno-myosarcoma of the kidney (Wilms' tumor) with arterial hypertension. Ann. intern. Med. **11**, 1613 (1938). — Broders A. C.: The microscopic grading of cancer. In Pack, G. T., and E. M. Livingstone. The treatment of cancer and allied diseases, p. 19. New York: Paul B. Hoeber 1940. — Brown, J. J. M.: Nephroblastoma. Brit. J. Urol. **23**, 358 (1951). — Bruce, J., and G. H. D. McNaught: Leiomyosarcoma of the kidney. Brit. J. Urol. **25**, 114 (1954). — Buckley, W.: Malignant transformation in a previously benign tubular adenoma of the kidney. Brit. J. Surg. **32**, 315 (1944).

Campbell, M.: Urology, p. 260. Philadelphia and London: W.-B. Saunders Company 1954. — Clinical pediatric urol. Philadelphia: W. B. Saunders Company 1951. — Carver, J. H.: Renal adenoma. Brit. J. Urol. **7**, 229 (1935). — Cystic disease of the right kidney in a infant. Brit. J. Urol. **21**, 228 (1949). — Childs, P., and W. B. Waterfall: Renal adenoma. A review with a report of two further cases. Brit. J. Urol. **25**, 187 (1953). — Chynn, K. U., and J. A. Evans: Nephrotomography in the differentiation of renal cyst from neoplasm: a review of 500 cases. J. Urol. (Baltimore) **83**, 21 (1960). — Clay, J.: Mixed tumour of the kidney in a patient aged 80. Brit. Med. J. **2**, 1083 (1930). — Clinton-Thomas, C. L.: A giant leiomyoma of the kidney. Brit. J. Surg. **43**, 497 (956). — Clinton-Thomas, C. L., and T. M. Robinson: Adenocarcinoma of the kidney in childhood; report of a case and a review of the literature. Brit. J. Urol. **28**, 132 (1956). — Colston, J. A. C., and J. A. Arcadi: Bilateral renal papillomas; transpelvic electro resection with preservation of kidney, contralateral nephrectomy; four-year survival. Trans. Amer. Ass. gen.-urin. Surg. **46**, 134 (1954). — Colvin jr., S. H.: Certain capsular and subcapsular mixed tumours of the kidney herein called "Capsuloma". J. Urol. (Baltimore) **48**, 585 (1942). — Conley, C. L., J. Kowal and J. d'Antonio: Polycythaemia associated with renal tumours. Bull. Johns Hopk. Hosp. **101**, 63 (1957). — Constance, T. J.: Bilateral rhabdomyoma of the kidney. J. Path. Bact. **59**, 492 (1947). — Creevy, C. D.: Confusing clinical manifestations of malignant renal neoplasms. Arch. intern. Med. **55**, 895 (1935). — Cristol, D. S., A. E. Bothe and P. J. Grotzinger: Renal adenoma; survey of reported clinical cases and another case report. J. Urol. (Baltimore) **64**, 58 (1950). — Culp, O. S., and F. W. Hartman: Mesoblastic nephroma in adults. A clinico-pathological study of Wilms' tumors and related renal neoplasms. J. Urol. (Baltimore) **60**, 552 (1948).

Damon, A., D. A. Holub, M. M. Melicow and A. C. Uson: Polycythaemia and renal carcinoma. Amer. J. Med. **25**, 182 (1958). — Daniel, W. E.: The hypertensive factor in Wilms' tumor. Sth. med. J. (Bgham, Ala). **32**, 1014 (1939). — Deming, C. L.: Renal tumors. J. Urol. (Baltimore) **55**, 571 (1946). — Deuticke, P.: Nierentumoren. Dtsch. Z. Chir. **231**, 767 (1931). — Donovan, H.: Nephrectomy for renal cancer; a posterior approach. With a note on the removal of growth from the vena cava. Brit. J. Urol. **27**, 371 (1955). — Dos Santos, R., C. Lamas and C. J. Pereira: Arteriografia dos membros. Med. Contemp. **47**, 93 (1929). — Doss, A. K., H. C. Thomas and T. B. Bond: Renal arteriography, its clinical value. Tex. St. J. Med. **38**, 277 (1942). — Dukes, C. E.: The pathology of essential haematuria and of renal tuberculosis. Trans. med. Soc. Lond. **65**, 391 (1948).

Eberth, C. J.: Myoma sarcomatodes renum. Vischows Arch. path. Anat. **55**, 518 (1872). — Esersky, G. L., S. H. Saffer, C.E. Panoff and J. Mendel: Wilms' tumour in the adult; review of literature and report of three additional cases. J. Urol. (Baltimore) **58**, 397 (1947). — Evans, J. A., W. Dubilier jr., and J. C. Monteith: Nephrotomography: A preliminary report. Amer. J. Roentgenol. **71**, 213 (1954). — Evans, J. A., J. C. Monteith and W. Dubilier jr.: Nephrotomography. Radiology **64**, 655 (1955). — Ewing, J.: Neoplastic disease, 4th ed., p. 799. Philadelphia and London: W. B. Saunders Company 1940.

Falkinburg, L. W., M. N. Kay and E. A. Sayer: Recurrence of nephroblastoma (Wilms' tumor) eight years after nephrectomy. J. Amer. med. Ass. **155**, 1228 (1954). — Farinas, P. L.: Retrograde abdominal aortography. Amer. J. Roentgenol. **46**, 641 (1941). — Fergusson, J. D.: Observations on familial polycystic disease of the kidney. Proc. roy. Soc. Med. **42**, 806 (1949). — Fey, B., R. Dossot and L. Quénu: Traité de technique chirurgicale, Tome VIII: Appareil urinaire et appareil genital de l'homme, 2nd ed., p. 48, p. 97 Paris: Masson & Cie. 1956. — Fish, G. W., and W. L. McLaughlin: Liposarcoma of kidney; report of a case presenting an unusual syndrome. J. Urol. (Baltimore) **55**, 28 (1946). — Fite, G. L.: Classification of tumours of the kidney. Arch. Path. (Chicago) **39**, 37 (1945). — Foot, N. C., G. A. Humphreys and W. F. Whitmore: Renal tumors; pathology and prognosis in 295 cases. J. Urol. (Baltimore) **66**, 190 (1951). — FRAZIER, T. H.: Multilocular cysts of the kidney. J. Urol. (Baltimore) **65**, 351 (1951).

GAHAGAN, H. Q., and W. K. REED: Squamous cell carcinoma of the renal pelvis; three case reports and review of the literature. J. Urol. (Baltimore) **63**, 139 (1949). — GALBRAITH, W. W.: Pedunculated vascular tumour of the ureter. Brit. J. Urol. **22**, 195 (1950). — GIBSON, T. E.: Lymphosarcoma of the kidney. J. Urol. (Baltimore) **60**, 838 (1948). — Operation on neoplasm of solitary kidney. Urologists Correspondence Club. March 17th 1953. — GILBERT, J. B.: Diagnosis and treatment of malignant renal tumours; historical data. J. Urol. (Baltimore) **39**, 223 (1938). — GLASER, S.: Simple renal cysts. Brit. J. Surg. **40**, 74 (1952). — GORDON-TAYLOR, G.: Gigantic benign tumour of kidney weighing 22 pounds. Nephrectomy; cure. Brit. J. Surg. **17**, 551 (1930). — GRAHAM, W. H.: The operative approach to the kidney. Modern trends in urology edit. E. W. Riches, p. 70. London: Butterworth & Co. 1953. — GRAWITZ, P.: Die sogenannten Lipome der Niere. Virchows Arch. path. Anat. **93**, 39 (1883). — GRIFFITHS, I. H.: A preliminary report on abdominal aortography in urology. Brit. J. Urol. **22**, 281 (1950). — Personal communication 1957. — GRIFFITHS, I. H., and A. C. THACKRAY: Parenchymal carcinoma of the kidney. Brit. J. Urol. **21**, 128 (1949). — GROSS, R. E.: The surgery of infancy and childhood, p. 588. Philadelphia: W. B. Saunders Company 1953. — GROSS, R. E., and E. B. D. NEUHAUSER: Treatment of mixed tumors of the kidney in childhood. Pediatrics **6**, 843 (1950). — GURNEY, C. W.: Erythremia in renal disease. Trans. Ass. Amer. Phycns **73**, 103 (1960).

HAMER, H. G., and W. N. WISHARD jr.: Osteogenic sarcoma involving the right kidney. J. Urol. (Baltimore) **60**, 10 (1948). — HAND, J. R., and A. C. BRODERS: Carcinoma of the kidney. J. Urol. (Baltimore) **28**, 199 (1932). — HARRISON, J. H., T. W. BOTSFORD and M. R. TUCKER: The use of the smear of the urinary sediment in the diagnosis and management of neoplasm of the kidney and bladder. Surg. Gynaec. Obstet. **92**, 129 (1951). — HARVEY, N. A.: Kidney tumours. J. Urol. (Baltimore) **57**, 669 (1947). — HARVEY, R. M.: Wilms' tumour; evaluation of treatment methods. Radiology **54**, 689 (1950). — HENDERSON, D. ST. C. L.: A case of persistent priapism secondary to a transitional cell carcinoma of the left-kidney. Brit. J. Urol. **22**, 223 (1950). — HENTHORNE, J. C.: Peripelvic lymphatic cyst of the kidney. Amer. J. clin. Path. 8, 28 (1938). — HERBUT, P. A.: Urological pathology, Vol. I., p. 603, 618. London: Henry Kimpton 1952. — HEWLETT, J. S., G. C. HOFFMANN, D. A. SENHAUSER and J. C. BATTLE: Hypernephroma with erythrocythaemia; report of a case and assay of the tumour for an erythropoietic-stimulating substance. New Engl. J. Med. **262**, 1058 (1960). — HICKS, W. K.: Benign tubular adenoma with malignant transformation. J. Urol. (Baltimore) **71**, 162 (1954). — HIGBEE, D. R., and D. M. ATKINS: Leiomyosarcoma in a double kidney. J. Urol. (Baltimore) **71**, 166 (1954). — HIGGINS, C. C.: Tumors of the renal pelvis. Review of fortyseven cases. Ann. Surg. **137**, 195 (1953). — HIGGINS, T. T., D. I. WILLIAMS and D. F. E. NASH: The urology of childhood, p. 194, 209. London: Butterworth & Co. 1951. — HILL, R. M.: Embryoma of the kidney in the adult. Brit. J. Urol. **18**, 53 (1946). — HILLAS SMITH and E. RICHES: Haemoglobin values in renal carcinoma. Lancet **1963**I, 1017. — HOVENANIAN, M. S.: Implantation of renal parenchymal carcinoma. J. Urol. (Baltimore) **64**, 188 (1950). — HOWARTH, V. S.: Renal hydatid disease. Brit. J. Surg. **38**, 350 (1950). — HOWELL, R. D.: Ureteral implantation of renal adenocarcinoma. J. Urol. (Baltimore) **66**, 561 (1951). — HOWES, W. E.: Kidney tumours. Radiology **42**, 319 (1944). — HÜSCH, P.: Über einen Fall von Myxom des Nierenbeckens. Z. Urol. **42**, 286 (1949). — HYMAN, A.: Clinical and surgical aspects of renal neoplasms. Surg. Gynec. Obstet. **41**, 298 (1929).

IMMERGUT, S., and Z. R. COTTLER: Intrapelvic fibroma. J. Urol. (Baltimore) **66**, 673 (1951). — ISRAEL, J.: Über einige neue Erfahrungen auf dem Gebiete der Nierenchirurgie. Dtsch. med. Wschr. **22**, 345 (1896).

JEPPESEN, F. B.: Lymphangioma of the ureter. J. Urol. (Baltimore) **70**, 410 (1953). — JOHNSON, W. F.: Carcinoma in a polycystic kidney. J. Urol. (Baltimore) **69**, 10 (1953).

KERR, J. A.: Gastric and renal leiomyosarcoma. Brit. J. Surg. **41**, 478 (1954). — KIRKMAN, H., and R. L. BACON: Malignant renal tumours in male hamsters (Cricetus auratus), treated with estrogen. Cancer Res. **10**, 122 (1950). — KLINGER, M. E.: Secondary tumours of the genito-urinary tract. J. Urol. (Baltimore) **65**, 144 (1951). — KRETSCHMER, H. L., and C. DOEHRING: Adenomas of the kidney. Surg. Gynec. Obstet. **48**, 629 (1929). — KÜSTER, E.: Die Chirurgie der Nieren. In Deutsche Chirurgie, S. 511. Stuttgart 1902.

LADD, W. E., and R. R. WHITE: Embryoma of the kidney (Wilms' tumor). J. Amer. med. Ass. **117**, 1858 (1941). — LANGLEY, G. F.: Pararenal teratoma in a boy aged nine years. Brit. J. Urol. **22**, 217 (1950). — LAWRENCE, J. H.: Polycythaemia: Physiology, diagnosis and treatment. New York: Grune & Stratton 1955. — LAZARUS, J. A., and F. FRIEDMANN: Leiomyosarcoma of the kidney. Amer. J. Surg. **87**, 251 (1954). — LEARY, T.: Crystalline estercholesterol and adult cortical renal tumors. Arch. Path. (Chicago) **50**, 151 (1950). —

Le Brun, H. T., H. S. Kellett and C. L. O. Macalister: Renal hamartoma. Brit. J. Urol. 27, 394 (1955). — Levi, W. M., and B. E. Ferrara: Spontaneous rupture of renal neoplasia: review and report of a case of ruptured renal cell carcinoma. Amer. Surg. 25, 257 (1959). — Lindblom, K.: Percutaneous puncture of renal cysts and tumours. Acta radiol. (Stockh.) 27, 66 (1946). — Lindblom, K., and S. I. Seldinger: Renal angiography as compared with renal puncture in the diagnosis of cysts and tumours. Trans. 10. Congr. Internat. Soc. Urol., Athens, 1, p. 331, 1955. — Lindgren, E.: Technique of abdominal aortography. Acta radiol. (Stockh.) 39, 205 (1953). — Lubarsch, O.: Die meist destruierend wachsenden und meist gewebsabweichenden Gewächse von bald ortsgleicher, bald ortsfremder Beschaffenheit. In F. Henke und O. Lubarsch (eds.), Handbuch der speziellen pathologischen Anatomie und Histologie, Bd. VI/1, S. 607. Berlin: Springer 1925. — Lucké, B.: Carcinoma in the leopard frog; its probable causation by a virus. J. exp. Med. 68, 457 (1938). — Lucké, B., and M. D. Schlumberger: Tumors of the kidney, renal pelvis and ureter. Washington, D. C.: Armed Forces Institute of Pathology 1957.

Macalpine, J. B.: Papillomatous disease of the renal pelvis. Brit. J. Surg. 35, 113 (1947). — Papilloma of the renal pelvis in dye workers; two cases, one of which shows bilateral growths. Brit. J. Surg. 35, 137 (1947). — Implantation of secondaries from a renal carcinoma ("Hypernephroma") within the ureteric lumen. Brit. J. Surg. 36, 164 (1948). — Tumours of the vesical end of the ureter. (Discussion.) Brit. J. Urol. 22, 330 (1950). — Masina, F.: Tumours of the vesical end of the ureter. Brit. J. Urol. 22, 320 (1950). — McDonald, D. F., and R. R. Lund: The role of the urine in vesical neoplasm. I. Experimental confirmation of the urogenous theory of pathogenesis. J. Urol. (Baltimore) 71, 560 (1954). — McDonald, J. R., and J. T. Priestley: Malignant tumours of the kidney. Surgical and prognostic significance of tumor thrombosis of the renal veins. Surg. Gynec. Obstet., 77 295 (1943). — Carcinoma of renal pelvis. Histopathologic study of seventy-five cases with special reference to prognosis. J. Urol. (Baltimore) 51, 245 (1944). — Melicow, M. M.: Classification of renal neoplasms. J. Urol. (Baltimore) 51, 333 (1944). — Messinger, W. J., and W. D. Jarman: Rhabdomyosarcoma of the kidney. Surgery 2, 26 (1937). — Miller, J. B., and J. J. Kaufman: Spontaneous rupture of the kidney by tumour. Brit. J. Urol. 35, 137 (1963). — Mintz, E. R.: Sarcoma of the kidney in adults. Ann. Surg. 105, 521 (1937). — Mogg, R. A.: Rare renal tumours with special reference to those occurring in children. Brit. J. Urol. 29, 287 (1957). — Moolten, S. E.: Hamartial nature of the tuberous sclerosis complex and its bearing on the tumor problem. Report of a case with tumor anomaly of the kidney and adenoma sebaceum. Arch. intern. Med. 69, 589 (1942). — Moore, G. E., and W. W. Walker: Metastatic hypernephroma of the thyroid gland with subtotal thyroidectomy, nephrectomy and resection of pulmonary metastases. Surgery 27, 929 (1950). — Morgan, G. S., J. V. Straumfjord and E. J. Hall: Angiomyolipoma of the kidney. J. Urol. (Baltimore) 65, 525 (1951). — Mortensen, H.: Transthoracic nephrectomy. J. Urol. (Baltimore) 60, 855 (1948). — Mortensen, H., and L. Murphy: Primary epithelial tumours of the ureter. A report of six cases and a review of the recent literature. Brit. J. Urol. 22, 103 (1950). — Muir, E. G., and A. J. B. Goldsmith: The prognosis of malignant renal tumours. Proc. roy. Soc. Med. 28, 31 (1935). — Murray, R. S., and G. C. Tresidder: Renal angiography. Brit. med. Bull. 13, 61 (1957).

Nagamatsu, G.: Dorso-lumbar approach to the kidney and adrenal with osteoplastic flap. J. Urol. (Baltimore) 63, 569 (1950). — Neibling, H. A., and W. Walters: Adenocarcinoma and tuberculosis in the same kidney; review of the literature and report of seven cases. J. Urol. (Baltimore) 59, 1022 (1948). — Nelson, O. A.: Arteriography of abdominal organs by aortic injection. Surg. Gynec. Obstet. 74, 655 (1942). — Newcomb, W. D.: The search for truth, with special reference to the frequency of gastric ulcer cancer and the origin of Grawitz tumours of the kidney. Proc. roy. Soc. Med. 30, 113 (1937). — Newman, B., and T. Reed: Liposarcoma of the kidney. J. Urol. (Baltimore) 62, 292 (1949). — Nicholson, G. W.: An embryonic tumour of the kidney in a foetus. J. Path. Bact. 34, 711 (1931). — Nightingale, H. J., and S. N. Lytle: Fibroma of the kidney with cyst. Brit. J. Surg. 25, 57 (1937). — Nitch, C. A. R.: Tumour of the kidney. Proc. roy. Soc. Med. 21, 907 (1927).

O'Conor, V. J., and J. R. Head: Transthoracic nephrectomy for Wilms' tumor. J. Urol. (Baltimore) 65, 193 (1951). — O'Flynn, D.: Multilocular cystic disease of kidney. Brit. J. Urol. 25, 41 (1953). — Omland, G.: Polycythaemia in renal carcinoma. Acta med. scand. 164, 451 (1959). — Owen, R. A. C.: A case of fibromyoma of the kidney. Brit. J. Surg. 40, 403 (1953).

Papanicolaou, G. N.: Cytology of the urine sediment in neoplasms of the urinary tract. J. Urol. (Baltimore) 57, 375 (1947). — Patch, F. S.: Three unusual primary kidney tumours. Brit. J. Urol. 9, 339 (1937). — Patton, J. F., and N. S. Bricker: Renal function studies in polycystic disease of the kidney. A preliminary report. Trans. Amer. Ass. gen.-urin. Surg. 45, 147 (1953). — Pemberton, J. de, and R. J. Bennett: Hypernephroma of the thyroid gland: A review of the literature and a report of two cases. Surg. Clin. N. Amer. 14, 593 (1934). — PERLMANN, S.: Virchows Arch. path. Anat. 268, 524 (1928). — PIERCE, E. C.: Percutaneous femoral artery catheterisation in man with special reference to aortography. Surg. Gynec. Obstet. 93, 56 (1951). — POWELL, T., R. SHACKMAN and H. D. JOHNSON: Multilocular cysts of the kidney. Brit. J. Urol. 23, 142 (1951). — POWER, S.: Cystadenoma of the kidney. Brit. J. Urol. 27, 285 (1955). — PRIESTLEY, J. T.: Survival following removal of malignant renal neoplasms. J. Amer. med. Ass. 113, 902 (1939). — Treatment of malignant renal tumours. Surg. Clin. N. Amer. 21, 1173 (1941). — PYRAH, L. N., and F. P. RAPER: Some uses of an isolated loop of ileum in genito-urinary surgery. Brit. J. Surg. 42, 337 (1954).

RAGINS, A. B., and H. C. ROLNICK: Mucus producing adenocarcinoma of the renal pelvis. J. Urol. (Baltimore) 63, 66 (1950). — RALL, J. E., and H. M. ODEL: Congenital polycystic disease of the kidney. Amer. J. med. Sci. 218, 399 (1949). — RAPPOPORT, A. E.: Haematuria due to papillary haemangioma of the renal pelvis. Arch. Path. (Chicago) 40, 84 (1945). — RICHES, E.: Polycythaemia in renal new growths. S. Afr. med. J. 37, 274 (1963). — RICHES, E.: On carcinoma of the kidney. Ann. roy. Coll. Surg. Engl. 32, 201 (1963). — RICHES, E. W.: A case of primary carcinoma of the ureter. Brit. J. Surg. 29, 392 (1941). — Modern trends in urology, 13, p. 133. London: Butterworth & Co. 1953. — Carcinoma of the kidney and renal tuberculosis. Trans. med. Soc. Lond. 70, 208 (1954). — Irradiation therapy in urology; the kidney. Brit. J. Urol. 26, 319 (1954). — The present status of renal angiography. Brit. J. Surg. 42, 462 (1955). — Radiotherapy in renal new growths. J. Fac. Radiol. (Lond.) 8, 19 (1956). — Factors in the prognosis of carcinoma of the kidney. J. Urol. (Baltimore) 79, 190 (1958). — RICHES, E. W., I. H. GRIFFITHS and A. C. THACKRAY: New growths of the kidney and ureter. Brit. J. Urol. 23, 297 (1951). (The B.A.U.S. series). — RICHES, E. W., and A. C. THACKRAY: "Essential Haematuria". Proc. roy. Soc. Med. 49, 696 (1956). — RICHES, E. W., and C. G. WHITESIDE: Abdominal aortography. Surgical Progress 1955, 1, 1. London: Butterworth & Co. 1955. — RICKHAM, P. P.: Bilateral Wilms' tumour. Brit. J. Surg. 44, 492 (1957). — ROBERTSON, T. D., and J. R. HAND: Primary intrarenal lipoma of surgical significance. J. Urol. (Baltimore) 46, 458 (1941). — ROBINSON, G. L., Perlmann's tumour of the kidney. Brit. J. Surg. 44, 620 (1957). — ROWLANDS, B. C.: Pyrexia as the sole presenting symptom of a hypernephroma. Brit. J. Urol. 23, 267 (1951). — RUIZ RIVAS, M.: Diagnostico radiologico. El neumorriñon. Technica original. Arch. esp. Urol. 4, 228 (1947/48). — RUSCHE, C.: Renal hamartoma. (Angiomyolipoma.) Report of three cases. J. Urol. (Baltimore) 67, 823 (1952).

SANDFORD, H. L.: Carcinoma of both kidneys. Report of a case with a review of the literature on multiple primary malignant tumours. Surg. Gynec. Obstet. 53, 360 (1931). — SCHOFIELD, T. L.: Hypernephroma; secondary deposits removed by operation. Brit. J. Urol. 24, 211 (1952). — SCHOLL, A. J.: Peripelvic lymphatic cysts of the kidney. Report of two cases. J. Amer. med. Ass. 836, 4 (1948). — SCHOONOVER, R.: Adenocarcinoma in the horseshoe kidney. Amer. J. Surg. 86, 417 (1953). — SCORER, C. G.: Cutaneous metastases from renal neoplasms. Brit. J. Urol. 23, 250 (1951). — SCOTT, L. S.: Bilateral Wilms' tumour. Brit. J. Surg. 42, 513 (1954). — Wilms' tumour: its treatment and prognosis. Brit. med. J. 1, 200 (1956). — SCOTT, W. W.: A review of primary carcinoma of the ureter. Presenting two cases. J. Urol. (Baltimore) 50, 45 (1943). — SCOTT, W. W., and H. L. BOYD: A study of the carcinogenic effect of betanaphthylamine on the normal and substituted isolated sigmoid loop bladder of dogs. J. Urol. (Baltimore) 70, 914 (1953). — SELDINGER, S. I.: Catheter replacement of the needle in percutaneous arteriography. Acta radiol. (Stockh.) 39, 368 (1953). — SHEACH, J. M.: Bilateral Wilms' tumour; a case report with a review of the literature. Brit. J. Urol. 25, 109 (1953). — SILVER, H. K.: Wilms' tumor. J. Pediat. 31, 643 (1947). — SIMPSON, G.: Carcinoma of the kidney. Brit. J. Surg. 21, 388 (1933). — SMYTH, M. J.: Silent hypernephromata. Brit. J. Surg. 27, 266 (1939). — SNELLING, M.: Results of the treatment of hypernephroma. J. Fac. Radiol. (Lond.) 1, 172 (1949). — SOUTHWOOD, W. F. W., and V. F. MARSHALL: A clinical evaluation of nephrotomography. Brit. J. Urol. 30, 127 (1958). — STEINER, R. E.: Venography in relation to the kidney. Brit. med. Bull. 13, 64 (1957). — SWAN, R. H. J., and H. BALME: Angioma of the kidney. Report of a case with an analysis of 26 previously reported cases. Brit. J. Surg. 23, 282 (1935). — SWIFT-JOLY, J.: Tumours of the renal pelvis and ureter. 5. Cong. Internat. Soc. Urol., London. Brit. J. Urol. 5, 327 (1933).

THACKRAY, A. C.: The pathology of renal tumours; modern trends in Urology (ed. E. W. Riches), 12, p. 117. London: Butterworth & Co. 1953. — Ten-year follow-up of cases of adenocarcinoma of the kidney. Proc. roy. Soc. Med. 50, 362 (1957). — TRESIDDER, G. C.: Ureterograms. Brit. J. Urol. 26, 240 (1954). — TRESIDDER, G. C., and R. P. WARREN: Primary neoplasms of the ureter. Brit. J. Urol. 26, 139 (1954).

UYS, C. J.: Tumours of the kidney in the Bantu races of South Africa. A pathological study based on 3707 consecutive autopsies. Brit. J. Urol. 28, 75 (1956).

Wade, H.: Observations on the role of excretion urography in the diagnosis of disease. Brit. Med. J. 1, 353 (1933). — Walsh, A.: Solitary cyst of the kidney and its relationship to renal tumour. Brit. J. Urol. 23, 377 (1951). — Waterfall, W. B.: Renal angioma causing severe haematuria. Brit. J. Urol. 22, 142 (1950). — Watson, E. M., H. R. Sauer and M. G. Sadugor: Manifestations of the lymphoblastomas in the genito-urinary tract. J. Urol. (Baltimore) 61, 626 (1949). — Watts, G. T.: Renal adenoma treated by partial nephrectomy. Brit. J. Urol. 27, 294 (1955). — Weisel, W., M. B. Dockerty and J. T. Priestley: Wilms' tumour of the kidney: a clinico pathological study of forty-four proved cases. J. Urol. (Baltimore) 50, 399 (1943). — Sarcoma of the kidney. J. Urol. (Baltimore) 50, 564 (1943). — Wells, C. A.: Irradiation therapy in urology: the kidney (Discussion). Brit. J. Urol. 26, 324 (1954). — Weyrauch, H. M., and M. M. Berger: Haemangioma of the kidney; report of a case simulating pyelo-ureteritis cystica; review of literature. Stanf. med. Bull. 9, 43 (1951). — Weyrauch, H. M., H. L. Wanless, J. L. Goebel and K. G. Scott: Biopsy of the kidney for suspected neoplasm. J. Urol. (Baltimore) 67, 60 (1952). — White, M.: Renal tumours in children. Brit. J. Urol. 23, 357 (1951). — Whitehouse, W. M., and I. Lampe: Osseous damage in irradiation of renal tumors in infancy and childhood. Amer. J. Roentgenol. 70, 721 (1953). — Whiteside, C. G.: Advances in the radiology of the urinary tract. Modern trends in urology (edit. E. W. Riches), 4, p. 20. London: Butterworth & Co. 1953. — Willis, R. A.: Pathology of tumours. St. Louis: C. V. Mosby Co. 1948. — Wilms, M.: Die Mischgeschwülste der Niere, S. 1. Leipzig: A. Georgi 1899. — Witten, D. M., L. F. Greene and J. L. Emmett: An evaluation of nephrotomography in urologic diagnosis. Amer. J. Roentgenol. 90, 115 (1963). — Wright, H. W. S.: A study of the surgical pathologi of hypernephromata; with special reference to their origin and symptomatology. Brit. J. Surg. 9, 338 (1922). — Wynn-Williams, D., and A. D. Morgan: Lymphangioma of the kidney. Brit. J. Surg. 37, 346 (1950).

Yates-Bell, J. G.: Rovsing's operation for polycystic kidney. Lancet 1, 126 (1957).

Addendum

Bixler, L. C., K. W. Stenstrom, and C. D. Creevy: Malignant tumours of the kidney: review of 117 cases. Radiology 42, 329 (1944). — Bloom, H. J. G.: Hormone treatment of renal tumours: Experimental and clinical observations. In: Tumours of the kidney and ureter (Ed. E. Riches), Chap. XXV, p. 311. Edinburgh: Livingstone 1964. — Bloom, H. J. G., C. E. Dukes, and B. C. V. Mitchley: Hormone dependent tumours of the kidney. 1. The oestrogen induced renal tumour of the Syrian hamster. Hormone treatment and possible relation to carcinoma of the kidney in man. Brit. J. Cancer. 17, 611 (1963). — Bloom H. J. G., and D. M. Wallace: Hormones and the kidney: possible therapeutic role of testosterone in a patient with regression of metastases from renal adenocarcinoma. Brit. med. J. 1964 II, 476. — Bodian, M., and L. L. R. White: British Empire Cancer Campaign reports for 1951, p. 133 and 1955, p. 184. — Bothe, A. E.: The effect of röntgen therapy upon tumours of the kidney. Amer. J. Röntgenol. 33, 529 (1935). — Boyland, E., C. E. Dukes, P. L. Grover, and B. C. V. Mitchley: The induction of renal tumours by feeding lead acetate to rats. Brit. J. Cancer 14, 283 (1962). — Case, R. A. M.: Comparison of mortality in selected countries. In: Tumours of the kidney and ureter (ed. E. Riches), chap. III, p. 28. Edinburgh: Livingstone 1964. — Emmett, J. L., S. R. Levine, and L. B. Woolner: Coexistence of renal cyst and tumor: incidence in 1007 cases. Brit. J. Urol. 35, 403 (1963). — Farber, S., G. J. D' Angio, A. Evans, and A. Mitus: Clinical studies on actinomycin D with special reference to Wilms tumour in children. Ann. N.Y. Acad. Sci. 89, 421 (1960). — Flocks, R. H., and M. C. Kadesky: Malignant neoplasms of the kidney: an analysis of 353 patients followed five years or more. J. Urol. (Baltimore) 79, 196 (1958). — Jenkins, G. D.: Regression of pulmonary metastasis follwing nephrectomy for hypernephroma: eight year follow-up. J. Urol. (Baltimore) 82, 37 (1959); — Final report: regression of pulmonary metastasis following nephrectomy for hypernephroma: 13 year follow up. J. Urol. (Baltimore) 94, 99 (1965). — Kerr, W. K., M. Barkin, I. A. D. Todd, and Z. Menczyk: A hypernephroma associated with elevated levels of bladder carcinogens in the urine: case report. Brit. J. Urol. 35, 263 (1963). — Kimball, K. G.: Amyloidosis in association with neoplastic disease. Ann. intern. med. 55, 958 (1961). — Koletsky, S., and G. E. Gustafson: Adenoma or renal cortex induced by X-irradiation in rats. Cancer Res. 15, 100 (1955). — Nairn, R. C., J. Philip, T. Ghose, I. B. Porteus, and J. E. Fothergill: Production of a precipitin against renal cancer. Brit. med. J. 1963 I, 1702. — Pačes, V., A. Placherova et Vl. Čapek: L'irradiation preoperatoire des tumeurs du rein chez l'adulte. Trans. XIII Congr. Internat. Soc. Urol., London 2, 461 (1964/65). — Pavone-Macaluso, M.: L'etiologie et la pathogéne des tumeurs expérimentales. Gaz. méd. Fr. 70, 1143 (1963). — Petkovic, S. D.: An anatomical classification of renal tumours in the adult as a basis for prognosis. J. Urol. (Baltimore) 81, 618 (1959). — Reboul, J., G. Bal-

LANGER, J. DELORME, I. TAVERNIER et M. GEINDRE: Association radiotherapie preoperatoire et chirurgie dans le traitment du cancer primitif du rein de l'adulte. Ann. Radiol. 5, 283 (1962). — RICHES, E.: The place of radiotherapy in the management of parenchymal carcinoma of the kidney. J. Urol. (Baltimore) 95, 313 (1966). — ROSEN, V. J., T. J. CASTANERA, D. J. KIMELDORF, and D. C. JONES: Renal neoplasms in the irradiated and non-irradiated Sprague-Dawley rat. Amer. J. Path. 38, 359 (1961). — TURNER-WARWICK, R. T., and A. D. THOMSON: Connective tissue and mixed tumours in adults. In: Tumours of the kidney and ureter (ed. E. RICHES), chap. XI, p. 99. Edinburgh: Livingstone 1964. — WATERS, C. A., L. G. LEWIS, and W. A. FRONTZ: Radiation therapy of renal cortical neoplasms with special reference to pre-operative irradiation. Sth med. J. (Bgham, Ala.) 27, 290 (1934). — WHARTON, L. R.: Preoperative irradiation of massive tumours of the kidney. A clinical and pathologic study. Arch. Surg. 30, 35 (1935). — WILLIAMS, D. I.: Urology in childhood, In: Handbuch der Urologie, Bd. XV, S. 175. Berlin-Göttingen-Heidelberg: Springer 1958; — The surgical treatment of nephroblastoma. In: Tumours of the kidney and ureter (ed. E. RICHES), chap. XX, p. 255, chap. XIX, p. 238. Edinburgh: Livingstone 1964). — ZOLLINGER. H. U.: Durch chronische Bleivergiftung erzeugte Nierenadenome und Carcinome bei Ratten. Virchows Arch. path. Anat. 323, 694 (1953).

Tumours of Bladder

By

ARTHUR JACOBS

With 47 Figures

With co-operation of

G.W. BLOMFIELD, K.M. GIRDWOOD, J.M. SCOTT and T. SYMINGTON

A. Aetiology

Since attention was drawn at the end of last century to the high incidence of bladder tumours in aniline dye workers, the aetiology of bladder cancer has been of especial interest and widespread research.

The subject has been extensively reviewed by POOLE-WILSON (1953) who mentions that it was REHN's original observation in 1895 which first brought suspicion on the chemicals employed in the dye manufacturing industry as possible carcinogenic agents in the bladder. Investigation revealed that in the manufacture of synthetic dyes from coal tar there were three main stages. First the production of crude coal tar compounds; secondly the conversion of these to intermediate compounds; and lastly the manufacture of aniline dyes. It was found that tumours developed mainly amongst men engaged in the second stage and to a lesser extent in the final stage of production. This fact was not easy to establish, as in the early days of aniline dye manufacture workers were less confined to set stages in the industry.

Aniline was the first chemical to be suspected of being a bladder carcinogen with the result that the condition is still often referred to as "aniline cancer". Further investigations, however, have revealed that pure aniline is not a causal factor (GROSS 1940, GOLDBLATT 1949, SCOTT 1952, BARSOTTI and VIGLIANI 1949 and 1952, DI MAIO 1949) and it is now considered that naphthylamine impurities in the benzene used for making aniline are actually responsible (WALPOLE WILLIAMS and ROBERTS 1952). α-Naphthylamine was also believed to be an agent causing bladder carcinoma. This view has been refuted and it is now thought that the by-product of 5% β-naphthylamine is the impurity which is the responsible factor. Experimental work on dogs has proved that α-naphthylamine is not a carcinogen (GEHRMAN, FOULGER and FLEMING 1949) whilst benzidine on the other hand is a proved carcinogenic agent (SCOTT 1952). In this connection it was thought that concomitant exposure to β-naphthylamine was a factor (GEHRMAN, FOULGER and FLEMING 1949) but clinical observation of cases occurring in men exposed to benzidine only, has clarified the point. Although administration of benzidine to experimental animals has failed to produce bladder tumours, BAKER (1950) has shown that a metabolite excreted in the urine of workers exposed to benzidine does act as a bladder carcinogen in mice.

Reports from several sources have confirmed the danger of β-naphthylamine as a urinary tract carcinogen in man. Experimental work with β-naphthylamine was at first inconsistent until it became clear that animals varied in their susceptibility to bladder cancer (SCHAR 1930, BERENBLUM and BONSER 1937). This variation in sensitivity between the species led to the speculation that differences in the metabolism of β-naphthylamine might result in varying concentrations of metabolites in different parts of the body. As both man and the dog produced tumours in the renal pelvis and the bladder only, it was suggested that the carcinogenic agent, for example a metabolite of β-naphthylamine might be maximally concentrated in the urine.

Examination of the urine of dogs and rats fed with β-naphthylamine resulted in isolation of 2-amino-1-naphthol (WILEY 1938, MANSON and YOUNG 1950). BONSER, CLAYSON and JULL (1951) then fed known amounts of β-naphthylamine to animals of different species and found a definite correlation between the amount of 2-amino-1-naphthol excreted in the urine and the susceptibility of the particular animal to develop bladder cancer. In dogs, who were the most susceptible, the concentration in the urine was 200 times that of the plasma. Further proof that 2-amino-1-naphthol was a carcinogen followed the experiments of introducing into bladders of mice wax pellets containing a 30% suspension of the metabolite. After 27—34 weeks, examination of the bladders revealed changes ranging from epithelial hyperplasia to transitional cell carcinoma with invasion of the bladder wall. Similar experiments using β-naphthylamine failed to produce any tumours. It is interesting to note that instillation of 20-methylcholanthrene by the same technique produced epithelial changes culminating in the development of a squamous carcinoma.

This evidence of the association of carcinogenic metabolites with occupational tumours and their close resemblance to spontaneous tumours both as regards histology and site in the urinary tract has awakened interest in the search for other metabolites as a cause of spontaneous bladder cancer in man. Experimental confirmation of the urogenous theory of pathogenesis was obtained by McDONALD and LUND (1954) who fed dogs and rabbits on β-naphthylamine after surgically isolating the dome half of the urinary bladder from contact with urine. Tumours developed only in the portion of the bladders receiving urine. This raises the interesting point as to the relative importance of the duration of contact between the agent and the epithelium, and the concentration of the metabolite. The increased efficacy of any water-borne agent through prolonged contact would explain the frequency with which tumours are found in the bladder compared with the upper urinary tract and also the differing incidence of bladder tumours in the sexes (male: female $= 3:1$) which could be due to the greater tendency to bladder neck obstruction and consequent stasis in the male. The high association of carcinoma in bladder diverticula also lends some weight to this idea. KENNAWAY (1930) has recorded that brewery workers who receive a daily issue of 2 pints of beer have a very low incidence of bladder tumours. Review of the incidence of bladder tumours throughout Great Britain shows that it is higher in those areas where the rainfall is least; the deduction is that low rainfall areas have a lower humidity and therefore the inhabitants tend to lose more fluid from the skin of the body and have a lower output of a more concentrated urine.

In searching for an actual carcinogen in man, WALLACE and his co-workers (1956) have found that the urine of patients with bladder tumours contains greater amounts of 3-hydroxyanthranilic acid and 3-hydroxykynurenine than normal controls. These two substances are end products of the metabolism of tryptophane, and the latter has caused bladder tumours when administered to rats. Moreover BOYLAND and WATSON (1956) by a modification of the BONSER-JULL technique confirmed that 3-hydroxyanthranilic acid and 3-hydroxykynurenine produced bladder tumours in mice. Further investigation revealed that these products are excreted in man in the form of a compound conjugated with sulphuric or glucuronic acid. These conjugated forms are not carcinogenic, but WALLACE has discovered that urine in man contains two enzyme systems, glucuronidase and sulphatase. His theory is that these enzyme systems may act on the conjugated metabolites and release potential carcinogens. This proposition gains weight from the fact that the urine in cases of bladder tumours contains a much higher level of glucuronidase compared with controls (BOYLAND, WALLACE and WILLIAMS 1955). The enzyme is also found in high concentration in the actual tumours. It is interesting to note too, that the urine continues to show a high level even after the bladder has been rendered tumour free.

If this theory on bladder tumour aetiology is accepted, there are two main methods whereby prophylaxis may be attempted and proof of the theory obtained.

1. By modifying the metabolism of tryptophane for example by some dietary measures.

2. By control of the activity of the urinary enzymes.

WALLACE et al are at present undertaking an interesting clinical experiment, administering orally to a series of bladder tumour cases who have a high urinary glucuronidase a substance called 1:4 saccharolactose, a compound which has the effect of inhibiting glucuronidase activity. A control group are given citric acid. Their results will be of great interest.

An immense amount of research has been undertaken in the case of bladder tumours and much is still under way. As each new discovery is made, further fields open up. At the moment the initiative appears to be with the biochemists, but the influence of changes in social habits including drinking, smoking, and diets, all require assessment. CLEMMESEN and NIELSEN (1956), for example, have found that the incidence of bladder tumours among Danish men doubled in the years 1948—1952 in comparison with 1943—1947. This increase was seen in the capital city only (the rural rise was slight by comparison) and it affected males only. In Britain, CASE (1956) has reported a steady increase in male mortality from bladder cancer. A concomitant rise has been observed in carcinoma of the lung, and in view of the considered association between smoking and this latter condition (DOLL and HILL 1956) the question has been raised whether tobacco smoking is a factor in the causation of bladder cancer. HAMTOFT and LINDHART (1956) have reported a higher incidence of cigarette smoking in Copenhagen compared with the rural areas. LILIENFELD, LEVIN and MOORE (1956) from an analysis of the records of the Roswell Park Memorial Institute, Buffalo, report statistically significant differences in the smoking habits of men with bladder cancer. More men with bladder cancer had been smokers compared with men with benign bladder disease, prostatic carcinoma or no disease; more of the bladder cancer patients had smoked cigarettes; and more of the cigarette smokers with bladder cancer had smoked for 30 or more years. These findings did not apply at all to women.

B. Symptomatology

Haematuria in the form of visible blood in the urine is the primary and outstanding symptom in about 75 per cent of patients suffering from bladder tumour (ASH 1940). The bleeding may be copious or slight, persistent or evanescent, the tendency towards more severe and prolonged attacks increasing as the growth progresses. Most often the blood is uniformly mixed with the urine, the diffusion taking place before voiding. Initial haematuria characterised by the appearance of a little blood at the beginning of urination is most likely to arise from a tumour in immediate proximity to the internal meatus; terminal haematuria with blood appearing only at the end of urination may occur from expression of blood from the tumour as the bladder contracts down on it.

Bladder irritability in the form of frequent and urgent urination associated with burning pain and often with pyuria, is the outstanding manifestation of tumour in some 30 per cent. The possibility of these symptoms denoting the presence of a bladder tumour is often demonstrated when cystoscopic examination, immediately after a visible haematuria was first observed by the patient, reveals an extensive growth which must obviously have been present over a lengthy period. Delay in the appearance of visible haematuria is most likely with non-papillary tumours, especially when their location is on the upper zones of the bladder. Non-epithelial tumours also show a less frequent incidence of bleeding as a presenting symptom.

Pain, unless from obstruction due to clot or from a tumour which encroaches on the vesical outlet, is not a common symptom until the diease becomes extensive and infection supervenes. The consequent cystitis and tumour necrosis then give rise to severe vesical tenesmus with pain in the suprapubic area and along the urethra, associated with a feeling of incomplete emptying. In advanced cases there may be pain in the rectum, back and thighs.

When the growth remains uncontrolled, symptoms of progressive renal failure due to dilatation and urosepsis appear and terminal uraemia is the outcome. This is the mode of death in over thirty per cent of those who succumb to bladder cancer.

C. Diagnosis

Grossly demonstrable blood in the urine or persistent blood cells in the urinary sediment with vesical irritability should be considered as indicating the possibility of bladder tumour, especially if occurring after 40 years of age. The diagnosis rests on cystoscopy, which gives an objective identification of the growth.

I. Cystoscopy

When after an inspection of the entire bladder surface, the presence of neoplasm is confirmed, details of its features should be observed and recorded. A small benign papilloma will be seen as a pale pink or grayish cluster of fine villi which wave about in the irrigating fluid. The attachment to the bladder wall is by a thin and comparatively long pedicle. Other papillomas are more squat with a raspberry-like surface appearance. Their villi are shorter but still delicate and a blood vessel can be seen to occupy each one. Though pedunculated, an abbreviated stalk may give the impression that the tumour is lying directly on the bladder wall. Hypertrophied vessels may often be seen coursing towards the root of the papilloma through the vesical mucosa which otherwise looks normal. These tumours are often single.

Papillary carcinomas or malignant papillomas are composed of closely packed stunted villi and therefore present a smoother surface likened to that of a cauliflower or marine sponge. A pedicle if present is short and thick; in the absence of one the size of the base may approach that of the tumour surface. If infiltration from pedicle or base has taken place, the surrounding mucosa may have an irregular granular look or may be oedematous. Coincidental with these changes, cystitis tends to occur with necrosis of the tumour and phosphatic encrustation of its surface.

A nodular tumour projects into the bladder as a solid reddish-yellow or brownish fleshy mass with an irregular or lobulated surface; or it may present as a flat plaque rou circular in ghlyshape with thick rounded margins. Later,

necrosis and ulceration produces a crater-like appearance in the central portion
on which calcareous material may be deposited or clots adhere. The margins
become increasingly rounded and everted.

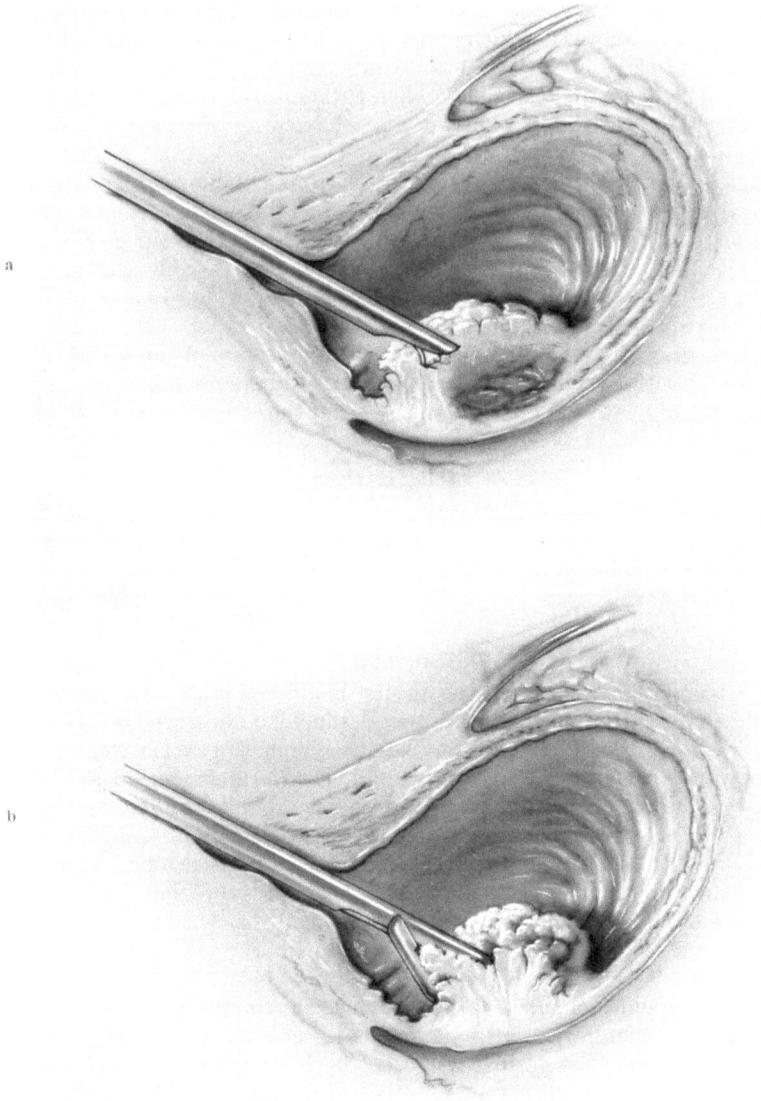

Fig. 1a and b. a Transurethral resection of a papillary tumour. b Tumour tissue in grasp of LOWSLEY's biopsy
forceps

Cystoscopy of a bladder tumour must include a study of the surrounding vesical
mucosa as well as a search for associated lesions which are not uncommonly present,
especially in male patients. Oedema or cystitis cystica may be due to infection

or result from tumour spread. Puckering of the mucosa indicates a submucosal neoplastic infiltration radiating from the main growth. Localised irregular mucosal elevations distal to a tumour may be due to surface manifestations of a submucosal lymphatic infiltration or to commencing new tumours (MASINA 1952).

The commonest coincidental lesion encountered with bladder tumour is a prostatic enlargement. Whether or not this gives rise to obstructive symptoms, a marked intra-vesical prostatic intrusion may prevent adequate visualisation of a tumour and cause difficulty in treating it by the transurethral route if the gland is not first resected or removed. The finding of a diverticular opening creates an additional problem in diagnosis for it then becomes necessary to ascertain if the sac also harbours a tumour to which that in the bladder is a satellite. If tumour protrudes through the opening or lies immediately within the neck it will be visible at least in part to cystoscopic view. It may however be entirely hidden within the diverticulum and in that event contrast cystography will be required and the detection of tumour aided by the demonstration of a filling defect. The sac may of course be the repository of a tumour without any growth in the bladder.

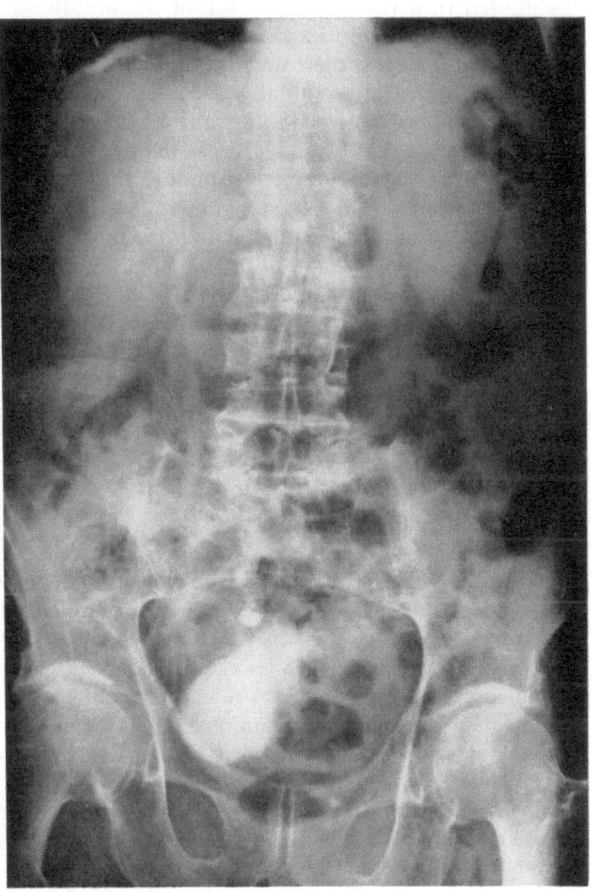

Fig. 2. Intravenous urogram showing gross filling defect on left side of bladder; there is no excretion from the left kidney

Cystoscopic examination should thus provide information on the size, site, and type of tumour, the nature of its base, the condition of the surrounding mucosa, the presence of multiple growths and of coexisting lesions. A simple cystoscopy may provide all this information and the needed data to determine the optimum method of therapeutic attack. In the presence of large tumours, or when there is infection and cystitis, more detailed knowledge will however be obtained if the cystoscopic inspection is made with the patient anaesthetised. In any event, complete relaxation under anaesthesia for a bimanual examination is required when there is any evidence pointing to infiltration and tissue for biopsy can at the same time be removed.

Bimanual examination helps to determine the depth of infiltration and if correlated with the histological findings, permits a substantially accurate staging of a tumour. If the tumour is deeply infiltrating and accessible to the finger, a thickened indurated mass will be felt on ordinary rectal examination and will indicate an incurable lesion. By extending the rectal examination to bimanual palpation with complete muscular relaxation, nearly the whole bladder can be made accessible to palpation and areas of thickening as distinct from actual tumour masses will be felt. It is recommended that the anal sphincter be dilated to permit the insertion of the middle as well as the index finger. The free hand is made to exert counter pressure on the suprapubic area so forcing the bladder downwards towards the fingers in the rectum. The prostate having been defined, each seminal vesicle and the bladder base is palpated. With firmer pressure both lateral walls and the entire posterior wall of the male bladder can be explored. If thickening or induration of an inferolateral ligament is felt on the same side as a tumour, this will indicate extravesical extension, for the normal inferolateral ligaments are not palpable. The anterior wall of the bladder behind the pubis is inaccessible to palpation and in the female that portion of the posterior wall in front of the cervix uteri cannot generally be felt.

If a tumour is impalpable or palpable but freely mobile and not infiltrating, depth of infiltration is likely to be limited to the submucosal or superficial muscular layers. The microscopic findings of the pedicle or from the bladder wall beneath, will differentiate between these two depths. When there is thickening of the bladder wall with resistance to palpation, the deep muscular layer is involved, whilst irregularity of the wall with a mass bigger than seen cystoscopically indicates perivesical spread.

II. Biopsy

When microscopic findings are considered desirable before deciding on the mode of treatment, the tissue should be obtained when cystoscopy and bimanual examination under anaesthesia are being carried out. When bladder changes are indefinite a specimen will be required for differential diagnosis and the nature of mucosal changes distal to a main tumour may remain equivocal without a histological examination. Tumour tissue for biopsy can be secured with special forceps such as LOWSLEY's or RICHES' which are designed for this purpose or a resectoscope or punch may be employed (Fig. 1). An excision of the intravesical projecting portion of the tumour should first be made and the specimen (No. 1) obtained labelled for histological study. Tissue from the tumour base, which should include the underlying muscularis, should next be excised and separately specified (No. 2). Specimen 1 should be used for grading and specimen 2 for staging.

An interpolation of the biopsy findings, in particular the extent of penetration with those of the bimanual pelvic examination is of inestimable value in assessing prognosis. JEWETT (1954) correlates the microscopic evidence of infiltration and that obtained by pelvic examination in the following way:

Bimanual Examination:	*Depth of infiltration:*
1. No thickening. No mass	Submucosal or superficial muscular
2. Thickening: No mass	Submucosal or superficial muscular
3. Mass, rubbery consistence, movable. Lateral ligaments negative	Deep muscular or perivesical (if not large papilloma)
4. Mass, stony consistence, movable. Lateral ligaments negative	Perivesical
5. Mass, stony consistence, movable. Ligaments thickened and indurated	Perivesical
6. Mass, stony consistence, fixed	Perivesical

JEWETT regards the prospects of cure as good for the first two groups, poor in the middle two, and very poor in the last two.

III. Urography

Excretion urography should always be carried out for it provides an indication of the renal function and demonstrates any dilatation in the upper urinary tract resulting from the presence of a bladder tumour. Should there be dilatation on one side associated with a tumour near the corresponding ureteric orifice, it generally means that the growth has infiltrated the wall of the ureter. When the ureter is complete obstructed there is usually no visualization of the kidney above, though its function may not have entirely ceased (Fig. 2). Evidence of an associated renal papillary tumour will be furnished by the excretory pyelogram show-ing a filling defect, but an inadequate visualiza-tion may call for retro-grade pyelography to con-firm the suspected renal lesion. The bladder tu-mour may however ob-scure the ureteric orifice and prevent its catheteri-sation.

The cystogram pro-vides a record of the size and position of a tumour and assists in assessing its stage. An intravesically projecting growth without

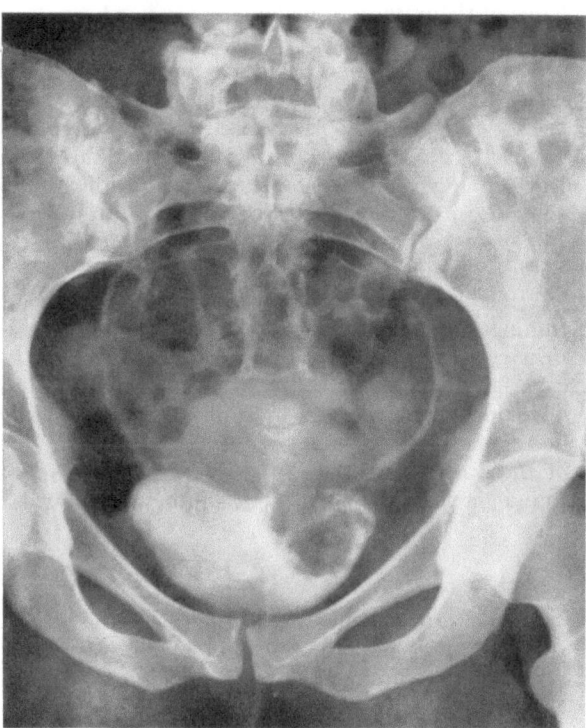

Fig. 3. Intravenous cystogram showing defect due to a papillary tumour

infiltration will cause a filling defect in the lumen of the viscus, whilst the wall of the bladder retains its normal outline (Fig. 3). An infiltrating tumour on the other hand, by causing a loss of elasticity, creates a defect in the outline which exhibits a flattened appearance or with deep infiltration, a bite-like deformity.

IV. Cytology of urinary sediments

The examination of smears of urinary deposits is being increasingly employed as an aid to diagnosis especially in connection with large scale screening of workers engaged in occupations associated with recognised bladder tumour hazards. Whilst repeated routine cystoscopic examinations over the years for thousands of employees who are symptom-free may be difficult to organize, regular examinations of urine can be easily arranged. Positive findings will indicate the need for cystoscopy and afford the opportunity of early diagnosis before the onset of symptoms.

Examination of wet unstained smears of urinary sediments were initiated
in some of the British and Continental dyestuffs factories more than thirty
years ago. If red blood cells were present, this was regarded as a criterion for
cystoscopy and enabled many pre-symptomatic diagnoses to be made on men
who had been exposed to carcinogens.

The work of PAPANICOLAOU and MARSHALL (1945) and of PAPANICOLAOU
(1947 and 1954) has shown that a study of the urinary sediments for exfoliated
cells can result in an identification of tumour and that the method is of diagnostic significance. The interpretation of the smears requires a thorough knowledge of the criteria of malignancy as has been described by PAPANICOLAOU (1954)
and is not easy. CRABBE et al. (1956) reported that out of 1800 men screened
during the years 1951 to 1956, positive smears were found in sixty-seven and
that 76.1% of that number (51) showed bladder tumours on first cystoscopy;
twenty-five had no R.B.Cs. or other manifestations of bladder tumour.
Five with positive smears showed no tumour on initial cystoscopy but developed
tumours in five to forty-eight months whilst another positive without a bladder
tumour was discovered to have a prostatic carcinoma. Six men with negative
PAPANICOLAOU smears, cystoscoped because of R.B.Cs. or other symptoms had
bladder tumours.

It is evident that in spite of the possibilities of false negative and positive
findings, the PAPANICOLAOU technique combined with an examination of the
urinary sediment for R.B.Cs. is a valuable aid in the detection of bladder tumours
especially when mass industrial investigations are required. The subsequent
development of tumours by some patients with malignant cells in the urine who
had no sign of tumour at the initial cystoscopy is of special interest. The origin
of these malignant cells might be from foci of cellular activity such as can be
found in normal-looking epithelium in tumour-bearing bladders.

D. Treatment

In spite of the advances in surgical and irradiation techniques over the last
10 to 15 years, the treatment of bladder tumours continues to be an uncertain
abstraction in urological practice. Surgery, electrosurgery, and irradiation alone
and in combination are all employed and can be applied in the following ways:

I. By transurethral diathermy or resection, sometimes supplemented by
interstitial irradiation from radon seeds inserted cystoscopically.

II. By transvesical diathermy or resection, alone or combined with interstitial irradiation from radium or radon, or from artificially produced sources
of radioactive isotopes such as gold[198], yttrium[90], tantalum[192], and cobalt[60].

III. By intracavitary irradiation from a solid central source of radium or
cobalt[60] or from a solution of bromine [82], or cobalt[60].

IV. By partial cystectomy.

V. By total cystectomy.

VI. By external irradiation.

The choice of method or of the integration of multiple measures will depend
on the assembled information acquired from the diagnostic investigations, in
particular on the location, size, number of tumours and the condition of the
surrounding mucosa as disclosed at cystoscopy; on the histological grade and
degree of spread revealed by biopsy and bimanual examination under anaesthesia; on the state of the upper urinary tract as demonstrated by urography
and on the general condition of the patient. The facilities available at different

centres, especially those for irradiation and the personal experience of the surgeon also inevitably influence procedure. In the following account of methods of treatment, the varieties of tumour that can be appropriately dealt with by the different techniques are indicated.

I. Transurethral diathermy and resection

The villous papilloma, single or multiple, with a narrow-based stalk which has not penetrated through the submucosa is the variety of bladder tumour most readily destroyed by cysto-diathermy. These relatively benign growths if small are effectively eliminated by the direct application of diathermy through a BUGBEE electrode inserted through an operating cystoscope. For the bigger papillomas, a ball electrode is preferable, the larger terminal facilitating a more rapid coagulation. A flexible ball electrode must be employed with the operating cystoscope but with a panendoscope a rigid one is used. This latter assembly is preferable for the fulguration of tumours which the operator can visualize with the foroblique lens system and is therefore signally suitable when the location is near the vesical outlet or on the anterior portion of the floor. For the sessile papillary tumours demonstrable through the foroblique lens system, removal with the resectoscope is, however, best of all. The projecting tumour tissue can be cut away with the loop and if penetration has extended through the submucosa, the resection can be deepened into the muscle layer. Bleeding points are controlled with the ball electrode which is also used to fulgurate the denuded area resulting from the resection.

Some workers consider that the vast majority of bladder tumours should be treated by transurethral resection. MILNER (1954) who holds this view endeavours to include the entire base in the resection. If the tumour infiltrates deeper than it is possible to cut, he makes a cystoscopic radon implantation with seeds of $1^1/_2$ millicuries placed at 1 cm. intervals. Whilst the majority of urologists who use radon seeds believe that a more uniform implant can be made through the open bladder, MILNER considers that interstitial irradiation is best when done cystoscopically with the bladder distended.

FRANKSSON (1950) suggests that the increased trauma caused by a resectoscope may be responsible for the higher incidence of recurrences at the bladder neck which he noted after the use of this instrument. In his series the recurrence rate was 33.3 per cent when treatment had been by resectoscope compared with 9.2 per cent following cystoscopic diathermy.

II. Transvesical diathermy and resection

When the size, location, or multiplicity of tumours make effective destruction by the perurethral route impracticable, the approach must be made through the open bladder. Thus tumours too large to be completely visualized in a single cystoscopic field and those situated well back in the bladder, especially on the vault and upper zones of the postero-lateral walls, may require to be dealt with in this way. So also may large multiple tumours located in different areas which are not of a high enough grade to warrant more radical surgery. With the bladder interior adequately exposed, pedunculated tumours or those which by traction can be made to produce a false pedicle of underlying mobile mucosa, can be cut across by diathermy needle through the base, which is then fulgurated (Fig. 4). With sessile papillary tumours the surface should be fulgurated at the outset, the dessicated tumour reamed off, and the process repeated until the base has been included in the resection. Immediate closure of the bladder is possible with the majority of cases thus dealt with. If this is effected with a three layer suture line and urethral catheter drainage maintained for five days, suprapubic leakage should seldom follow.

As a palliative for large infiltrating tumours, papillary, nodular, or ulcerative, similar diathermy treatment by the transurethral or transvesical route can be

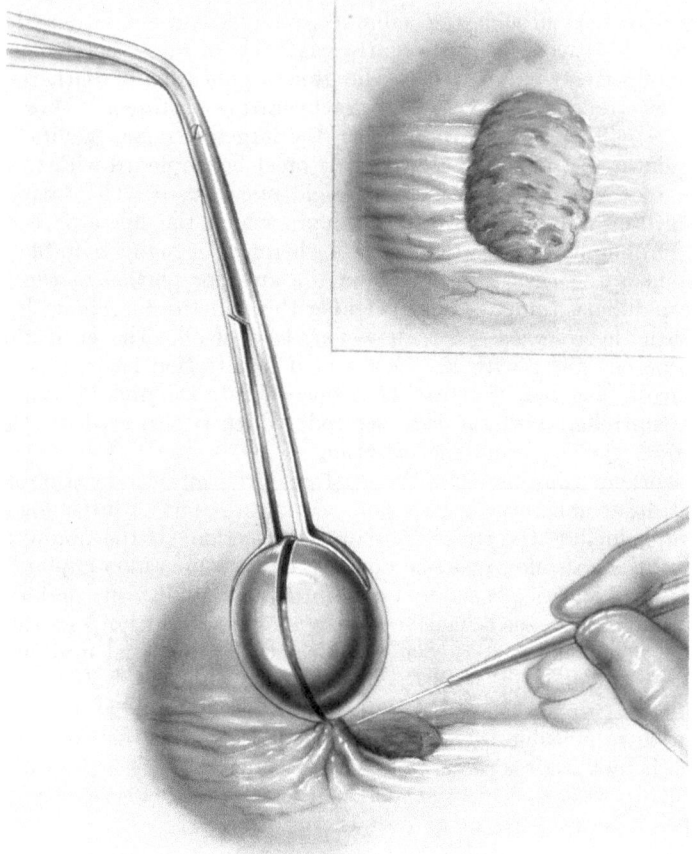

Fig. 4. Removal by diathermy of a papillary tumour which is encased within a FERGUSON forceps. The insert illustrates the resultant tumour-bearing area, the floor of which is formed by the muscle layer

employed with benefit. The removal of intravesical tumour projections relieves frequency by increasing bladder capacity, strangury is diminished especially if the carcinomatous mass is in the vicinity of the bladder neck and bleeding is lessened.

The modern operating cystoscope, panendoscope, and resectoscope reduce the need of a transvesical approach to a bladder tumour when the objective is its destruction by fulguration or resection. These instruments should not however, be determinedly used when circumstances favour optimum access by open operation. If this is followed by bladder closure the post-operative hospital stay need only be increased by a short period of days.

III. Irradiation

Radiotherapy can be applied to bladder tumours by interstitial, intracavitary, and external irradiation techniques. The subject is fully dealt with by POOLE-WILSON, BOLAND and DOBBIE (Radiotherapy, vol. V/2) but in order to complete the overall picture of therapeutic measures, a description of these forms of treatment is included in this section. The objective of radiotherapy is to take advantage of the basic fact that irradiation is destructive to tumour tissue, some varieties being highly sensitive and others very resistant. The more resistant a tumour is to irradiation, the greater is the amount required for its destruction. This has to be achieved by means which will cause minimal damage to the surrounding normal tissues.

Interstitial irradiation provides a method of destroying a tumour without adversely affecting the surrounding normal tissue. The naturally occurring radiating elements of radium and radon have been used for this purpose over a long period. Artificially produced radioactive isotopes such as cobalt[60], tantalum[192], gold[198], and yttrium[90] have been developed in recent years and attempts have been made with these media to improve the technical application of the radioactive source. As the effect of interstitial irradiation is localised, the method must be employed selectively. Especially suitable for this form of therapy are (a) the solitary broad-based papillary transitional cell carcinoma on the lower zones of the bladder having a base not greater than 5 cm. in diameter, which is not penetrating beyond the superficial muscle layer; (b) two or three contiguous tumours with these characteristics which jointly occupy an area up to 5 cm. in diameter; (c) two separately placed tumours neither of which exceeds a 5 cm. diameter. Deeply infiltrating tumours are unlikely to be effectively influenced nor will those with associated unstable areas at distal sites where future tumour development is probable.

The mode of using the alternative agents is largely influenced by their half-life which for radon is 3 days, 18 hours. Radon seeds are now in large measure used in preference to the radium element as their short half-life obviates any need for their subsequent withdrawal (JACOBS 1952). Unless the area to be irradiated, which should include a 1.0 cm. uninvolved margin, is very small, an optimum distribution of seeds aiming to deliver a cancerocidal dose to the tumour area is most likely to be achieved through the open bladder. Projecting tumour tissue should be removed by electro-excision. When infiltration is present the depth of excision should extend through the muscle layers and if necessary even into the extravesical fat. The seeds are then inserted just below the resultant surface level of the tumour-bearing area as a single layer implant. In order to deliver a uniform and accurate dose level which should be between 7000 r. and 7500 r., it is best to use a maximum number of seeds of comparatively low strengh in preference to a smaller number, each of greater strength. Seeds 5 mm. in length of 0.6 mc. strength, with an 0.5 mm. gold screenage are very suitable. An area of 3 cm. diameter would call for a total dose of 10.9 mc., made up of 18 seeds, 12 of which would be placed equidistant from each other round the periphery and 6 as an inner circle, 1 cm. within the outer implant (Fig. 5a). A uniform distribution over larger areas, which in properly selected cases should, as stated, seldom exceed a 5 cm. diameter, is aided by the employment of malleable metal circular guides of appropriate size. Complete closure of the bladder in three layers after this technique, and continuous urethral catheter drainage for five days, will generally result in primary healing.

YATES-BELL and HENRIQUES (1957) prefer to use radio-active gold grains (Au[198]) which have an isotope half-life of 2.7 days. The grain is cylindrical, measures 2 mm. by 0.5 mm. and is screened with platinum. The activity of each grain at the time of implantation is 4.5 millicuries and a tumour dose of

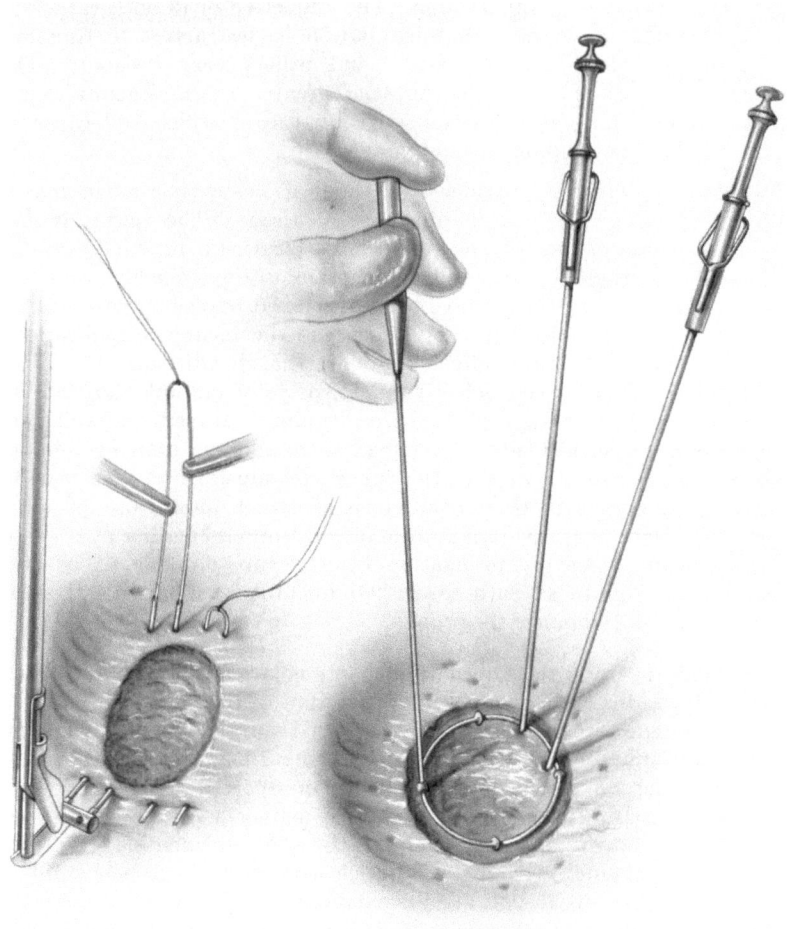

b   a

Fig. 5 a and b. a Irradiation of an area of 3 cm. diameter with radon. Twelve seeds have been distributed as an outer circle; an inner circle is being inserted within a guide held firmly against the bladder wall. b Irradiation with radioactive tantalum wire

between 6000 and 7500 rontgens is aimed at. The "gun" designed by Professor SMITHERS for interstitial irradiation, which can be loaded with fifteen grains, is used to introduce the element. This may be done through the open bladder, the projecting tumour tissue being removed by electroexcision before making the implant; or the grains may be entered by suprapubic puncture with a cysto-scope in position to guide the distribution. With this latter approach it is recommended that transurethral resection of the protruding growth be carried out a few days earlier. The needle is advanced into the partially distended bladder through a half inch incision above the pubis and guided into the tumour area. The magazine containing the fifteen gold grains is loaded into the needle and the gun screwed on. The operator uses one hand to direct the needle to the desired position and the other to manipulate the cystoscope whilst the assistant fires the gun on instruction as the successive punctures are made. In the female the needle may be passed through the urethra alongside the cystoscope. Urethral catheter drainage is maintained for forty-eight hours.

YATES-BELL considers that a more uniform dosage is obtained with less trauma and greater speed than with radium or radon.

Radioactive tantalum wire is another artificially produced isotope which is suitable for interstitial irradiation in the bladder. When radioactive, the metal emits both γ and β rays and is therefore sheathed in a thin platinum covering to protect the tissues from overdosage of β irradiation. With a half life of 112 days, the same material can be used over a relatively long period and can be reactivated in the atomic pile as required. It is thus possible to have an adequate supply of the wire constantly available at an economic cost. WALLACE (1957) considers that with its use a better control of dosage can be obtained than with any of the short half-life isotopes which are allowed to remain as permanent implants. With radon, accuracy of dosage depends on the positioning of the radioactive source whilst with tantalum, which is removed when the estimated required radiation has been achieved, the dosage can be varied at will by allowing the wire to remain for longer or shorter periods. The tantalum insertion is made through the open bladder but the technique of withdrawal permits a complete bladder closure. The wire with its platinum sheath, is easily flexible. It is bent into a hair pin shape with a loop at the closed end and the limbs, one a few mm. shorter than the other, 1 cm. apart. The tumour bearing area is prepared in the same way as for a radon implant. Should this leave a large mucosal defect, it is an advantage to close it by suture in order to reduce the healing time and consequent infection liable to follow the irradiation. The insertion of the wire is made by threading the open ends of the hairpin into the points of a double shafted hollow needle mounted on a boomerang holder. The double shafted needle is directed through the area about 5 mm. below the surface level so that the hollow points emerge facing the internal urethral meatus. On withdrawing the loaded needle, the wire is brought into position with the loop directed towards the internal meatus. One or more hairpins may be required according to the size of the area to be irradiated (Fig. 5 b). Each is connected by a loose suture passed through the loop to a urethral catheter and the bladder closed. When the requisite dosage has been delivered, the catheter is removed; traction on the attached sutures effects an easy withdrawal of the tantalum wire through the urethra.

Intracavitary irradiation can be supplied from a solid central source or from a radioactive solution. Because any deflection of a solid source away from the centre of the bladder could profoundly affect dosage and cause a necrosis of one wall with failure to destroy tumour on the other, the medium has to be very carefully placed. MILLEN (1953) using a solid centre of cobalt[60] which has a half-life of about five years, devised a method of obtaining a fixed and accurate positioning by introducing it into a distended balloon attached to the end of a metal urethral catheter. As the tumour-lethal dose of the radiation extends to a depth of only 6 mm., POOLE-WILSON (1954) considers that this method of therapy is not indicated when there is deep penetration nor when papillomata project beyond 5 mm. into the bladder.

A fluid source contained within an indwelling balloon catheter has the advantage of giving a homogeneous distribution of radiation and avoiding the eccentricity of dosage liable with a solid source. WILDBOLZ and PORETTI (1955) as well as MULLER (1955), using a cobalt[60] solution have reported some encouraging results especially when the malignant tumours dealt with were relatively superficial. A major risk associated with cobalt[60] solution is that of the balloon rupturing. With an isotope of such a long half-life, subsequent absorption could result in the patient developing leukaemia or a sarcomatous bone lesion; and in the event of accidental spilling of the solution, the area where this occurred would remain a radiation hazard for years. Because of these dangers, WALLACE (1957) has tried the effect of alternative media, first with radioactive sodium chloride which has a half-life of only fourteen hours and subsequently with radioactive bromine in the form of ammonium bromide solution the half-life of which is 35 hours. This isotope has a higher proportion of γ to β radiation and a greater degree of penetration than sodium chloride. WALLACE points out that a bulky bladder tumour may by its mere volume displace the bladder wall away from the surface of the balloon and that this could result in the base of a tumour being moved outside the zone of effective irradiation. This drawback could be overcome by instilling directly into the bladder without an intervening container, a radioactive solution, which would make direct contact with the tumour and penetrate by way of the interpapillary spaces to its centre. The tumour, irradiated from all aspects of its surface, would thus receive an amount of radiation considerably in excess of that arriving on the normal mucosa. ELLIS (1955) has employed this technique using radioactive gold colloidal solution which has a half life of 2 days, 17 hours, and is a pure β ray emitter. The bladder muscle level is not therefore reached by the irradiation which is in consequence only applicable to small tumours, essentially papillary in type in the mucosal stage. If incomplete tumour regression occurs, it may be possible to destroy the residual lesions by diathermy, which can be applied without risk, as the underlying blood supply will not have been jeopardized by penetrating γ irradiation. The two methods combined could thus prove useful against multiple scattered superficial papillomata. WALLACE (1957) believes that yttrium[90], an isotope which is mainly a β emitter may prove a more optimum medium for this technique of intracavitary irradiation as it has a considerably higher energy than colloidal gold.

IV. Partial cystectomy

The operation of partial cystectomy or segmental resection involves removal of the tumour bearing area of the bladder wall in its full thickness. The solitary, sessile, broadbased growth of average or high grade malignancy, with well defined edges, is the variety most aptly dealt with in this way, providing the area occupied is within a 6 cm. limit and on a site which favours segmental resection. Involvement of the superior wall or vault or the upper zones of the posterolateral walls particularly lends itself to the operation, especially when the bladder is a large one. Given these circumstances, wide resection, including the removal of an ample free margin is possible and should be followed within a period of a few months by a restoration of bladder capacity and normal function. There have been reports on the use of ileo-cystoplasty to circumvent an inadequate urinary reservoir occurring after extensive segmental resection (PYRAH and RAPER 1955).

Some sessile tumours, particularly those with vague ill-defined margins, may have an invisible microscopic focus distal to the obvious limits of the growth and MASINA (1954) has emphasized the futility of removing the visible tumour-bearing area in such instances without including the microscopic focus. This also applies to the nodular tumour, often of limited dimensions, which is associated with early submucosal neoplastic changes in areas distal to the main growth.

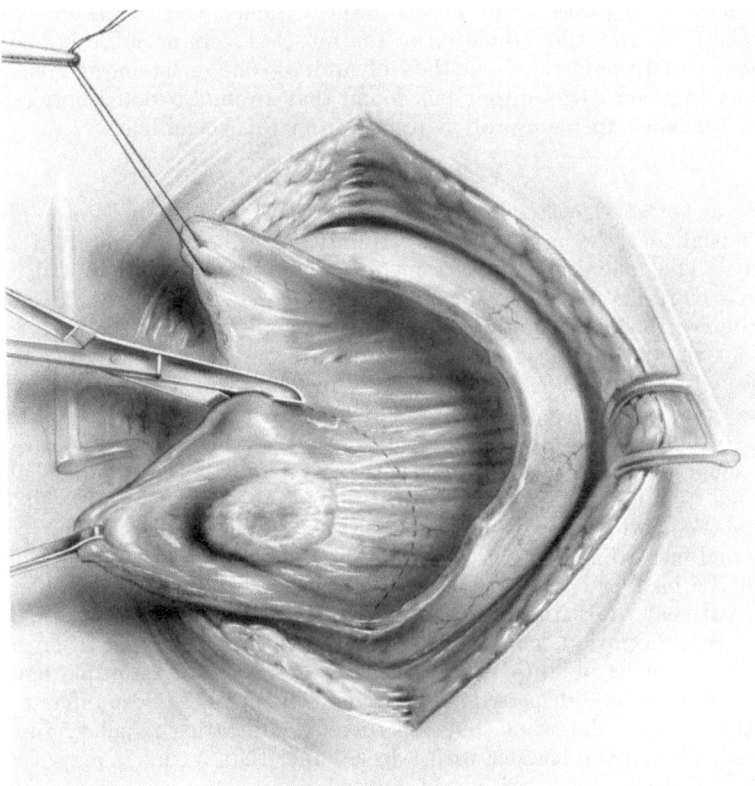

Fig. 6. Partial cystectomy; the location of the tumour permits an ample segmental resection without division of the ureter

A factor which militates against good final results after partial cystectomy is that tumours on the superior zones, for which the operation is most often performed, tend to be silent for protracted periods and therefore of long-standing and often deeply infiltrating at the time of diagnosis. A high proportion have in consequence already developed pelvic glandular metastases before coming to operation. Evidence of enlarged lymph nodes in the region of the aortic bifurcation, the common iliac arteries, and along the external and internal iliac vessels, should be looked for and the peritoneum opened for this purpose at the outset of the operation. If glandular or other distal secondaries are found, removal of the tumour-bearing segment may still, on occasion, be a worth-while palliative as a means of mitigating bladder symptoms.

In carrying out partial cystectomy it is best to mobilize the bladder before opening it, to an extent that will permit removal of the affected site along with a free margin, which should if possible extend to 3 cm. all round the tumour

(Fig. 6). Except for well-defined tumours of limited size situated on the vault or on the upper zones of the lateral, posterior, or anterior walls, this will necessitate tying off the superior vesical pedicle, sometimes on each side. After the increased mobilization which follows this step, the need or otherwise to divide the ureter will become apparent; the line of incision should never be limited in order to avoid doing so. If division is required, reimplantation should preferably be through a separate opening in the bladder rather than at the line of its reconstruction. A direct anastomosis of the trimmed end of the ureter to the bladder mucosa and approximation of the muscle layers over the implant after the manner of LEADBETTER's method of uretero-colic anastomosis gives a very proficient re-union. The author has found that an unobstructed ureteral outlet without tendency to regurgitation results from this technique.

V. Total cystectomy

The role of total cystectomy in the treatment of bladder cancer is highly controversial and the procedure is certainly less frequently resorted to than formerly. The magnitude of the operation and its association with a post-operative mortality of 12 to 15 per cent, even in experienced hands (JACOBS and STIRLING 1952 and RICHES 1956) partly account for this. Perhaps the chief reason for the decline in its use however, has been the gradual realisation that patients with highly malignant infiltrating tumours which cannot be controlled by conservative measures, are not cured of the malignancy even after the entire bladder had been removed. Cystectomy does nevertheless, have a place in the management of bladder tumours. There are certain varieties which although not of high grade malignancy, are difficult or impossible to get rid of by any other method, which are potentially curable by cystectomy. The papillary transitional cell carcinoma with multiple large deposits scattered over different areas of the bladder is in this category. If the cystectomy is performed as the initial treatment whilst the malignant process is still confined within the bladder and before penetration has passed beyond the superficial muscle layer, good final results can be obtained. This likewise applies to the multiple papillomas with an extensive and dispersed distribution. The operation also offers potential curability for the solid sessile but superficially infiltrating papillary tumour on the lower zones of the bladder with a base larger than 5 cm. It may also prove suitable for the nodular tumour of limited dimensions when associated with early neoplastic changes in other areas too widespread to be dealt with by segmental resection. Infiltrative lesions on the bladder base especially when approximating on to the outlet can, on occasions, be advantageously dealt with in this way, the palliative effects in the circumstances being sometimes an improvement on those obtained by other methods.

In the male, a cystectomy generally includes a removal of the prostate and seminal vesicles and in the female, hysterectomy is often advisable (Fig. 7). Some workers (HIGGINS 1952, LEADBETTER and COOPER 1950, WHITMORE and MARSHALL 1956) advocate the additional step of a pelvic lymphatic gland block dissection. A removal of the pelvic connective tissue as with the MILLIN-MASINA technique (1949) will include the regional lymph nodes in the adipose tissue behind the symphysis and the lymph glands along the base of the bladder located in the region of the ureters. A planned dissection of the aortic and iliac glands however adds greatly to the operation, is time consuming, and can be associated with fairly profuse bleeding. FRANKSSON (1950) has shown that if these groups are involved by tumour, coexistent metastases are likely to be present in other parts. This extension to the operation does not therefore seem worth while.

Consideration of cystectomy must also take into account the consequent necessity of urinary diversion. This is generally accomplished by transplanting the ureters into the pelvic colon at the time of the cystectomy though some prefer a two-stage operation. The risks of subsequent upper urinary tract damage and blood chemical imbalance are put forward as further reasons against cystectomy. Though renal damage and electrolyte absorption can result in morbidity and mortality, these sequelae have been much reduced in recent years by the introduction of improved techniques. If the uretero-colic anastomosis functions well, continued renal integrity will generally follow and electrolyte imbalance of a degree that cannot be controlled by suitable therapy will seldom arise

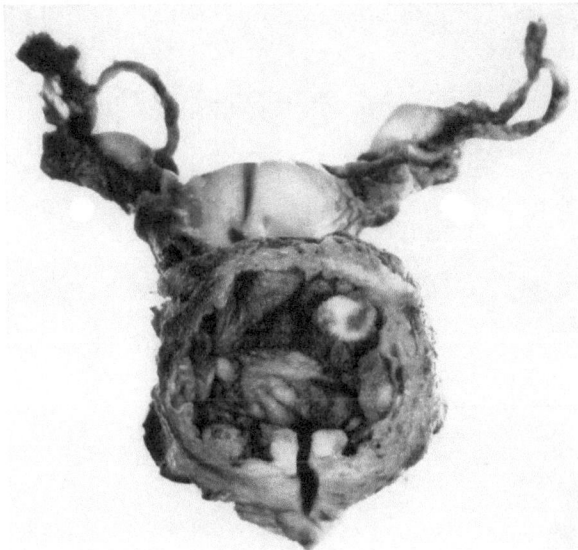

Fig. 7. Uterus and broad ligaments removed in continuity with the bladder. Widespread tumours of both nodular and papillary varieties can be seen

(JACOBS 1956). The alternatives to uretero-colic anastomosis are transplanting the ureters into an isolated segment of ileum with the establishment of an ileostomy (Fig. 8); uretero-colic anastomosis with exclusion of the urine receiving segment by division of the bowel above the site of implantation and the formation of a colostomy out of the cut end of the colon proximal to the line of division; and uretero-cutaneous anastomosis. In assessing the need or justification for these alternatives and the artificial stomata that must go with them, it should be borne in mind that though ureteric implantation into an isolated bowel segment or to the skin will generally be followed by a minimal disturbance of electrolyte balance (IRVINE et al 1956) it will not ensure against upper urinary tract damage. BRICKER et al (1954) reported pyelo-nephritic changes in 30 per cent of 65 cases of bladder substitution with isolated ileal segments followed for more than six months and CORDONNIER (1955) found progressive dilatation occurring in 17 per cent of 41 cases followed up to two years. HUMPHREY (1956) has observed a high rate of renal morbidity after cutaneous ureterostomy.

A method of diverting the urine into an excluded loop of bowel and avoiding an artificial stoma in an unnatural site has been described by LOWSLEY and JOHNSON (1955). The rectum is divided, the lower segment to be used as bladder is shut off and the ureters transplanted into it. The proximal bowel is mobilized

and drawn through the perineum under the external sphincter, to open immediately anterior to the anus. Two separate openings are thus established, one
for faeces and one for urine, under the control of a single muscle. Good functional
results have been reported by JOHNSON (1956) after performing the procedure
eight times for bladder cancer.

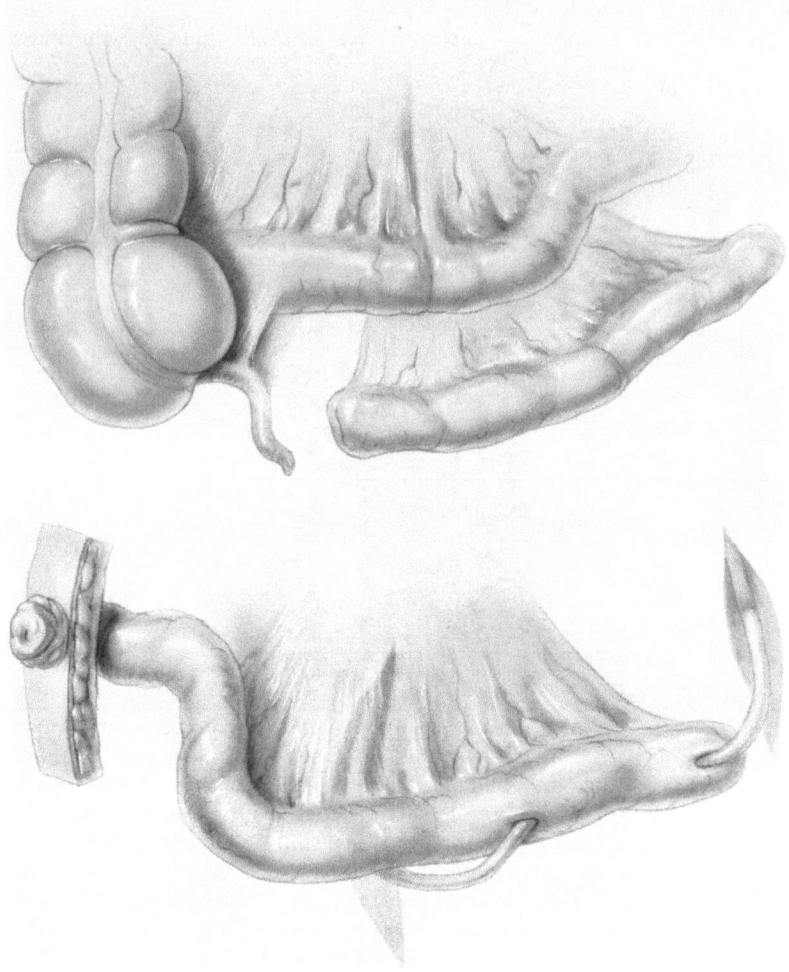

Fig. 8a and b. a Loop of small intestine isolated from the terminal ileum; the continuity of the ileum has been
restored by end-to-end anastomosis. b The ureters have been implanted into the loop the terminal end of which
has been passed through the abdominal wall to form an ileostomy

VI. External irradiation

(Contributed by G W BLOMFIELD)

Implantation of radioactive seeds or radium needles for malignant tumours
in the bladder will deliver a concentrated dose well suited to a small area, but
lesions involving a large portion of the bladder and multiple lesions are

increasingly difficult to cover in proportion to their size. When such lesions are beyond the stage range of interstitial radium or local excision, external irradiation by deep X-ray therapy will cover a much larger volume of tissue and may be the only line of treatment practicable. Total cystectomy may be the only alternative to this form of therapy and some cases unsuitable or too advanced for total cystectomy may yet be treated radically or palliatively by external irradiation.

1. General principles of external irradiation

Deep X-ray therapy for bladder tumours has been in vogue for many years, but the general lack of success in the past with comparatively few cases controlled, resulted in very few cases being relegated to this form of treatment: in fact, it was often given more as a palliative procedure to cases beyond remedy. Causes of failure have largely been due to the following factors:

1. Unsuitable selection of cases.

2. Inaccurate localisation of the tumour, combined with imperfect planning of the treatment.

3. Difficulty in reaching a high local dose without scattering the rays to adjacent and vulnerable tissues.

4. Superficial reactions limiting the depth dose.

The last two difficulties have been overcome by the employment of more appropriate apparatus, and there can be little doubt that much of the technical difficulties in the past were consequent upon limitations of the radiation available. By using a large number of beams directed with great accuracy, or by using a rotating beam, it is possible to make the best of orthodox irradiation in the 200 to 400 kilovolt range but there is no real substitute for the high energy irradiation which is now becoming available at all the principal radiotherapy centres. The high energy irradiation at the two million volt level upwards has the following characteristics which are so valuable:

1. Build up of dose beneath the skin with avoidance of troublesome skin reactions.

2. Greater penetration.

3. Reduction of scatter.

It should therefore be used in preference to the lower energy irradiation whenever possible.

2. Indications for external irradiation

External irradiation is still a mere alternative to more localised forms of therapy and the choice of this method implies rejection of methods which might be radical or equally successful by attacking the tumour locally. It does not replace diathermy, local excision, partial cystectomy or radon implant when these methods are appropriate and likely to be successful. The following clinical and pathological characteristics of a tumour would bring it towards this category[1]:

a) General pattern of tumour

Extensive or multiple papillary tumours covering a large portion of the mucosa.

b) Histology

Possibly the histology in itself has little bearing upon sensitivity, but undifferentiated and anaplastic growths are generally less amenable to other

---

[1] This follows headings taken from "The classification of tumours of the bladder" in Broadsheet No. 1 issued by the Institute of Urology

methods. Practically no tumour can be judged insensitive until irradiation is tried and some which are fixed and beyond the scope of surgery may yet resolve under X-ray therapy.

c) Direct continuity spread

A growth which has spread over a large area but not penetrated beyond the bladder wall will most likely be unresponsive to other methods of treatment excepting total cystectomy. It is here that one has to consider external irradiation as a useful alternative bearing in mind that total cystectomy may yet be carried out if irradiation fails.

Spread to the adjoining organs or fixation to the pelvic wall most likely brings the growth to the inoperable stage where irradiation is the only line of treatment left; palliative or radical X-ray therapy is then the only remaining line of defence.

d) Lymphatic spread

This is generally a difficult point to settle. In th eearly stages of lymph node invasion there will be no clinical evidence of lymph gland involvement. External irradiation can be extended to cover some of the lymphatic field and may control early and limited involvement of adjacent nodes. If the growth is anaplastic and there is any evidence of spread beyond the bladder the paravesical glands and the internal iliac group should be included in the treated area. Glands outside the true pelvis are unlikely to be controlled for long.

e) Spread by blood stream

Bone metastases are not uncommon and they will respond in a palliative way. Metastases to liver, brain, lung and other organs are seldom worthy of treatment and generally indicate widespread dissemination beyond the range of therapy.

3. Method of treatment

From an anatomical standpoint there are two main methods of delivering the radiation:

1. Treatment to a relatively small volume covering the affected area of the bladder and leaving adjacent structures and glands at a low dose.

2. Treatment of the whole bladder including the paravesical tissue and the regional lymph nodes in the immediate vicinity of the irradiated field.

Clearly the clinical and pathological findings will influence the choice of local or extensive irradiation but it must be remembered that treatment over a wide area results in a more limited dosage and may therefore limit delivery of a radical dose to a lesion which is only of local extent.

4. Localisation of the growth

Accurate localisation of the known growth is essential. Bladders vary in position though, in the male, it will generally be found that the base of the bladder is level with the upper margin of the symphysis pubis or slightly below it. A small quantity (about 50 ccs.) of opaque medium is injected into the bladder per urethram to obtain localising X-ray photographs. One phial of pyelectan mixed with forty cubic centimetres of a $12^1/_2\%$ solution of sodium iodide forms a very satisfactory mixture for this purpose and is

preferable to using more
concentrated solutions of
sodium iodide alone. It is
also less irritating to the
bladder. Radiographs are
then taken in the antero-
posterior and lateral pla-
nes. These will accurately
locate the position, depth
and spatial relationships
of the bladder. The posi-
tion of the primary lesion
relative to the bladder
should be known by
cystoscopic examination
and may show quite well
as a filling defect. Figs. 9
and 10 are taken from
two such localising radiographs. The shadow seen
in relation to the pubis is
from a bag of bolus ma-
terial (beads of sago or
ground rice) placed on the
abdomen in order to pre-
vent over exposure and
fogging of the plate. The
bladder is outlined by the
contrast medium which in
this case shows an ob-
vious filling defect, from
a large tumour on the
left side of the bladder.

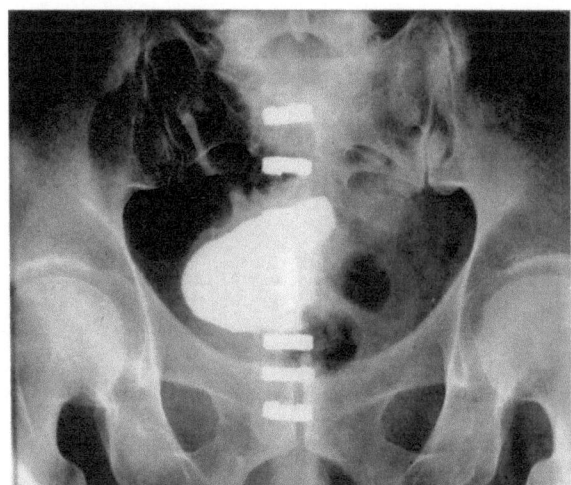

Fig. 9. Antero-posterior X-ray of the pelvis showing contrast medium in the bladder. A large filling defect is present on the left side. This locates the tumour vertically and laterally

After obtaining the
localising X-rays the contour of the pelvis at the
level of the lesion, is outlined on paper, and the
X-ray fields can then be
accurately planned. Iso-
dose charts for a typical
case are shown in Figs. 11,
12 and 13. These are
drawn up for 200 Kilo-
volt, 2 million volt and
8 million volt X-ray
therapy and show the
advantages of high energy
irradiation. It can be seen

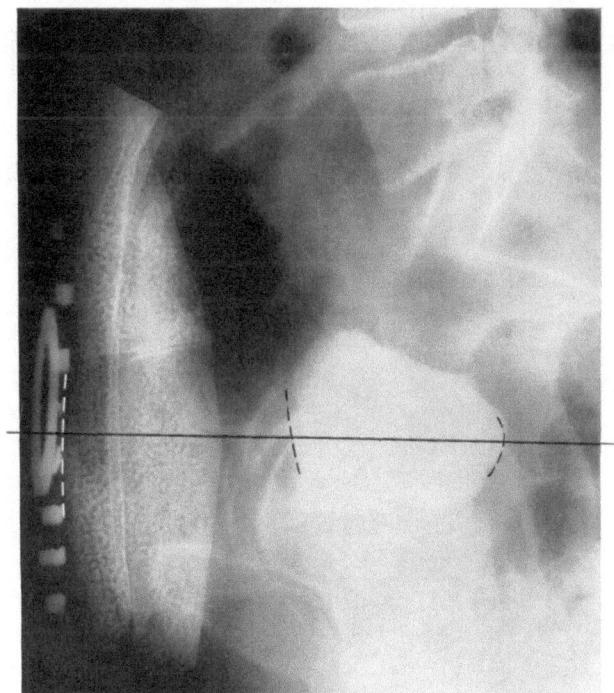

Fig. 10. Lateral X-ray corresponding to Fig. 1. This is taken to locate the depth of the tumour below the symphysis pubis. A ring is shown (edgewise) on the anterior abdominal wall. The shadow below this is due to packing material placed in a position to prevent fogging of the film

that the distribution of dosage at the higher energy is confined more accurately
to the tumour area; the pelvic skeleton suffers less irradiation and the skin
escapes injurious reactions.

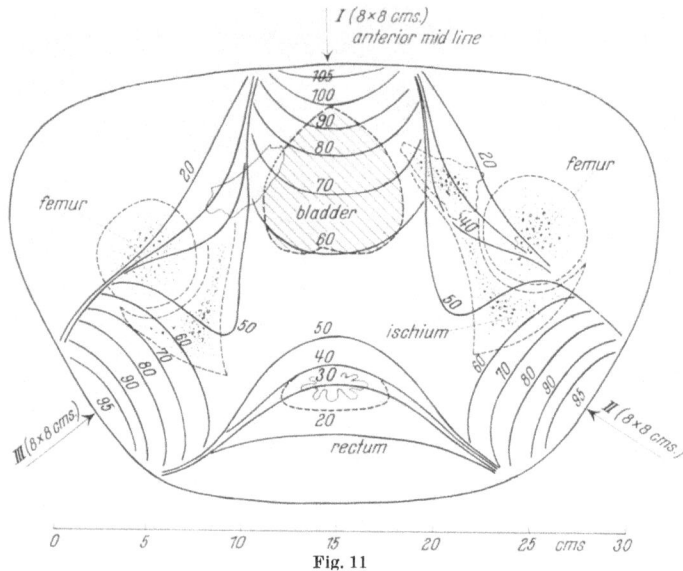

Fig. 11

Figs. 11, 12 and 13. In these three figures the isodose curves are shown for the same case when using the same x-ray fields at 200 kilovolts, 2 million volts and 8 million volts. The contour lines show the x-ray dosage levels and illustrate the improved dosage distribution with high energy radiation. This is reflected clinically by the absence of skin reactions and a better resolution of the growth. (Photograph by courtesy of British Journal of Urology, from "Irradiation Therapy in Urology" G. W. BLOMFIELD, 1954.)

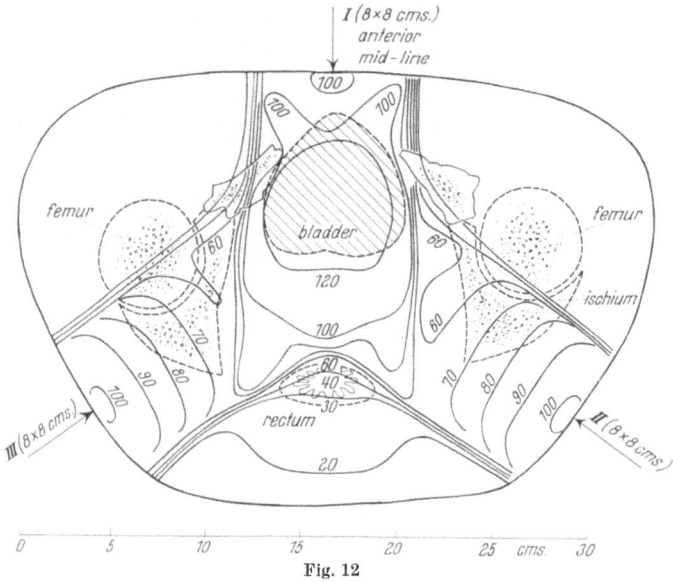

Fig. 12

5. X-ray dosage

Using orthodox X-rays and multiple beamed fields (say 5 instead of 3) a depth dose of 4000 to 4500 roentgens can be delivered in five weeks—at some risk of causing rather severe skin reactions. The treatment is generally an ordeal and is better given slowly to avoid severe side effects, nausea, vomiting, stranguary and diarrhoea. Diarrhoea and bowel reactions indicate that treatment is being carried out too rapidly for the tissues to tolerate and a reduction of the daily dose is then necessary. Urgency of micturition and strangury commonly appear towards the end of the course. Alkaline diuretics may relieve this, but a look-out should be kept for cystitis and infections which should be promptly treated.

With high energy X-rays or cobalt teletherapy a tumour dose of 5000 to 6000 roentgens is possible in 5 to 6 weeks; the depth dose must be carefully ascertained and there is no warning from skin reactions at the surface as to what s happening at depth.

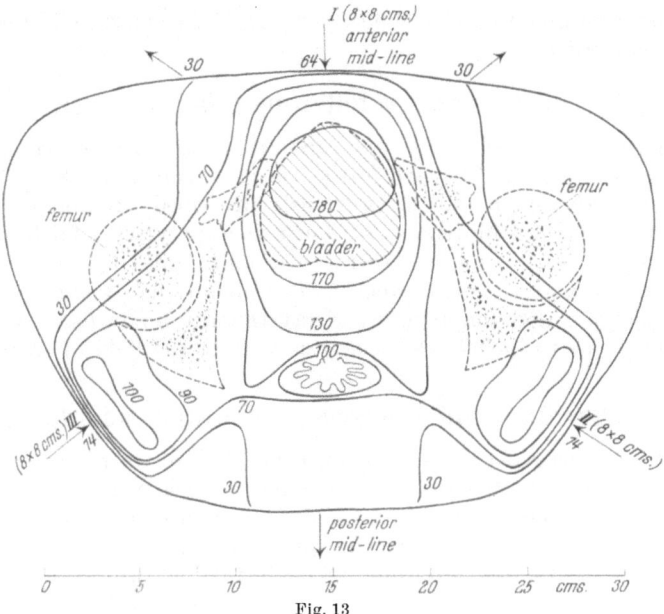

Fig. 13

6. Results

Nothing is more difficult to analyse than the relative results in advanced cases of cancer by various methods of treatment. The results of X-ray therapy are vitiated by reason of the more advanced cases being treated by this method. The following table, taken from the Sheffield Cancer follow up gives death and survival up to five years after X-ray treatment for 1950, 1951 and up to four years for 1952. It should be emphasised that these are cases of advanced carcinoma, and the 1951/52 figures included patients sent from further afield when it became known that special treatment was available at Sheffield for this type of case. The figures show, however, that there is a possible salvage of twenty to thirty per cent by X-ray therapy alone for the advanced case.

Carcinoma of bladder — SHEFFIELD

Year	No. of cases	1 Yr.	Dead			
			2 Yr.	3 Yr.	4 Yr.	5 Yr.
1950	26	9	4	1	1	11 alive
1951	34	16	7	3	1	7 alive
1952	66	32	14	6	14 alive	

7*

7. Palliative treatment

When growths of the bladder are quite beyond radical surgery and too extensive or widespread for any reasonable hope of cure by X-ray therapy, a single field of X-rays placed over the bladder and covering the growth will be sufficient for controlling haemorrhage and for effective palliation in the majority of cases. Preferably it should be high energy irradiation and the treatment should be given sufficient time to judge the response. In the most advanced cases it may be found desirable to desist after a short trial of two or three weeks but in others it may be worth while going on to more radical treatment according to the response. Relief of haematuria is the rule rather than the exception.

8. Recurrences following other measures

Recurrences after local surgery or radon treatment and recurrences of growth on an unstable mucosa, if beyond further surgery, can be dealt with by external irradiation in the same way as previously described. Radon to a small area should not preclude a full course of X-ray therapy after a fairly long interval.

Recurrences in the surgical scar are usually from implantation of growth seedlings or carcinoma cells during or after operation. It is not an uncommon sequel to surgery and consequently X-ray therapy following cystotomy should be so planned as to include the operation scar in the field of treatment.

9. Treatment of recurrences following radiotherapy

It is difficult to give radiotherapy over the same area without risk of exceeding the tolerance dose to the tissues so irradiated. The main chance of cure lies with the original course of treatment and a full radical course of external irradiation cannot be repeated to cover the same tissues.

Previous radiotherapy does not preclude further surgery unless tissues are badly scarred or fibrosed. This scarring and fibrosis can be kept at a minimum if high energy irradiation is properly used but is likely to be present after a full course of orthodox-X-ray therapy at lower voltages. Partial or total cystectomy may yet remain as possible lines of defence when radical X-ray therapy has failed to cure the primary tumour.

10. Follow up

A regular follow up of patients having external irradiation is as desirable as for any others, but it is wise to avoid cystoscopy for two or three months after treatment because reactions will take at least two months to settle down. Cystoscopic appearances are difficult to interpret until all reactions have cleared.

11. Complications

The chief complications which follow irradiation to the bladder are:
1. Contracted and irritable bladder.
2. Haemorrhagic mucosa.
3. Ulceration of bowel.

Contraction and fibrosis are more likely to occur in association with a septic bladder. Haemorrhage may result from a severely damaged mucosa when reactions have been excessive. It is uncommon but sepsis again can aggravate this condition. It is, however, a condition which follows destruction of the normal, healthy and very regenerative mucosa. A similar condition has been met with as a complication following the irradiation of the bladder mucosa by a central source or by irradiation delivered from radioactive solutions introduced into the bladder. Usually the haemorrhage will cease with catheter drainage and attention to bladder sepsis. If not, a total cystectomy may be the only remedy. It is not necessarily associated with persistence of growth which may indeed be quite absent when haemorrhage appears.

Ulceration of bowel is a rare complication if treatment is properly planned. It is due to overdosage of the bowel and the rectum is the most likely area to be involved. It is avoided by careful attention to the positioning of the X-ray fields and by keeping the bowel dose below "tolerance" level.

12. Availability of apparatus

Almost all the principal radiotherapy centres now have access to high energy irradiation. Apparatus of this type gives radiation equivalent to X-rays of two million volts and upwards. The teletherapy from radioactive cobalt is equivalent to an X-ray machine of two to three million volt energy and equally good. Linear accelerators of four million volts and upwards are perhaps even better, but it will be seen from the isodose charts (Figs. 11, 12 and 13) that irradiation from any of the "high energy" machines will give an adequate dosage distribution. The higher energy has certain biological advantages, and in addition to better dosage distribution it is less damaging to skeletal tissues. On a long term basis it is no more expensive than orthodox irradiation but the capital cost is considerable.

E. Discussion on methods of treatment

The means at our disposal for the treatment of patients suffering from bladder tumours have been described. It is generally accepted that the superficial histologically benign papilloma is best treated by electroexcision or electrocoagulation which should be applied by the transurethral route unless the size and position of the growth make a suprapubic approach preferable. A five year and more survival can thereafter be expected for over 85 per cent of cases. Yet during the five year period a reappearance of neoplasm is likely in over 50 per cent and this will prove to be histologically malignant in 15 to 20 per cent (NICHOLS and MARSHALL 1956). The treatment though highly effective can thus never be regarded as curative and follow-up cystoscopic examinations over a life-time are necessary. Treatment on similar lines is also of merit for carcinoma of low grade and superficial stage, that is when invasion has not penetrated beyond the halfway level of the muscle layer. High grade, deeply infiltrating tumours are not, however, satisfactorily controlled by this method which in the circumstances has but little to offer in terms of survival. Thus in an analysis of patients treated by electroexcision and electrocoagulation, NICHOLS and MARSHALL (1956) reported a survival of five years and longer for 82.6 per cent of 69 patients with low grade superficial tumours; for 76.5 per cent of 17 patients with high grade superficial tumours; and 16.7 per cent of 18 patients with high grade deeply infiltrating tumours. Because palliation often follows, with good prospects of retaining bladder function at a low operative risk, the treatment is particularly attractive for patients of advanced years or whose general condition precludes major surgical procedures.

The operation of partial cystectomy or segmental resection is a useful and satisfactory procedure in selected cases. Although it may be applied to any part of the bladder, the technique becomes increasingly difficult when the tumour involves the trigonal area or neck and division of a ureter with its re-implantation site is in consequence required. In the female, a resection which encroaches on the internal meatus is liable to result in incontinence. CONWAY and BRODERS (1942) demonstrated microscopic extension 1 cm. beyond the periphery of the primary tumour in 19 of 40 specimens studied, the distance in some being up to 3 cm. Removal of such an adequate free margin with the tumour-bearing area is most readily achieved when segmental resection is performed for growths in the upper zones of the bladder. As with other methods of treatment, the long term results are directly related to the grade and degree of penetration. JEWETT and CASON (1948) in a study of 55 cases after segmental resection found no subsequent deaths from bladder cancer in 14 in which infiltration "had not reached the half way level".

The evolution of total cystectomy as a method of treating bladder cancer can be attributed to dissatisfaction with the results following conservative measures especially when used against infiltrating tumours and those of widespread distribution. It has however become increasingly apparent that cystectomy has not enhanced the control of high grade high stage tumours; a majority of the patients with this variety of growth fail to survive for even two years after the operation (JACOBS and STIRLING 1952). Neither radical total cystectomy with extensive lymph node dissection nor pelvic exenteration appear to improve the end results (WHITMORE and MARSHALL 1956). The enquiry instituted by DE GIRONCOLI (1955) showed that many surgons who formerly practised cystectomy had abandoned the procedure because of high mortality and poor late results. A greater number however, considered that there was still a definite role for cystectomy on a selective basis. The consensus of opinion amongst this group of surgeons was that the best prospects from cystectomy followed when the operation was performed for a generalised involvement of the bladder by low grade, low stage tumours; for persistent development of superficial multiple recurrences after conservative therapy had failed; and for tumours of large size with wide surface involvement. In addition it could provide symptomatic relief and temporary palliation in selected patients suffering from high grade high stage tumours.

When palliation is the sole possible objective, urinary diversion by ureterocolic anastomosis can by itself give much relief if vesical strangury is a pronounced feature. A reduction of haematuria can also be brought about by simultaneous ligation of the internal iliac arteries.

Irradiation therapies, both interstitial and intracavitary, as with other treatments, give best results when used against low grade superficial tumours. Some of the new techniques of application using the modern sources of radiation show evidence of providing a more accurate distribution and dosage but have furnished no dramatic change in the outlook of cure. It is likewise with external radiation. Megavoltage machines whether γ or X-ray enable the required dosage to be given more easily and rapidly and with less scatter. Painful reactions associated with conventional intensive deep therapy are thereby avoided or much reduced and there is in consequence less constitutional upset. But though some superficial tumours and an occasional infiltrating one may be destroyed, external radiation in general does not cure bladder cancer. Furthermore the successes have to be offset by the frequent sequelae of contracted bladder, vesical irritability and pain.

With a plethora of treatments ranging between such a minor procedure as cystodiathermy and extensive operations like cystectomy or even pelvic exenteration, the optimum method applicable to a particular patient can on occasions prove a difficult problem. It may be said that any therapeutic programme for bladder cancer should be based on the principle that superficial localised growths do not require more than complete local destruction or simple extirpation, whilst tumours that are deep and no longer localised cannot ordinarily be eradicated by conservative measures. A comparison of results on similar cancers has not yet proved whether local destruction by electroexcision and electrocoagulation with or without irradiation is as satisfactory as surgical removal by segmental resection or cystectomy. The assessment by MOSTOFI (1956) suggests that better long-term results follow surgical exstirpation.

The conclusion must be that after the relevant and efficient use of any of the competing methods, the final outcome of a bladder cancer is thereafter largely predetermined by the grade and stage of the growth rather than by the type of treatment administered to it.

F. Secondary tumours of the bladder

Secondary tumours, which are usually carcinomas, may reach the bladder by direct extension from the cervix. rectum, pelvic colon, prostate or seminal vesicles: by implantation from a primary focus in the renal pelvis or ureter; or rarely by metastasis from distant sites such as the stomach and adrenals.

Carcinoma of the cervix is the commonest extravesical malignant lesion to affect the bladder. Clinical manifestations are similar to those arising from primary bladder tumours. In the earlier stages of the disease before actual invasion has occurred, cystoscopy will show an elevation of the bladder base due to extra-vesical pressure which may be marked enough to produce distinct lateral sulci on the adjacent bladder wall. With the start of actual invasion, local areas of submucosal haemorrhage and oedema can be seen and are followed by nodular projections which eventually ulcerate. If survival is long enough, a vesico-vaginal fistula will occur. This complication was noted in sixty-two of 683 cases studied by GRAVES, KICKHAM, and NATHANSON (1936), who also stated that BEHNEY found vesico-vaginal fistulae in 22.3 per cent of 166 cases of carcinoma of the cervix at post-mortem examination. The spread of the malignant process with extension into the perimetrial tissue is often associated with dilatation in the upper urinary tract.

Carcinoma of the rectum or sigmoid colon is the other extravesical lesion most likely to be responsible for a secondary bladder cancer by direct extension. This may take place as a late recurrence subsequent to the removal of the affected bowel; OPPENHEIMER (1943) reported nine instances of invasion of the bladder and prostate after abdomino-perineal resection. Extension of the cancer is liable to result in a vesicocolic or vesicorectal fistula though in the female patient a rectovaginal fistula may be a sequel. Occlusion of one or both ureters by compression or actual invasion is common in these late stages of the disease, when death in most instances is due to urinary tract obstruction and infection.

The treatment of advanced invasive tumours arising from cervix or bowel is usually palliative and symptomatic. Transurethral resection will relieve symptoms resulting from a bulky intravesical tumour intrusion and diathermy will diminish bleeding. The constant soaking which follows the development of a malignant vesico-vaginal fistula can only be relieved by a urinary diversion which should generally be accomplished by uretero-colic anastomosis. When, however, the malignant process has spread to the posterior vaginal wall and threatens to extend into the rectum, cutaneous ureterostomies should be established or the ureters should be transplanted into an isolated loop of ileum and an ileostomy stoma formed; if as occasionally happens the vesico-vaginal fistula is associated with a recto-vaginal fistula, a colostomy will also be required. Radical surgery in the form of pelvic exenteration (BRUNSCHWIG 1948) can sometimes be offered as an alternative to palliative measures but the inherent malignancy of these extensive growths adversely affects the prospects of survival, and success following the operation is likely to be transitory.

Retrovesical sarcoma is an extravesical tumour which is generally considered to be unassociated with any particular organ though LAZARUS (1946) recorded two cases of primary sarcoma arising in the seminal vesicles. The tumour is located behind the bladder and usually between the seminal vesicles which lie on its posterior aspect. It is above the level of the prostate, which it tends to push downwards whilst displacing the rectum backwards and the bladder forwards. In most of the recorded cases the sarcomas have been of the small round cell variety. Invasion of the bladder wall with the tumour cells extending between the muscle fibres is late; the cystoscopic appearances then resemble a sarcoma arising in the bladder itself. Before this stage is reached a large mass may have developed which projects above the pelvic brim. The encroachment on the bladder reduces the capacity and may cause difficulty in voiding or retention. In a case reported by LAZARUS (1946), regression of the tumour took place following a partial excision and radon seed implantation. PYRAH (1954) has described three cases of retrovesical tumours which were respectively a retrovesical adenoma of the prostate, a retrovesical fibroma, and a large, recurrent retrotrigonal adenoma of the prostate. Surgical removal was accomplished in each.

I. Carcinoma of the renal pelvis or ureter

The association of a papillary carcinoma of the bladder with a similar lesion in the upper urinary tract has generally been considered to result from implantation of tumour cells carried in the urine. This view is questioned by some who believe that tumour cell implantation would be resisted by normal bladder mucosa and that the tumours are in fact primary in the bladder as well as in the upper urinary tract. The multiple focal involvement is attributed to a tumour diathesis, possibly activated by a carcinogenic agent in the urine such as has been described by WALLACE (1956). Support for this latter view is forthcoming from evidence of papillary tumours in the pelvi-calyceal system becoming manifest many years after a bladder tumour is first encountered. HELLSTRÖM (1956) reported a case of papillomatosis, first in the bladder, later in the urethra, and six years after the detection of the initial bladder lesion, in the upper calyx of the right kidney. Vesical recurrences continued after nephroureterectomy though the urethral papillomatosis was abolished by resecting the urethral mucosa. The author has also encountered a case of papillary carcinoma of the renal pelvis and ureter which became apparent ten years after the initial bladder neoplasia, which was characterised by new tumour formations throughout the ensuing decade. Intravenous pyelography at the outset had shown no evidence of renal tumour. On the other hand, SCOTT (1951) reported on a papillomatosis persistently recurring over a period of ten months after the destruction of three initial pedunculated growths, which was cured by nephroureterectomy and segmental cystectomy after investigation had revealed a papillary carcinoma of the ureter. The patient was well after ten years. If the causative factor in this instance was a carcinogenic agent in the urine, its excretion must have been confined to the one kidney.

II. Metastases from distant sites

Metastases to the bladder from primary tumours in distant organs are rare. HERBUT (1952) mentions a case of gastric carcinoma in a thirty-eight year old man, metastasizing to the bladder after five years. It was reported by HERMANN (1929) who in an analysis of his cases of KRUKENBERG tumours found ten instances of carcinoma of the stomach in four of which there was metastases to the bladder; one case of carcinoma of the caecum with metastasis to the bladder and one of carcinoma of the gall-bladder with metastasis to the urinary bladder. There was evidence of lymphatic spread in five out of the six in the series. HERBUT (1952) also quotes a case of a pedunculated tumour of the bladder in a three year old boy which had the same histology as a WILM's tumour removed five months earlier; and another unusual secondary bladder tumour which was removed from a sixty-five year old man who at autopsy eleven months later was found to have a melanotic sarcoma in both adrenals. Metastasis to the subcutaneous tissues also occurred.

In a report on a malignant tumour of the bladder secondary to an adenocarcinoma of the breast which manifested itself seven years after a left radical mastectomy, GUNEM and BALAT (1956) tabulated the primary focus of eighty cases found in the literature. The order of frequency of the focus of origin was as follows:

Gastric carcinoma	25	Renal cell carcinoma of the kidney	2
Malignant melanoma	18	Carcinoma of ovary	2
Carcinoma of female breast	16	Carcinoma of cervix	1
Lung carcinoma	6	Carcinoma of uterus	1
Rectosigmoid carcinoma	3	Carcinoma of caecum	1
Malignant tumour of testis	2	Carcinoma of gall bladder	1
Carcinoma of pancreas	2		

III. Metastasis

Extension of bladder tumours to other organs occurs by direct growth, by implants, and by lymphatic or haematogenous metastasis. Metastasis is chiefly found in the paravesical, iliac and inguinal lymph glands, the vertebral column, the liver, the lungs and the peritoneum. In FRANKSSON's (1950) series of 67 autopsy cases, metastasis was demonstrable in 33. By excluding eight with benign tumours, this 49.3 per cent incidence was increased to 54.4 per cent, a figure fairly consistent with LEADBETTER and COLSTON's (1937) 57.1 per cent of 98 autopsy cases and JEWETT and STRONG's (1946) 50.0 per cent of 104 autopsy cases. FRANKSSON considered that "a high grade tumour which has penetrated deeply presumably has manifest or incipient metastasis."

A rare form of metastasis is a secondary growth in the penis. In a review of the literature PAQUIN and ROLAND (1956) found that the bladder was the primary site of the lesion in 21 out of 64 patients. They considered that the mechanism was either by direct extension or by retrograde venous or lymphatic transport. The appearance of the deposit in the penis anticipated the death of the patient by a few weeks or months in all but three patients. The following is an account of a case of your author, not included in the review of PAQUIN and ROLAND (1956).

The patient, first seen in 1951, when 57 years of age had two sessile infiltrating tumours, one on the right infero-lateral wall of 2.5 cm. diameter and the other on the left postero-lateral wall of 3 cm. diameter. Treatment was by

transvesical electro-excision and interstitial irradiation with radon seeds. The
growths proved to be an active anaplastic transitional cell carcinoma with
infiltration of the deep muscular layers. During the ensuing twenty-two months,
new tumour formations were encountered and treated by transurethral diathermy.
Because of widespread involvement, further attempts at control by conservative

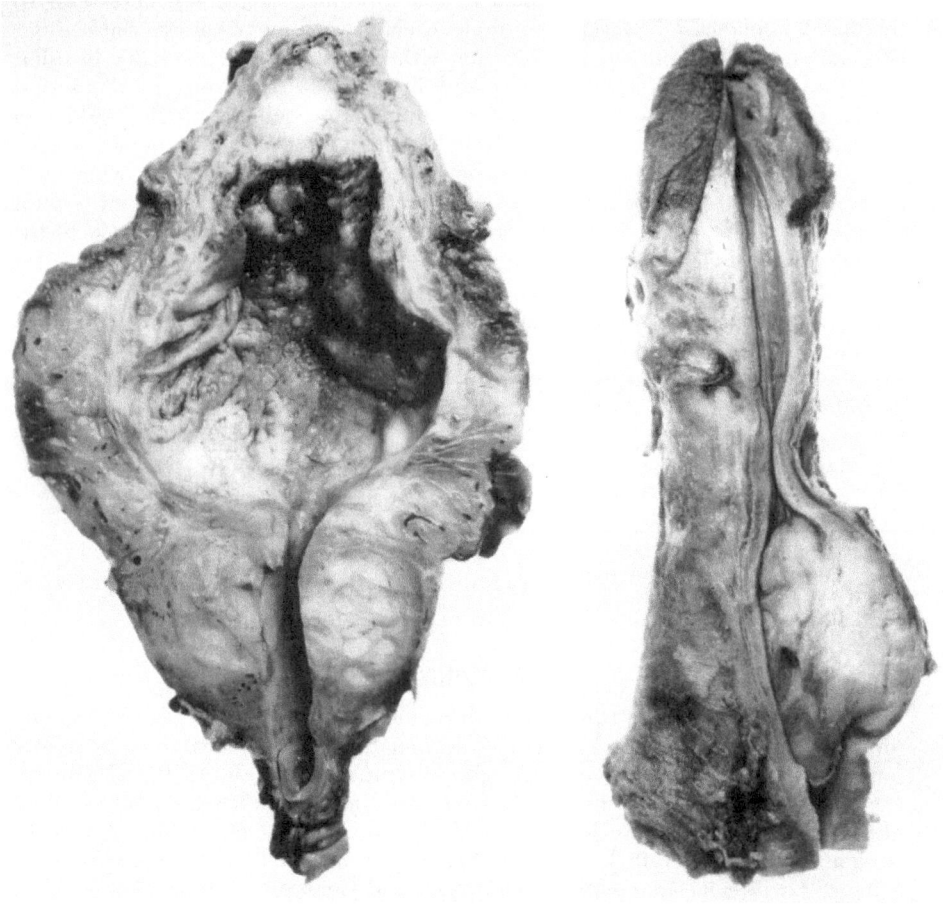

Fig. 14   Fig. 15

Fig. 14. Cystectomy specimen showing extensive tumour involvement twenty-two months after interstitial
irradiation with radon

Fig. 15. Penis with tumour in corpus spongiosum removed five months after cystectomy (same case as Fig. 14).
The urethral epithelium is intact

therapy were then deemed futile and on Juli 1953 a one-stage cystectomy was
carried out (Fig. 14). Five months later a tumour mass appeared on the dorsal
surface of the proximal penis and was dealt with by amputation of the penis and
orchidectomy. The tumour was located in the corpus spongiosum, normal
transitional urethral epithelium covering it (Fig. 15). Histologically it proved
to be an anaplastic carcinoma similar to the primary bladder growth. The
patient remained well and active until 1958, when evidence of metastases appeared. He died in August of that year.

G. Non-epithelial tumours of the bladder

Bladder tumours of mesenchymal derivation are rare. The relative incidence of authentic malignant cases is more accurately reflected in the lower of the figures (0.3 per cent) which have been quoted by Professor SYMINGTON, whilst that of benign mesenchymal tumours is even less.

The clinical manifestations of benign bladder tumours of connective tissue origin depend on their location and on associated pathological changes. Frequency and dysuria are generally the dominating symptoms, haematuria with the exception of angiomas, being inconspicuous in the absence of ulceration. When a tumour has reached large dimensions, a mass in the pelvis may be

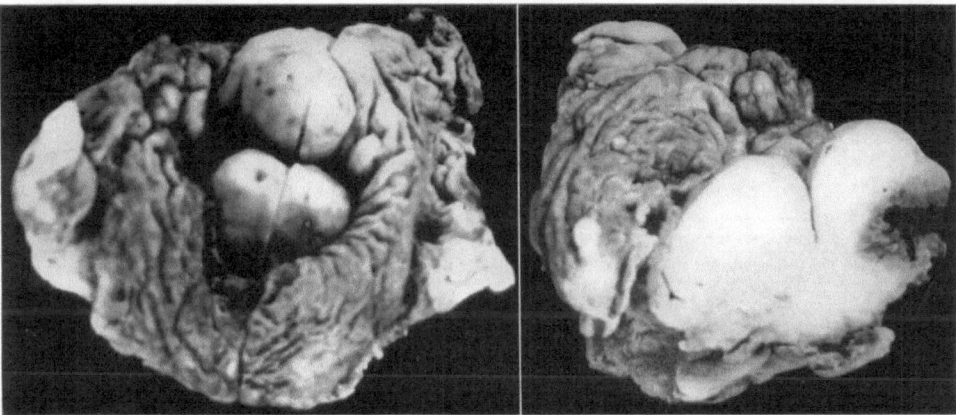

Fig. 16   Fig. 17

Figs. 16 to 21. From JACOBS, A., and SYMINGTON, T. Primary lymphosarcoma of the urinary bladder. Brit. J. Urol. 25, 2, 119. Reproduced by permission of the editor and publishers

Fig. 16. The bladder is opened to show two apparently separate tumours. The bladder wall is irregularly thickened and the two rounded growths project into its lumen. × 2/3

Fig. 17. The growth is seen on section. It is found to be a single tumour united over the lower half. Part of the growth, to the right, is necrotic. The portion of the bladder wall adherent to uterus is seen at the bottom. The growth does not extend through the whole thickness of the bladder (Fig. 21). A small nodule is seen projecting from the bladder wall immediately above the main mass. × 1

disclosed on combined abdominal and rectal or pelvic examination. Angioma is of special clinical interest, for it may be confused with varicosities or simple dilatation of vessels, quite commonly encountered on the bladder floor. A true angioma is one of the rarest of all bladder tumours and is seen cystoscopically as a bluish-black, irregular-shaped sessile mass most commonly located on the dome. The age incidence ranges from early childhood to old age.

The treatment of benign mesenchymal tumours depends on their size and location. Very small lesions can be dealt with by cystoscopic resection and fulguration; for the larger tumours a partial cystectomy is best.

Malignant mesenchymal tumours have been divided into seven groups in the pathological section, the classification being based on the histogenesis of the growths. All must be viewed as highly malignant or potentially so, but as with carcinomas, the degree of malignancy is largely influenced by the histological grading, depth of growth and degree of spread. Regardless of the specific variety of malignancy, the symptoms tend to be similar. Painless haematuria is usually the outstanding symptom in adults though it is less common in children in whom the initial complaint is more likely to be dysuria which can culminate in retention. The other dominating clinical manifestations are painful urination associated with suprapubic discomfort, frequency and dribbling incontinence.

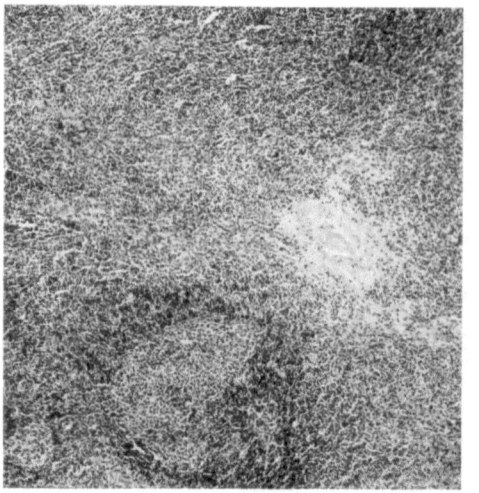

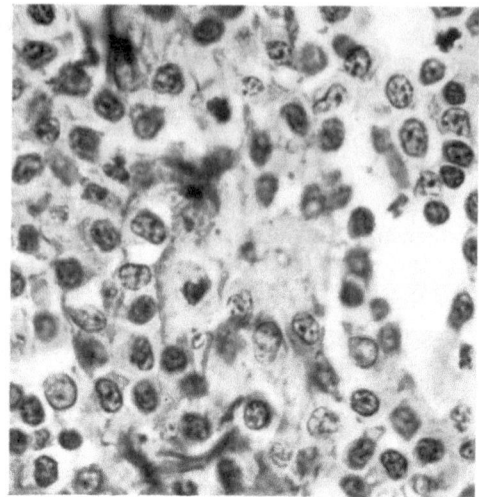

Fig. 18   Fig. 19

Fig. 18. The growth consists of a diffuse mass of mature and immature lymphocytes. A well-formed lymph
follicle with germ centre is seen to the left of the field Haematoxylin and Eosin. × 100

Fig. 19. The cells shown in the main mass in Fig. 18 consist mainly of immature lymphocytes (lymphoblasts) and
a few lymphocytes. Mitoses are seen in a few cells. Haemotoxylin and van Gieson. × 500

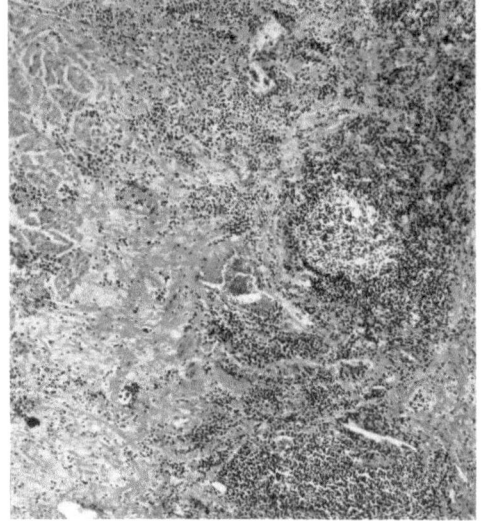

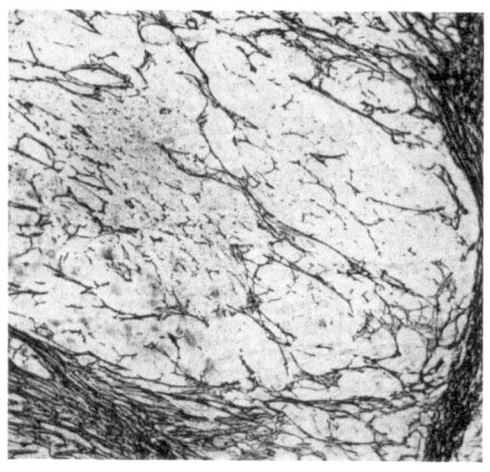

Fig. 20   Fig. 21

Fig. 20. The bladder muscle (seen below) is invaded by tumour cells in which a well-formed lymph follicle is
present. Haematoxylin and Eosin. × 100

Fig. 21. Sections taken from the base of the tumour where it is invading bladder muscle. The reticulin of
the bladder is compressed to the periphery. No reticulin fibrils are seen between individual tumour cells.
Reticulin × 57

A pelvic mass may be finally palpable, especially in the female patient. Priapism
has been described as a symptom (FEGGETTER 1937).

The results of treatment vary considerably. Whilst the final outcome of
bladder sarcoma is too often fatal and treatment therefore largely a palliative
problem, this is not invariably so. Our own case of lymphosarcoma, which was
treated by cystectomy and hysterectomy (Figs. 16 to 21) with uretero-colic an-

astomosis, carried out in 1950, is today, at the age of 69 years, well and active. Surgical extirpation when possible is the treatment of choice. If the lesion is discovered early this may be accomplished by a wide partial cystectomy but if the growth is too widespread to permit an adequate segmental removal, total cystectomy should be carried out.

H. Pathology

(Contributed by T. SYMINGTON, J. M. SCOTT, K. M. GIRDWOOD)

Introduction

Tumours of bladder are either epithelial or mesenchymal in origin and both groups are dealt with below in some detail. It is customary to subdivide epithelial growths into simple (papilloma) and malignant varieties, and to deal with them in that order. It is proposed to reverse the order, in this instance, and to deal first with malignant epithelial tumours and then discuss the simple growths. The main reason for this departure in procedure is that malignant epithelial tumours constitute the major part of this work, and such growths can only be studied with any thoroughness in cystectomy specimens, where simple epithelial tumours are seldom encountered.

I. Epithelial tumours

1. Malignant epithelial tumours

a) Sex, age and site incidence

Sex. Primary epithelial tumours of bladder show a marked predominance in male subjects as shown in the following table (from KRETSCHMER et al 1934).

Series	Male %	Female %
KRETSCHMER et al. (1934)	76.25	23.75
ALBARRAN (1903)	87.4	12.6
ULTZMANN (quoted by VERHOO- GEN) (1921)	67	33

Age. Epithelial tumours are rare before 20 years; the peak recorded by the American Registry (KRETSCHMER *et al.*) was between 60—64 years. The age distribution in their 921 cases was as follows:

160 patients between 60—65 years
153 patients between 55—59 years
569 patients between 50—69 years
5 patients before 30 years
10 patients after 80 years

Site. Personal observations agree with those of KRETSCHMER and his colleagues that multiple and single tumours occur in the following sites in the order of descreasing frequency: lateral wall, trigone, posterior wall, bladder neck, vault and anterior wall. Growths on the lateral wall tend to involve one or other ureteral orifice and tumours on the lateral wall, trigone and base constitute three-quarters of the cases. The high incidence of hydroureter, hydronephrosis and acute ascending pyelonephritis is accordingly not unexpected.

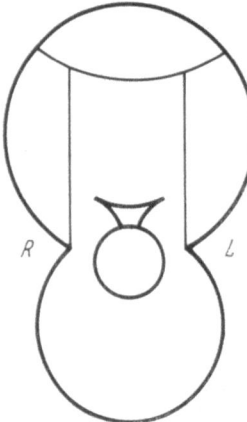

Fig. 22. The site and size of tumour indicated by clinical assessment, bimanual examination and cystoscopic findings, are recorded

A reliable diagnosis and assessment of epithelial tumours of the urinary bladder is possible only if suitable biopsies are supplied. From the pathologist's point of view the most satisfactory specimen for this purpose is a total cystectomy. A complete assessment of the growth is possible; the nature of the tumour, its size, shape and site in the bladder, its extension into the bladder wall and its relationship to other pelvic organs can be obtained. Such a specimen can give all the information required and, from the study of such material, it is possible to build up an experience which is applicable to the more difficult task of assessing the significance as well as the limitations of biopsy material.

Nevertheless, it is not sufficient for the pathologist merely to receive the excised bladder. It is imperative that he should have full clinical details and for this purpose the diagram illustrated in Fig. 22 should accompany the specimen.

b) Preparation of the specimen

It is not unnatural for the surgeon to wish to confirm his clinical impression by opening the bladder. Nevertheless, this natural curiosity should be curbed until after the bladder has been injected with a suitable fixative. The method used here is to employ Wentworth's No. 1. fixative, since the bladder tumour can then retained as a museum specimen. When the specimen and accompanying clinical form are received in the pathology department, both ureters and urethra are ligated and the bladder wall palpated to site the tumour. A full cystoscopic chart, if supplied, is of great help in this respect. If the growth is on the lateral or posterior walls, a large bore intravenous needle, attached to a syringe containing Wentworth's No. 1 fixative, is passed through the anterior

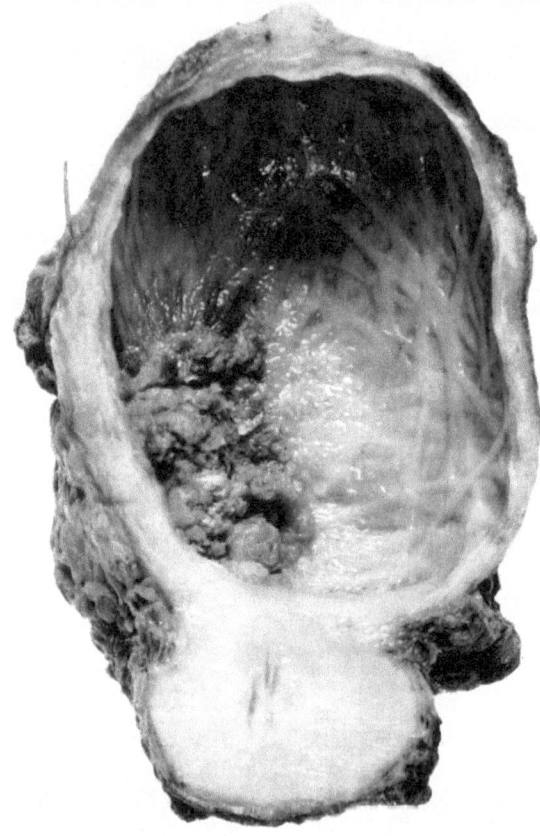

Fig. 23. The anterior wall of bladder has been removed to expose a single, irregular ulcerated tumour lying in the right lateral wall. The growth involves the right ureteric orifice. × 1.5

wall of the bladder. If the tumour is on the anterior wall, the injection is made
posteriorly. The fixative is injected until the bladder is tense and distended, when
it is immersed in Wentworth's No. 1 solution for 24 hours at least. Subsequent
treatment will depend on the site of the growth and whether it is single or

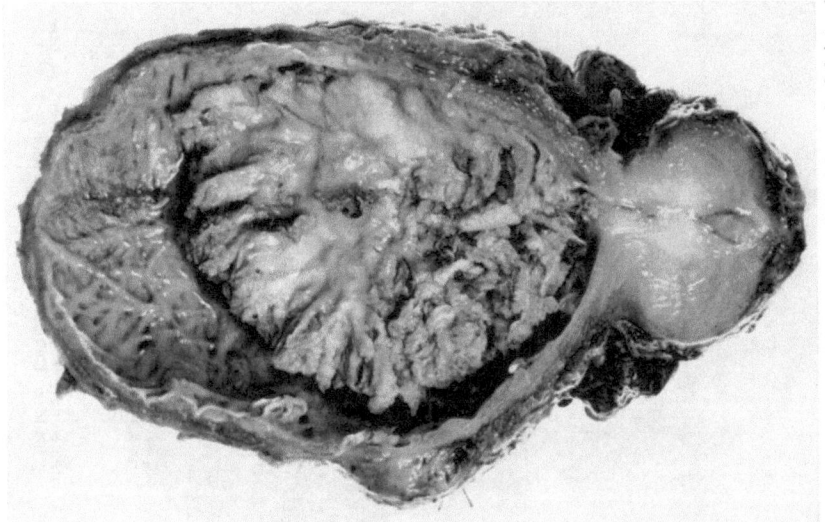

Fig. 25. A large papillary growth projects from the left lateral
wall. It occupies most of the lumen of the bladder

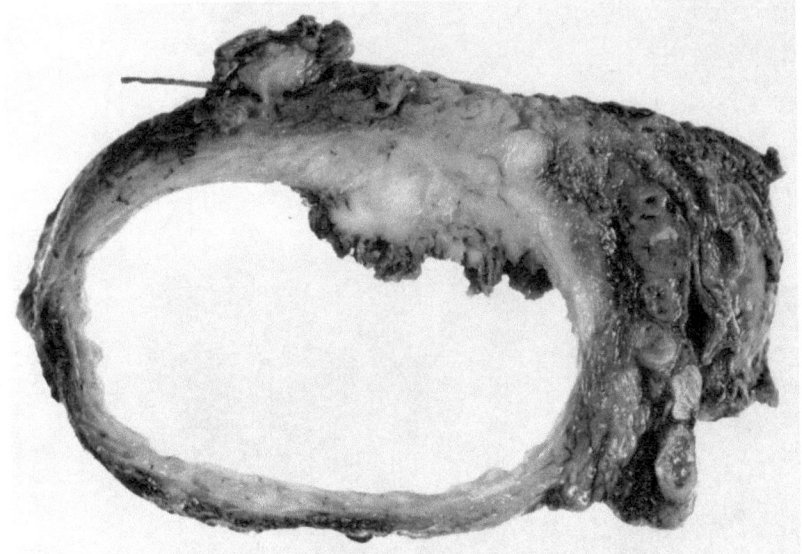

Fig. 24. The posterior wall of bladder (Fig. 23) has been removed.
The section has passed through the base of the growth and shows
the white tumour mass in the wall of the bladder

multiple. Generally the bladder is opened by a thin slice through the anterior
wall. This may be sufficient to demonstrate the growth (Fig. 23) or merely expose
its upper surface when a second cut brings the tumour into view. A posterior
slice is now removed through the tumour to expose the base of the growth which
can be identified and described and the extent of invasion of the bladder wall
noted (Fig. 24). Serial photographic records are taken at each stage and, if

required, the remainder of bladder, showing the tumour and invasion of wall, can be retained as a museum specimen. In addition, tissue from each tumour is taken for histological examination. Sections are made of the tumour and adjacent

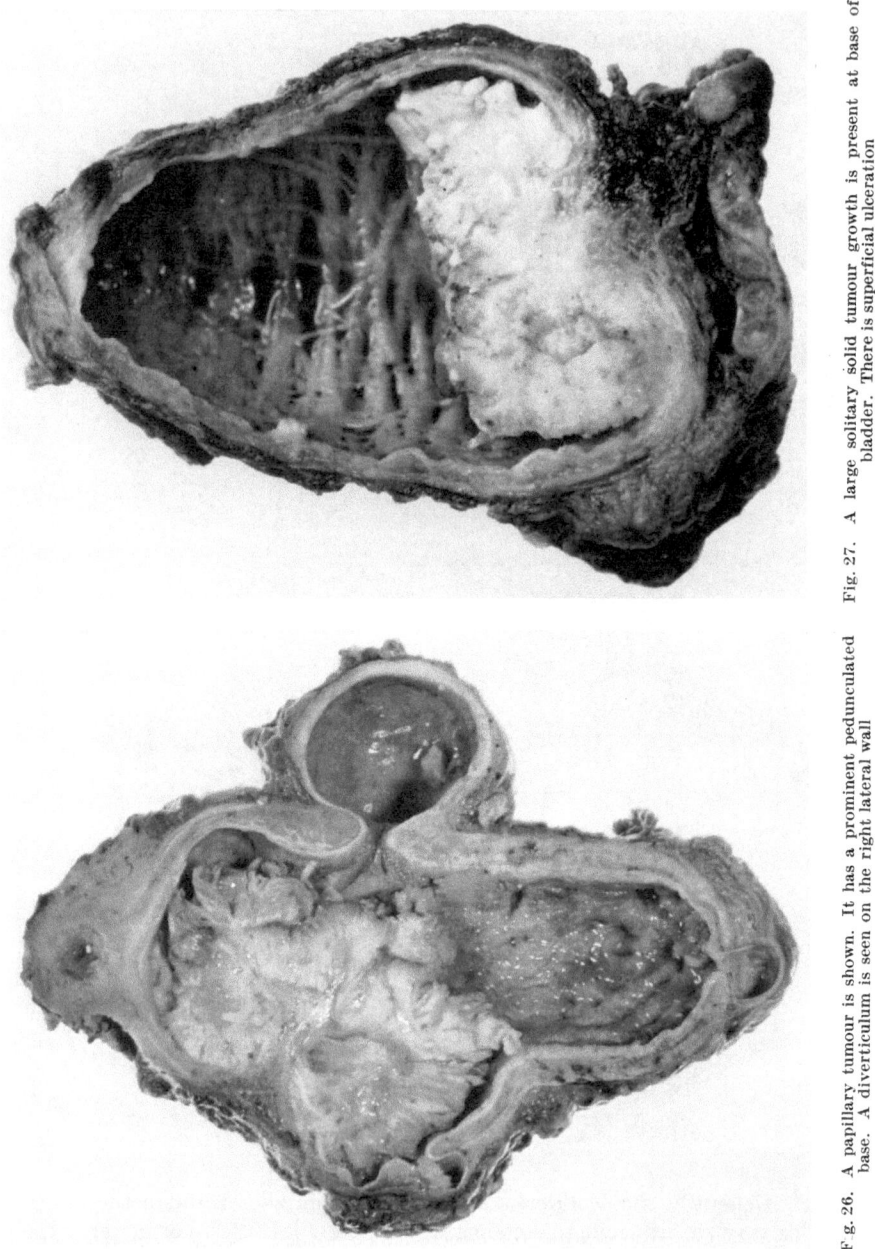

Fig. 27.   A large solitary solid tumour growth is present at base of bladder. There is superficial ulceration

F g. 26.   A papillary tumour is shown. It has a prominent pedunculated base. A diverticulum is seen on the right lateral wall

bladder wall to determine the extent of spread, while blocks are made of the bladder mucosa, particularly in relation to the tumour. All material is fixed in 10% neutral formalin, processed in the usual way and stained by haematoxylin and eosin.

c) Morbid anatomy

The following description is based on examination of 50 cases of complete or partial cystectomy for bladder carcinoma. It covers almost all forms of primary epithelial tumour seen in bladder with the exception of

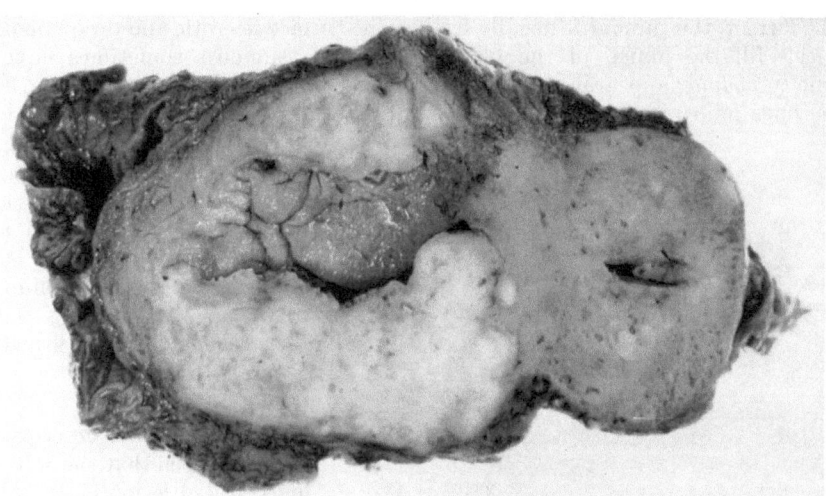

Fig. 29. Large, solid, ulcerated tumour growths are shown on the left and right lateral walls of the bladder

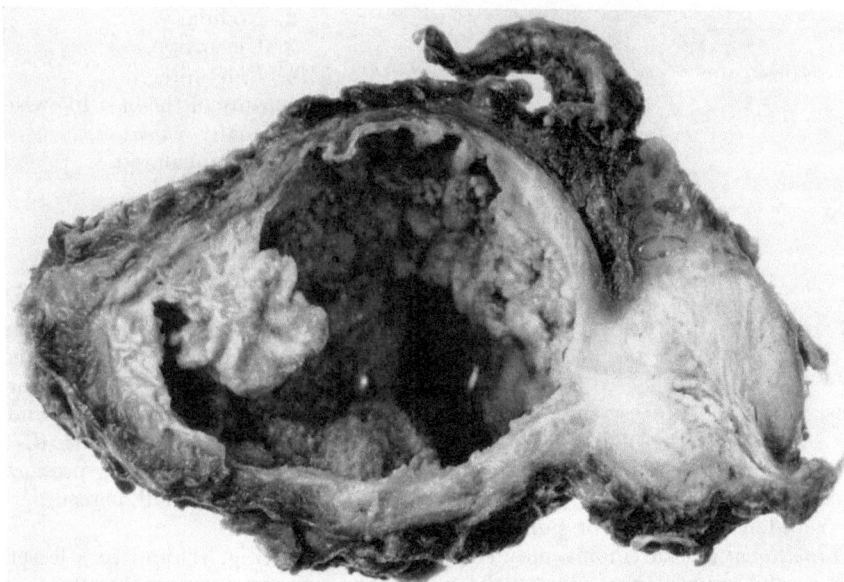

Fig. 28. Multiple tumours are shown. The growth arising from the left superior angle lis papillary and has a pedunculated base. Numerous small sessile growths are shown on the left posterior wall

simple papilloma, which will be considered during the discussion on biopsy specimens.

Primary epithelial tumours of bladder may be single or multiple. *Single* tumours vary greatly in size from a few centimeters to one which almost fills the lumen of the bladder (Fig. 25). The common type has a *papillary* arrangement, but it is not uncommon to see a large *ulcerated* growth with raised rolled edges

and haemorrhagic centre (Figs. 23 and 24). The combination of a papillary and
ulcerative growth occurs. The *base* of the growth may be *pedunculated*, the
pedicle being thin or broad, and in some instances appears as a massive stalk
(Fig. 26). In the ulcerative type and in some papillary growths the base is *solid*
(Fig. 27). *Multiple* tumours likewise vary in appearance. They are commonly
papillary in type (Fig. 28) but solid *ulcerative* growths are not uncommon (Fig. 29).
In the former, the tumour is usually friable, the surface necrotic and desquamated
particles fill the lumen of the bladder. Multiple tumours sometimes have a
distinctly *nodular* appearance when their surface is smooth and rounded (Fig. 30).
Sometimes multiple growths do not conform to any of the above patterns but
have an *indefinite* arrange-
ment. The *base* of *multiple
tumours* has a variable appearance. Most of them are
pedunculated, but some are
sessile, and it is not uncommon
to find both occurring together. The ulcerated growths
generally tend to have a *solid*
base.

From the above description
it can be seen that the following types of lesion occur:
1. Papillary.
2. Nodular.
3. Ulcerative.
4. Indefinite.
The nature of the base likewise
shows many variations:
a) Pedunculated.
b) Solid.
c) Sessile.

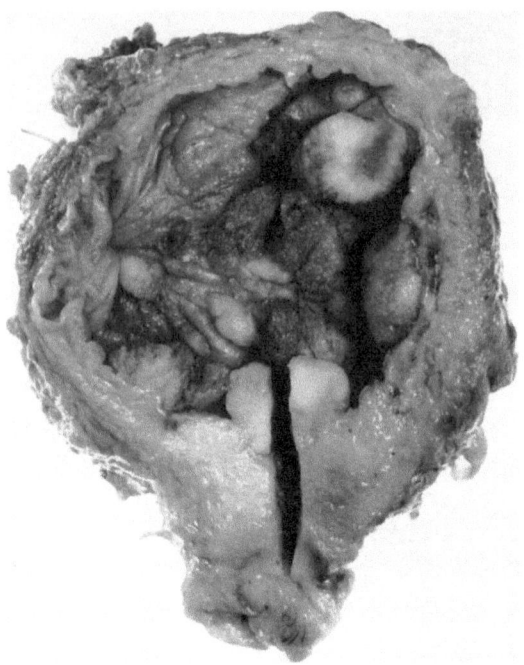

Fig. 30. Multiple nodular growths are present throughout the
bladder. Their surface is smooth and rounded

d) Microscopic appearance

As would be expected from
the varied macroscopic pic-
tures described above, the
microscopic appearances show considerable variations. "Every grade and
every sequence from solitary highly organized non-invasive papilloma to disorderly invasive carcinoma is seen" — WILLIS (1948). The great majority present
as variations of transitional cell carcinoma and only a very small percentage
occur as adeno-carcinoma or pure squamous epithelioma.

Transitional cell carcinoma may exist as a papillary (Fig. 31) and to a lesser
extent a solid growth (Fig. 32) and the cells of the tumour show varying degrees
of pleomorphism and cellular mitosis. When the growth is differentiated, the
cells resemble those of transitional epithelium; pleomorphism and mitosis are
scanty, and the lesion is localized to the mucosa (Fig. 33). When the tumour
is anaplastic the cells show a marked degree of pleomorphism and mitotic figures
are marked (Fig. 34); the lesion may be localized to mucosa but usually has
spread into and in many instances through the bladder wall (Fig. 35). As might
be expected all grades of pleomorphism and mitosis exist between the differen-

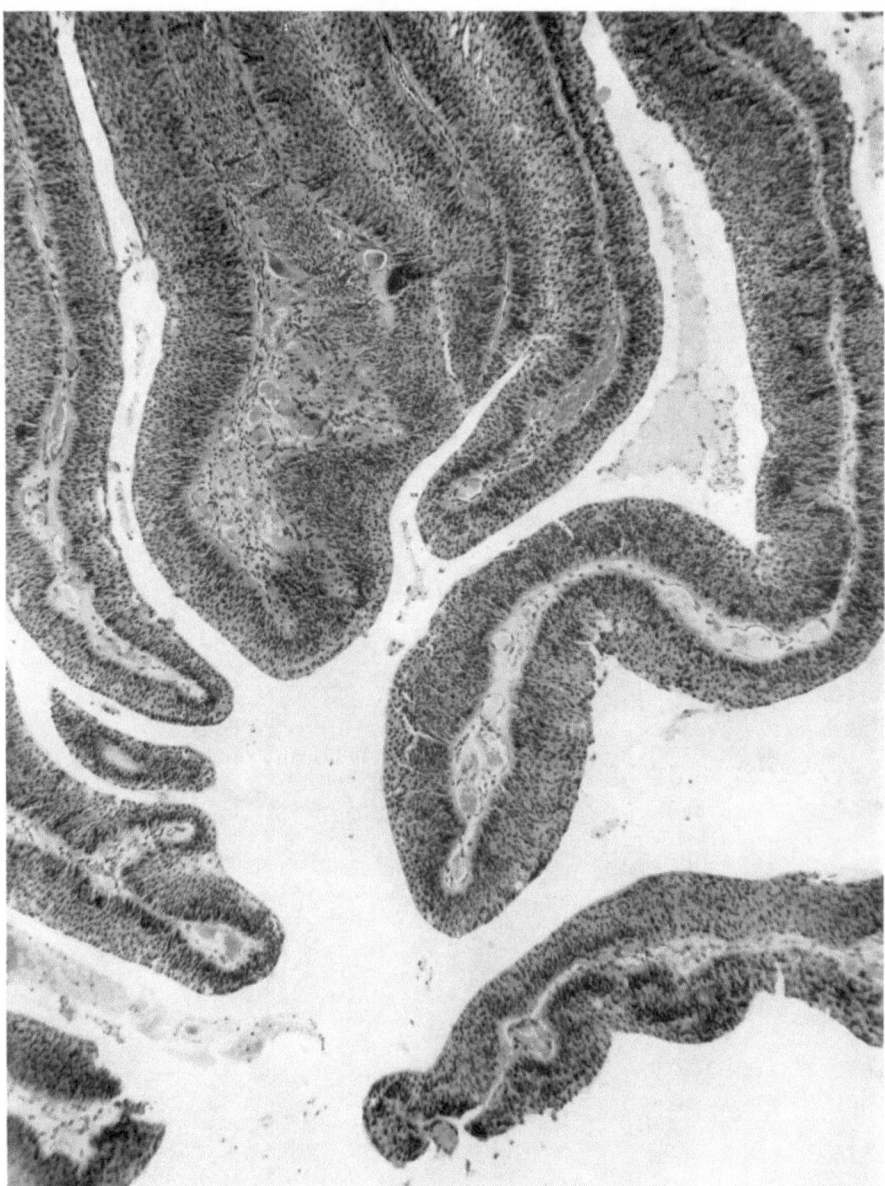

Figs. 31 and 32. Show respectively a typical papillary and solid type of bladder tumour. Fig. 31, Haematoxylin and Eosin. × 100

tiated and anaplastic groups, and although many histological patterns exist, it should be remembered that they are but variants of transitional cell carcinoma.

Metaplastic changes sometimes occur in transitional cell carcinoma. The major part of the growth retains the appearance of a transitional cell carcinoma and, to a variable degree, metaplastic changes occur either to squamous epithelium (Fig. 36) or to the formation of pseudo-glandular structures (Fig. 37). The significance of these changes will be discussed later. In some instances a pure squamous epithelioma occurs following leukoplakia (Figs. 38 and 39) and in

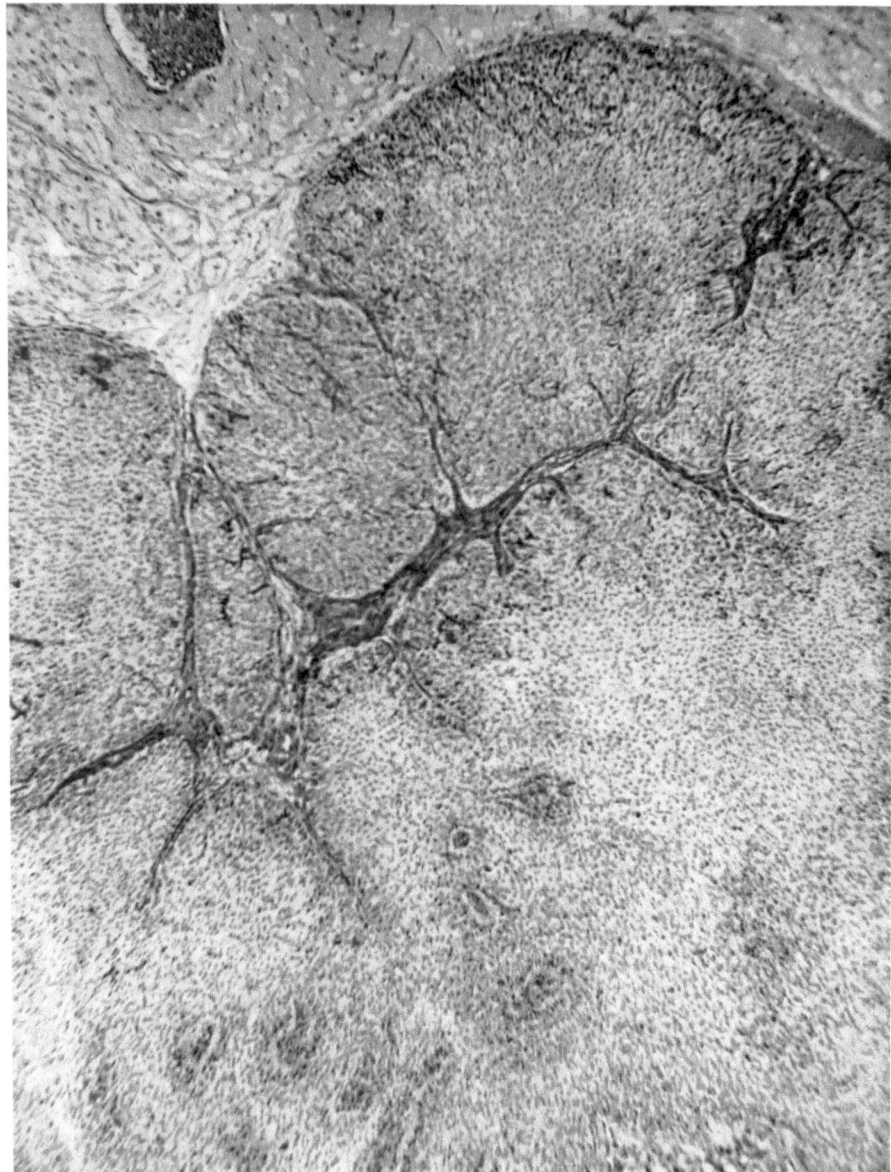

Fig. 32. Haematoxylin and Eosin. × 100

a smaller percentage of cases primary mucoid adenocarcinoma is encountered
(Fig. 40). It may be possible to show transition of mucosa from normal to frank
adenocarcinoma. The clear mucoid cells become heaped up and produce the
appearance of cystitis glandularis and adenocarcinoma.

e) Changes in bladder mucosa adjacent to tumour

The appearance varies considerably. In well-differentiated tumours, the
adjacent mucosa may be normal, while with anaplastic growths it is extremely

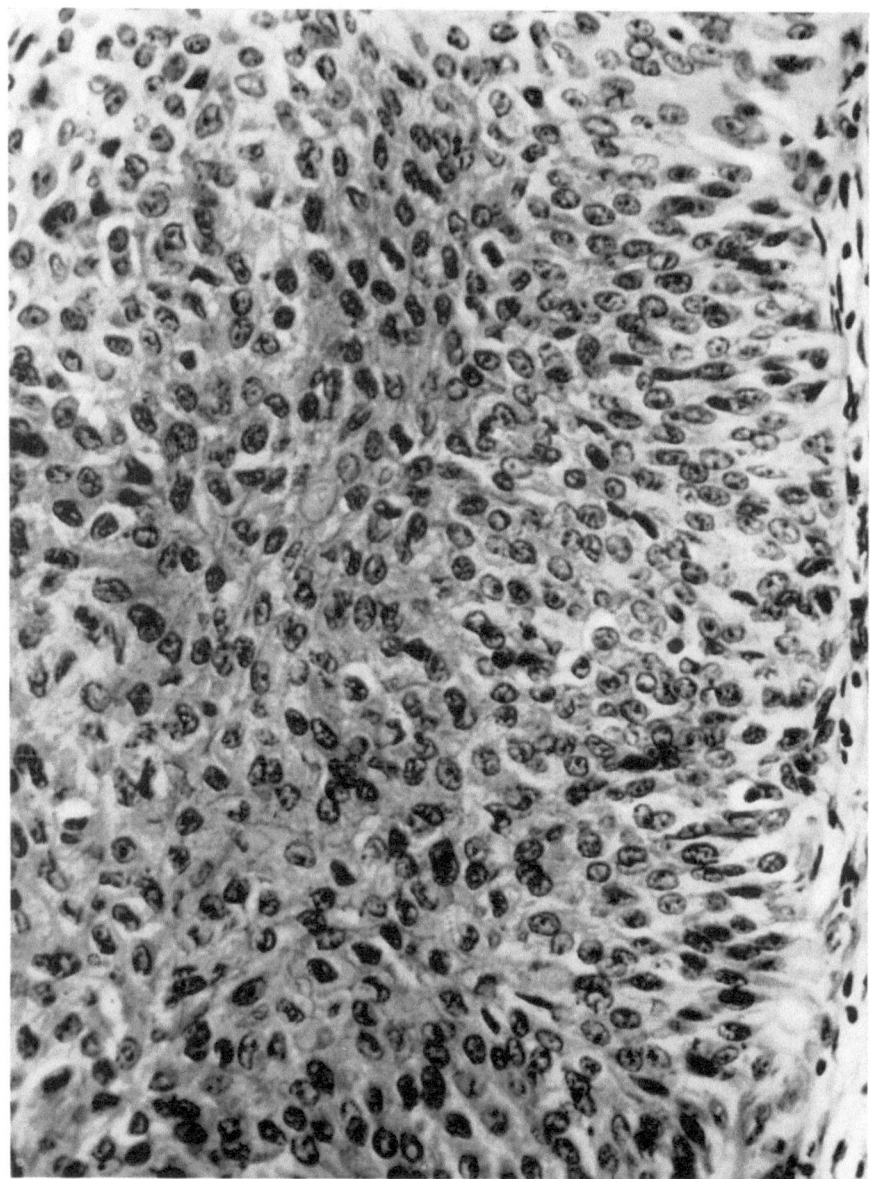

Fig. 33. A portion of well differentiated transitional cell carcinoma of bladder is shown. The cells resemble those of transitional epithelium; pleomorphism and mitotic activity are scanty. Haematoxylin and Eosin. × 500

active; pleomorphism of the cells is marked and mitotic figures prominent. The appearance is that of carcinoma *in-situ*. At times the bladder mucosa is completely denuded and in such instances the lining consists of very vascular granulation tissue. It is not uncommon to find a hyperplastic mucosa, part of which is ulcerated and lined by granulation tissue. Occasionally, the bladder mucosa has the appearance of hyperplastic squamous epithelium, in which the cells have a clear cytoplasm. The appearance resembles that seen in cervix (Fig. 38). On occasions the mucosa shows transitional changes. Areas of clear cells become

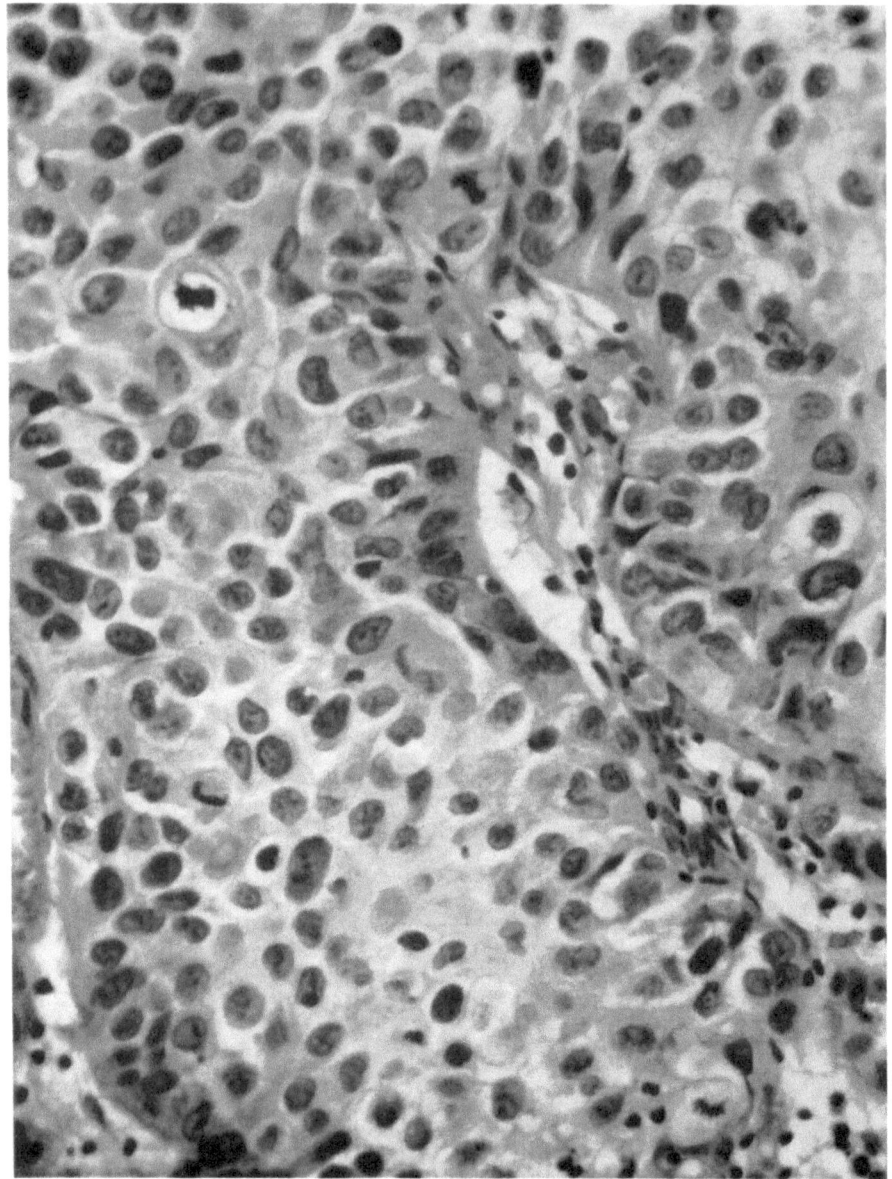

Fig. 34. This is a more anaplastic transitional cell carcinoma. Pleomorphism is marked and many cells show mitotic activity. Haematoxylin and Eosin. × 500

heaped up and assume glandular formations which eventually merge with frank adenocarcinoma. This transitional change from normal to malignant bladder mucosa is not infrequently observed, even in tumours of low grade malignancy.

Other changes observed are the presence of Brunn's nests, while lymph follicles, with or without germ centres and scattered collections of lymphocytes in the submucosa can be seen in a large number of cystectomy specimens.

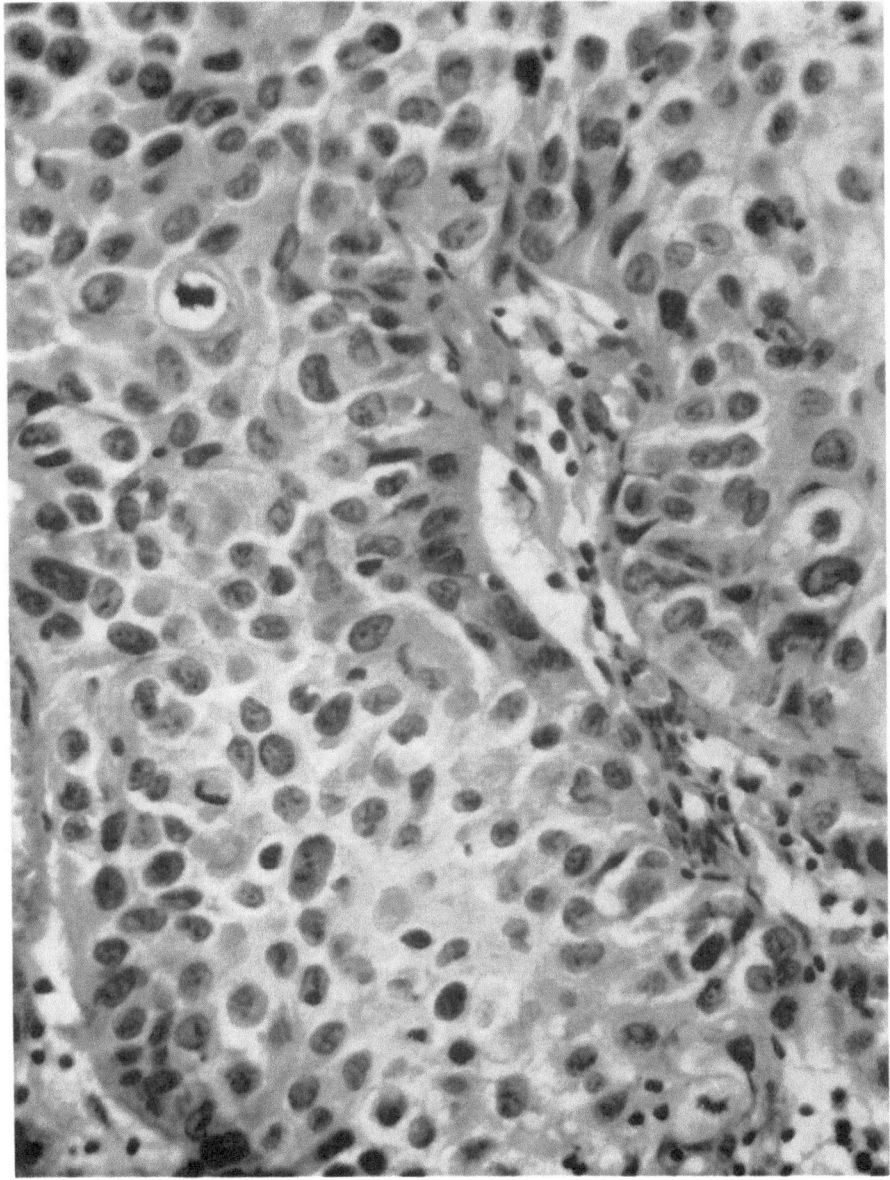

Fig. 35. An anaplastic papillary transitional cell carcinoma showing lymphatic invasion. The growth has spread through the muscle coat. Haematoxylin and Eosin. × 100

f) Spread of tumours

In our series of 63 post-mortems, the vesical neoplasm had remained localized to the mucous membrane for long periods without exhibiting further invasive tendencies even though, in parts, the tumour was morphologically active. When bladder carcinoma is invasive metastases occur to the regional lymph nodes or by the blood stream. Invasion of veins in the bladder wall by tumour cells is a common finding in such cases and may be the beginning of

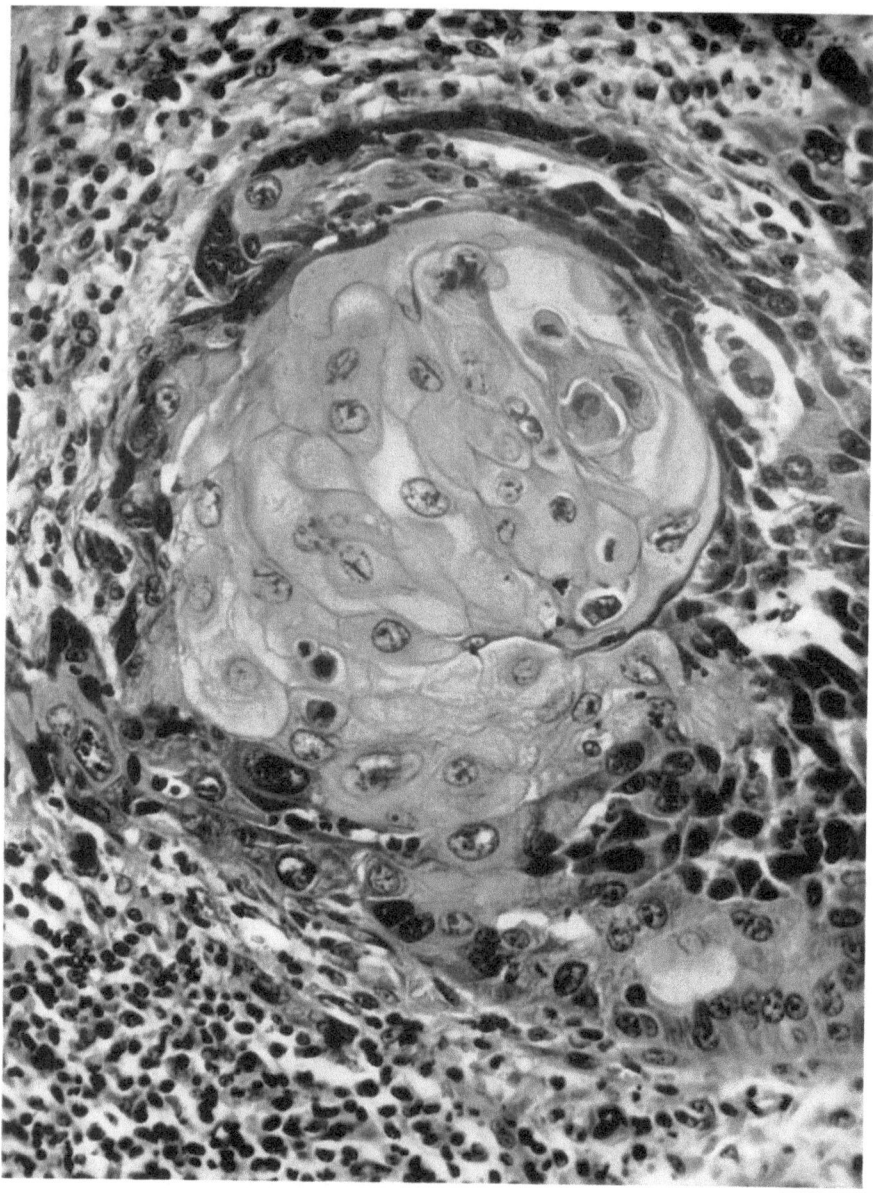

Fig. 36. An area of squamous metaplasia is shown. The predominant appearance of the growth was, however, one of transitional cell carcinoma. Haematoxylin and Eosin. × 500

haematogenous spread. However, it should be appreciated that clumps of tumour cells may be found frequently in capillaries in the base of a growth. When the tumour cells are well differentiated such a finding may have little prognostic significance, and may result from manual trauma during operative procedure.

Haematogenous spread usually occurs late in the disease and took place in 14 cases (22%) in our series. Osteogenic metastases, to vertebrae and pelvic

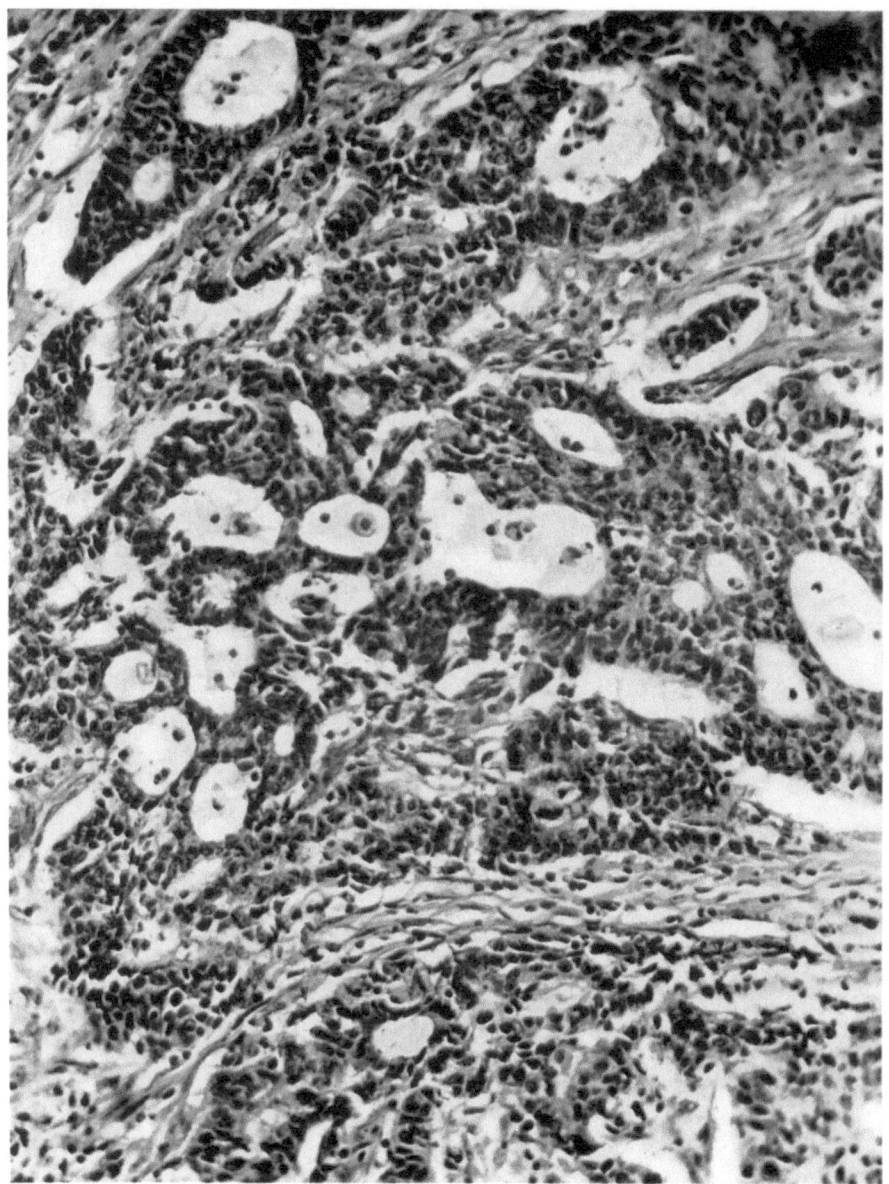

Fig. 37. Pseudo-glandular structures can be seen in a transitional cell carcinoma. Haematoxylin and Eosin. × 250

bones occurred in 4.8% of cases. The organs most frequently affected were liver, kidney, lungs and adrenals.

Extra-vesical spread occurred in 19 cases (30%).

g) Cause of death in bladder tumours

In our series, when the patient died 1 to 9 days following total or sub-total cystectomy with bilateral transplantation of ureters, the cause of death was

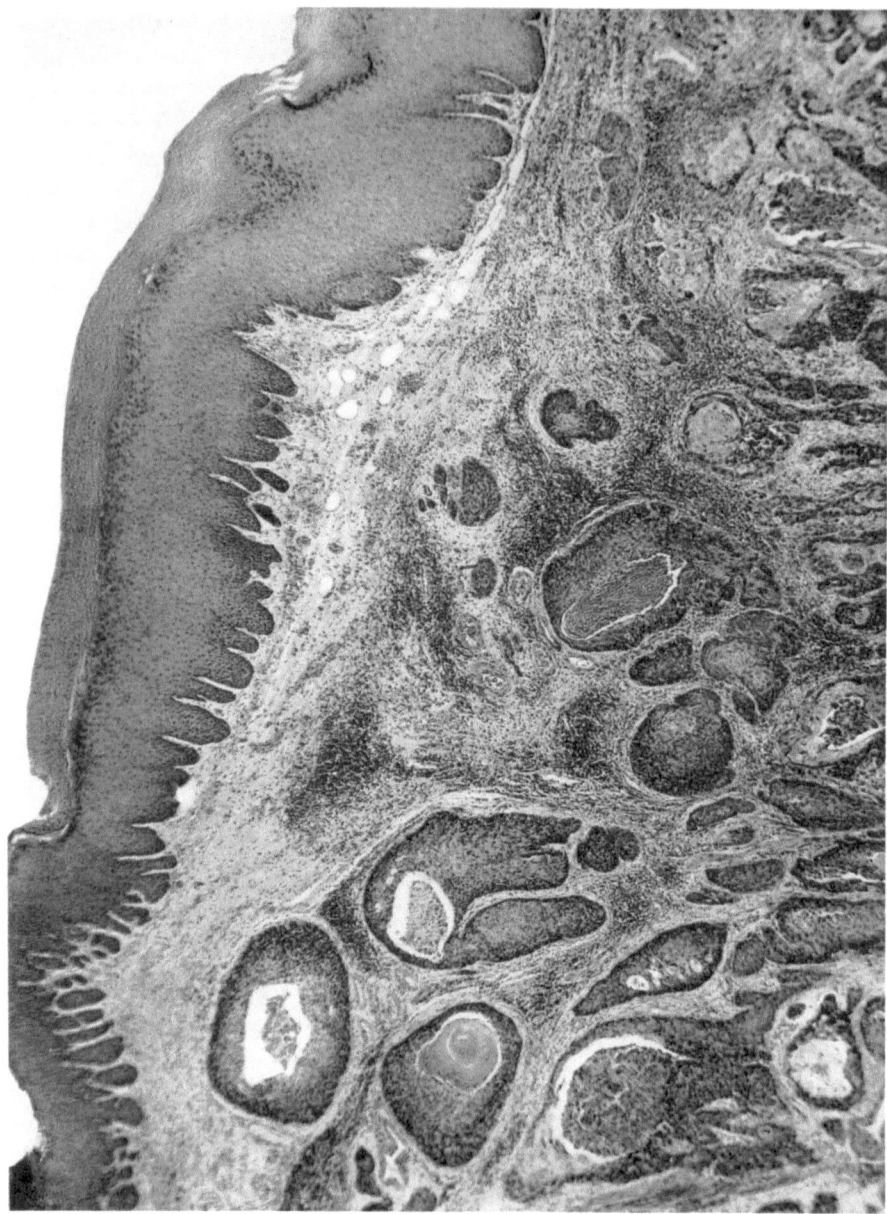

Figs. 38 and 39. Squamous epithelioma of bladder. The lining of the bladder (top) is squamous epithelium. Tumour growth, with prominent epithelial pearls, is seen in the submucosa (Fig. 38). A high power view of one of the epithelial pearls is shown (Fig. 39). (Fig. 38, Haematoxylin and Eosin. × 60. Fig. 39, Haematoxylin and Eosin. × 500)

found to be acute generalized peritonitis, haemorrhage, acute intestinal obstruction or pulmonary complications. When death occurred 8 months to 2 years following operation, an upper urinary tract infection invariably was found. When the patient had not been subjected to operation, a combination of metastases, toxaemia and upper urinary tract infection was observed.

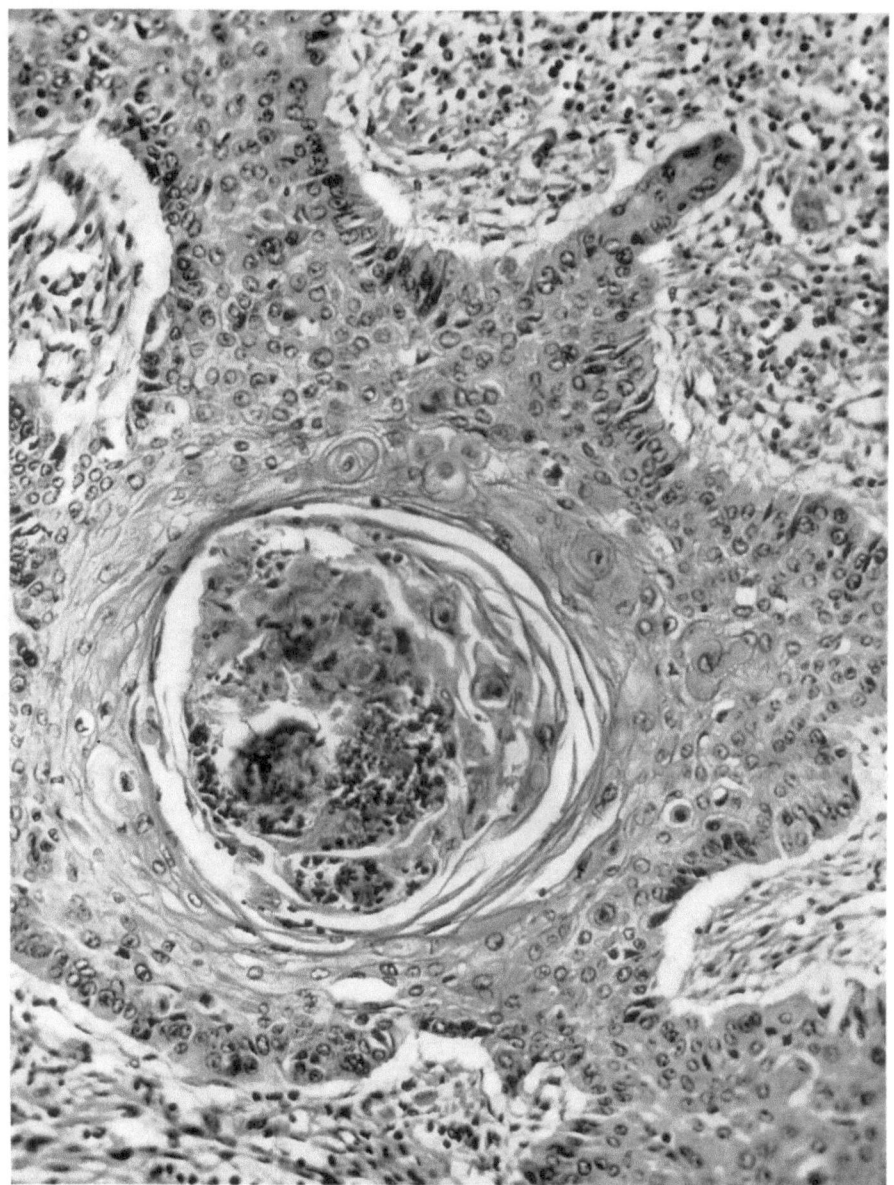

Fig. 39

h) Classification of epithelial tumours of bladder

It is obvious that transitional cell carcinoma of bladder can present with a wide variety of histological patterns. At one end of the series is a well-differentiated growth composed of transitional epithelium, which differs only from a simple papilloma in that the fronds are thicker and the epithelium more hyperplastic. The cells show some degree of pleomorphism and occasional mitoses are seen. At the other extreme, the growth is anaplastic and, in extreme cases, bears no resemblance to transitional epithelium; pleomorphism and mitosis are

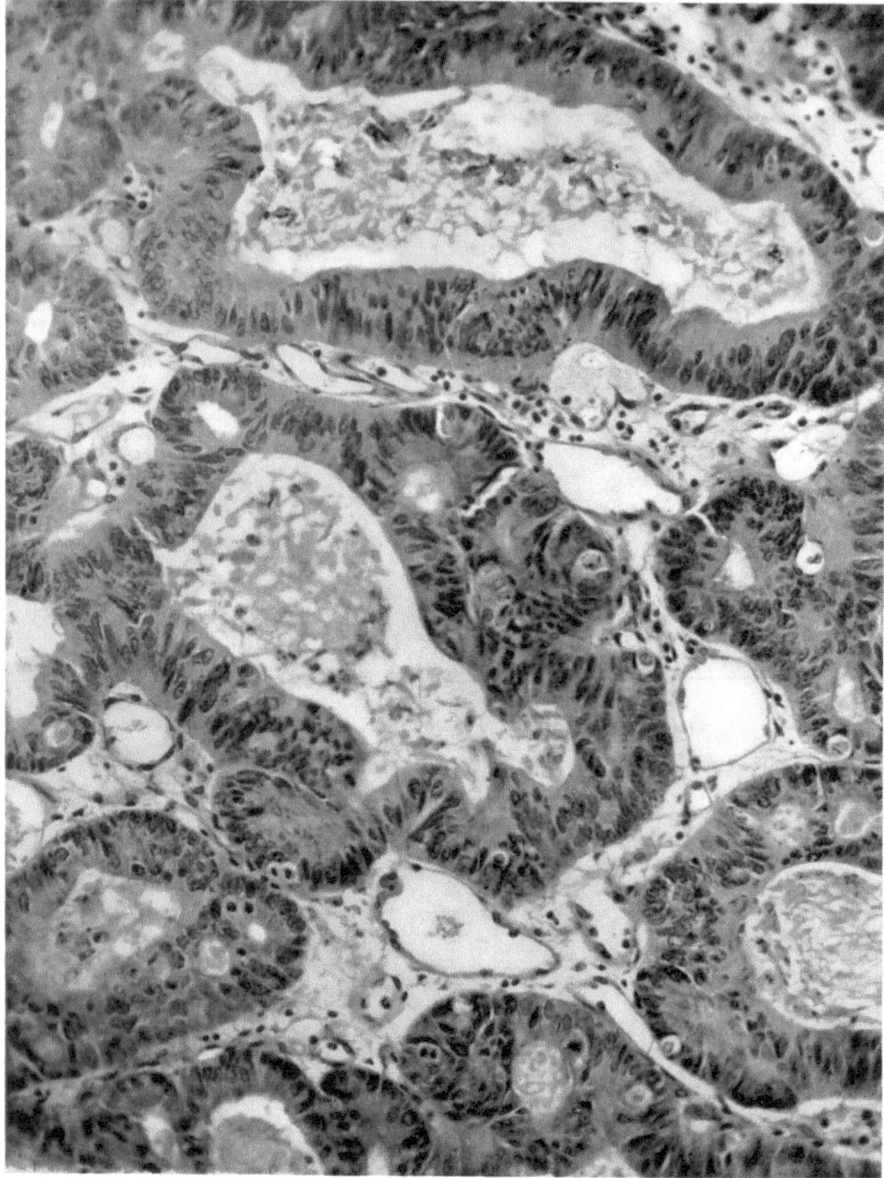

Fig. 40. The typical appearance of adenocarcinoma of bladder is evident. Haematoxylin and Eosin. × 300

marked. It is not unnatural that a large number of tumours fall into the wide range between the two extremes.

Growths which can be identified as squamous epithelioma and adeno-carcinoma are found and once more both groups show variable degrees of differentiation and anaplasia. Serving as a link between transitional cell carcinoma on one hand and squamous epithelioma and adeno-carcinoma on the other, are a large proportion of transitional call carcinoma with squamous and glandular metaplasia.

In addition to the variety of histological appearances, various stages of invasion of the bladder wall are encountered. The growth may be localized to the mucosa or tumour cells may be present in the lymphatics or blood vessels of the submucosa, muscularis or extra-vesical tissue. Again distant metastases may have arisen.

Thus, it becomes obvious that in any attempt to classify bladder tumours, two factors will predominate. 1. The histological type of the growth, and 2. the stage of invasion of bladder wall or extra-vesical tissues. Either one or both factors form the basis of most bladder tumour classifications, yet the large number of classifications in use is in itself an indication of the difficulties and problems involved. In order to arrive at some uniformity, it is proposed to review in some detail the more common classifications in use and, if possible, find a common factor in them.

α) DUKES and MASINA (1949) stress the importance of close co-operation between the surgeon and pathologist and maintain that a complete diagnosis and classification of bladder tumours should include details of the following features:

1. Gross characters of tumour. First notice whether lesions are single or multiple Then record size, shape, surface, surroundings and exact situation of lesions.

2. Histology of tumour.

a) *Non-malignant tumours*—simple transitional cell papilloma.

b) *Malignant tumours.*

 i) papillary transitional cell carcinoma.

 ii) solid (non-papillary) transitional cell carcinoma.

 iii) transitional cell carcinoma with metaplasia, squamous or glandular.

 iv) pure squamous carcinoma.

 v) pure adeno-carcinoma.

 vi) anaplastic spheroidal-cell carcinoma simplex.

The authors disagree with the term "Grade I carcinoma" for a simple transitional cell papilloma and believe that " if the epithelial cells of a papillary tumour closely resemble normal transitional epithelium and are non-invasive in character, the tumour should be called benign papilloma." They reserve the term "carcinoma" for unmistakable evidence of malignancy as shown by invasiveness or irregularity in size and shape of neoplastic cells or atypical nuclear structure.

Papillary and solid transitional cell carcinoma have appearances similar to those already described (p. 29). Most examples of carcinoma of bladder in their earliest phases grow in a papillary pattern and this arrangement may be retained throughout by growths of a relatively benign character. In order to bridge the gap between differentiated and dedifferentiated papillary and solid growths, DUKES and MASINA subdivide each into low, average and high grades, depending on the degree of pleomorphism and frequency of mitotic figures. Since most of the growths contain papillary and solid areas, they grade the tumour according to the predominance of papillary or solid areas. On occasions metaplastic changes occur in papillary or solid transitional cell carcinoma, either to squamous or glandular epithelium. They find these changes occur "most frequently in rapidly growing tumours of a relatively high grade of malignancy."

Pure squamous carcinoma (Fig. 38) and adenocarcinoma (Fig. 40) are relatively rare and since the prognosis is different from that of transitional cell carcinoma with squamous or glandular metaplasia, they are placed in different categories. Undifferentiated growths consisting of spheroidal or polygonal cells are classified as anaplastic spheroidal cell carcinoma simplex (2. b vi) (see

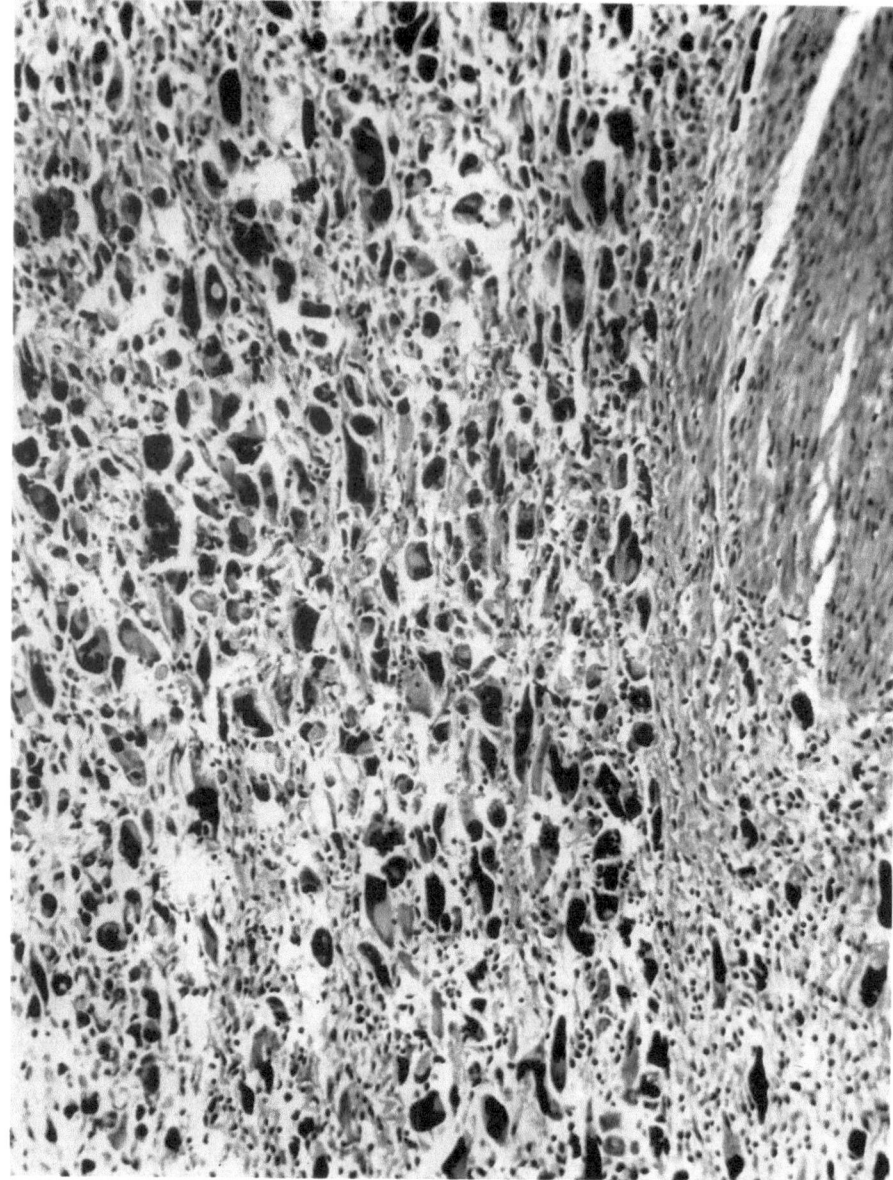

Figs. 41 and 42. Anaplastic transitional cell carcinoma of bladder. Note the very marked cellular pleomorphism. The growth could be mistaken for a sarcoma. This was the appearance of the tumour six months after partial cystectomy and irradiation. However, the microscopic appearance of the tumour at the time of cystectomy (Fig. 41) is that of an anaplastic transitional cell carcinoma. (Fig. 41, Haematoxylin and Eosin. × 100)

Figs. 41 and 42). "They are very invasive, grow rapidly, tend to metastasize early and consequently have a very bad prognosis. They vary in histology and sometimes resemble sarcomas."

3. *Extent of spread.* The four stages in the spread of tumours of bladder are shown in Fig. 43. When tumours are confined to the submucosa and muscle coat of the bladder, without having given rise to metastases, they are described

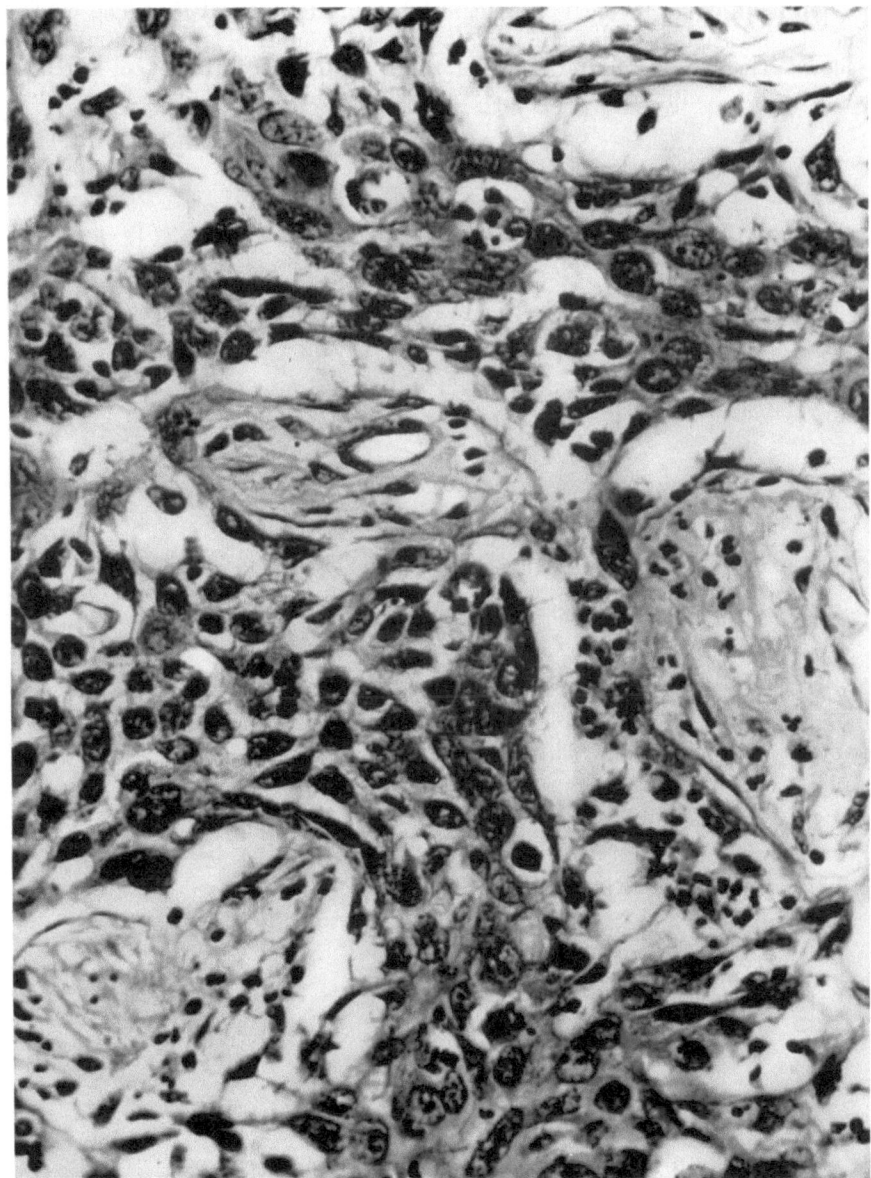

Fig. 42. Haematoxylin and Eosin. × 500

as in Stage 1. If the growth is confined to the mucous membrane and submucosa only, it is in Stage 1 A; if it has spread to bladder muscle — Stage 1 B. "Stage 2 is used to describe tumours which have spread by direct continuity into the perivesical fat or adjacent tissues but have not caused any lymphatic metastases. When metastases occur in regional lymphatics, in nodes capable of being removed *en bloc*, the 3rd stage is reached. Evidence of still further spread or of remote metastases would place the tumour in Stage 4."

DUKES and MASINA find that tumours "staged" in this way show a "fairly close correlation between gross character, histology and extent of spread."

β) **Institute of Urology histological grading (Dukes 1955).** The outline of
the classification recommended by the Institute of Urology (London) is shown
below. The general pattern of the tumour, direct continuity of spread and
evidence of lymphatic and blood spread, is similar to that recommended by

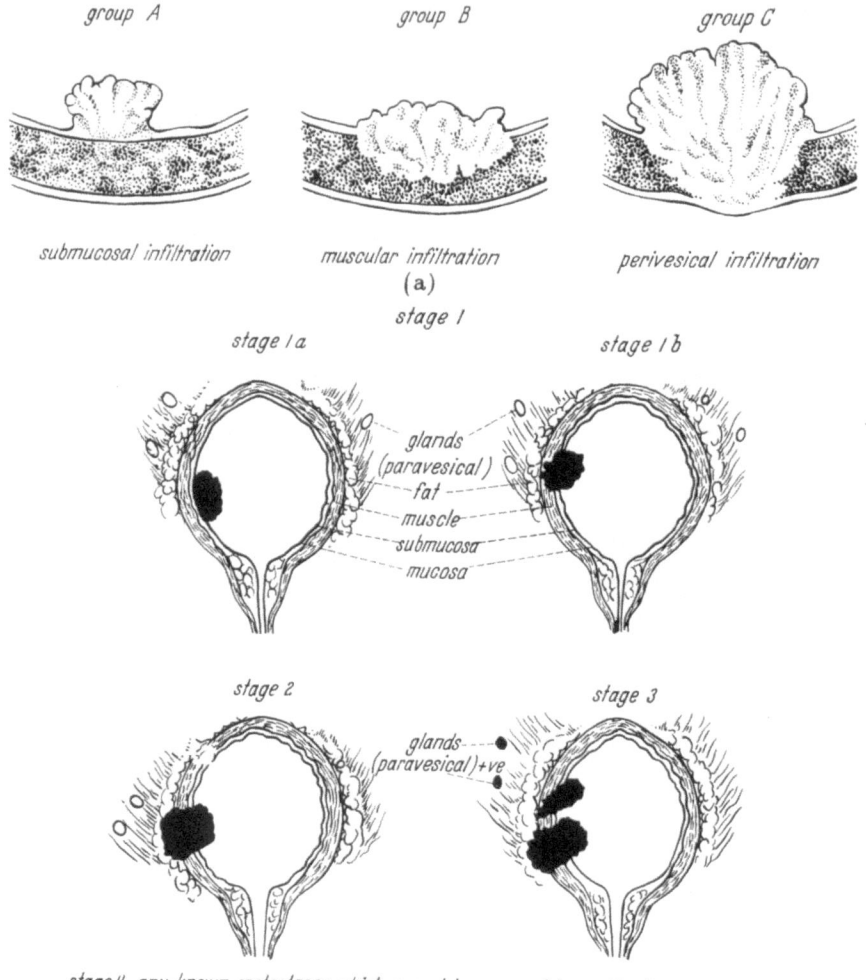

Fig. 43. Demonstrates the stages of invasion of bladder tumour. From Dukes and Masina (1949)

Dukes and Masina. The histological grading is simplified. Transitional cell
carcinoma may be papillary or solid and graded differentiated or anaplastic.
Included in the anaplastic group are transitional cell carcinoma with squamous
or glandular metaplasia, as well as spheroidal cell carcinoma simplex. The small
percentage of cases presenting as pure squamous cell carcinoma or adenocarcinoma are retained in a separate sub-group.

Outline of Classification

I. General pattern of tumour:
    1. Papillary
    2. Solid
    3. Papillary and solid

II. Histology:
    1. Papilloma (simple and benign)
    2. Differentiated transitional cell carcinoma
    3. Undifferentiated or anaplastic transitional cell carcinoma
    4. Squamous cell carcinoma
    5. Adenocarcinoma
    6. Other varieties of malignant tumour
III. Direct continuity spread:
    1. Within mucosa only
    2. Into bladder muscle but not beyond this
    3. Into perivesical tissue
    4. To adjoining organs
    5. Fixation to wall of pelvis
IV. Lymphatic spread:
V. Spread by blood stream.

γ) **Classification of bladder carcinoma-registry of American Urological Association** (KRETSCHMER et al. 1934). In this very comprehensive review involving 902 cases of bladder tumours, the usual attention is paid to site of tumour and whether multiple or single. The microscopic classification is based on the BRODER's system, which is a simple morphological classification depending on the extent to which the epithelium of the tumour is differentiated as shown below.

Grade 1. Three-quarters of the growth shows a differentiated epithelium and one-quarter is undifferentiated.
Grade 2. Both differentiated and undifferentiated elements are equal.
Grade 3. Three-quarters undifferentiated.
Grade 4. No differentiation.

In addition to BRODER's grading, the epithelial tumours are subdivided as follows:

    1. Papillary carcinoma.
    2. Infiltrating carcinoma.
    3. Unusual types of epithelial tumours.
      a) Adenocarcinoma.
      b) Colloid carcinoma.
      c) Adenoma malignum (intestinal origin).

The last group constitutes only 17 of the 902 epithelial tumours. Papillary or infiltrating carcinoma was diagnosed in 882 cases (approximately half the number of cases were infiltrating) and in each instance a definite grade was designed on the basis of BRODER's classification as shown below:

Grade 1. . .	172	19%
Grade 2. . .	293	32%
Grade 3. . .	299	33%
Grade 4. . .	101	11%
Not stated .	36	

δ) **Squier Urological Clinic Classification** (MELICOW 1955). This system takes into account cell variation and tissue penetration and the pathological diagnosis is given in the form of a complete sentence, e.g., "Well-differentiated, non-infiltrating, papillary carcinoma of urinary bladder Grade 1" or "poorly differentiated, infiltrating carcinoma of urinary bladder with invasion of muscularis Grade 3 B". The classification is shown below:

1. *Urothelial* (from lining transitional epithelium).
    A. *Papillary.*
      1. Papilloma.
      2. Papillary carcinoma. Penetration
        A (submucosa), B (muscularis), C (perivesical), or D (distant).

B. *Nonpapillary or solid.*
    1. Intraurothelial carcinoma (*in situ* or BOWEN's disease).
    2. Infiltrating carcinoma. Usual grades 2, 3, 4. A, B, C, D.
2. *Glandular* (from glandules normally present in vertex or subtrigone or from glandular rests) usually solid ± mucin in urine.
    1. Adenoma.
    2. Adenocarcinoma, Grade 1, 2, 3, 4. A, B, C, D.
3. *Metaplastic.*
    A. Toward squamous cell type.
        1. Papillary: papilloma or carcinoma, rare.
        2. Nonpapillary or solid (?preceded by leukoplakia).
    B. Towards glandular formation (? preceded by cystitis glandularis).
        Adenocarcinoma Grade 1, 2, 3, 4. A, B, C, D.

In a series of 914 cases of primary tumour of bladder, 95 per cent were of "urothelial" origin. 81 per cent of them were papillary, 7 per cent non-papillary, 11.5 per cent metaplastic (squamous cell and glandular) and 0.5 per cent were glandular (of urachal or subtrigonal origin). Of the papillary tumours 12.4 per cent were papilloma, 25.2 per cent Grade 1, 24.7 per cent Grade 2, 20.2 per cent Grade 3 and 17.5 per cent Grade 4.

ε) **Classification based essentially on depth of penetration of bladder wall** (JEWETT & STRONG 1946, MARSHALL 1952). JEWETT and STRONG in a review of 107 autopsy cases, found a close relationship between depth of penetration of vesical wall by primary neoplasm and theoretical prognosis, and on this basis they separated cases into three groups.

Group A. Tumour cell confined to mucosa.
Group B. Infiltration into but not through muscularis.
Croup C. Extension through muscle coat.

MARSHALL has enlarged this classification based on the extent of invasion. The muscularis is the focal point; growths which have not extended to the muscle are called *superficial*, while those in or through the muscle are referred to as showing *deep* penetration, as shown below:

Stage 0. Tumour limited to mucosa (papilloma and intraepithelial carcinomas).
Stage A. Tumour involves submucosa.
Stage B.   B1. Growth present in superficial part of muscle.
          B2. Deep penetration.
Stage C. Extension to perivesical fat or even capsule of adjacent organs but not further.
Stage D. The growth is beyond the limits of the bladder and perivesical fat. It is divided into D1 and D2.

D1 refers to lesions within the pelvis, including invasion of pelvic walls or recti muscles below level of umbilicus. This stage includes lesions below the sacral promontory, medial to the psoas and below not beyond the perineum externally. D2 — here the lesion extends beyond the limits of the pelvis and so includes distant metastases and metastases in lymph nodes external to the inguinal ligament and in nodes above the sacral promontory.

All available pre-operative findings, such as cystoscopy, bimanual palpation and X-ray, were used to judge the clinical stage, and the biopsy report relegated to minor roles. MARSHALL found that in 104 patients with superficial or deep invasion, the stage was diagnosed correctly in 81 per cent and wrongly in 19 per cent. A dual classification, on a basis similar to that described by DUKES and MASINA and MELICOW results by adding the following histological grades to the stages described above.

Originally MARSHALL used a histological classification along the following lines:

1. Papilloma — includes simple papilloma and papilloma with atypical cells.
2. Carcinoma Grade 1—4   Grades 1 and 2 — low grade carcinoma.
                           Grades 3 and 4 — high grade carcinoma.

ζ) CRAIK's classification (1957). Although this classification is a combination of a histological grade with the stage of invasion, it has the great advantage of expressing both factors as a single unit. Full details of CRAIK's classification are given below.

Grade I. Active recurrent type of transitional cell papilloma. The histological features are similar to those of the simple papilloma; there may be a few large cells with hyperchromatic nuclei and mitotic figures are sometimes more frequent. Tumours which are multiple or have recurred should be regarded as "not simple" however benign the histological structure. Invasion and metastases do not occur.

Grade II. Papillary transitional cell carcinoma. This may be single or multiple. it may be pedunculated or sessile, and it may vary in form from a fine localized and pedunculated papilloma to a huge, apparently solid, cauliflower mass. The epithelium covering the fronds tends to be thick and there is considerable cellular pleomorphism. In many areas the cells have lost their transitional orientation and are spheroidal or polygonal and not spindle-shaped. Areas of squamous metaplasia may occur, often being associated with infection. The nuclei are distinctly hyperchromatic and mitoses are frequent, many being tripolar or even multipolar. There may be commencing invasion of the stroma of the fronds or of the submucosa of the bladder wall but *not* of the muscle, lymphatics or blood vessels. The tumour often recurs after extirpation.

Grade III. Transitional cell carcinoma. This is an invasive tumour the essentia-feature being the histological demonstration of *invasion of muscle.* The supere ficial region may be a papilloma and when this is so the histological features arr indistinguishable from a Grade II (or even a Grade I) papilloma. The bladdel wall is invaded by thick columns of cells which may be branched giving a solid papillary structure. The cells, for the most part, retain their transitional orienta-tion having their long axes arranged at right angles to the edge of the columns. Mitotic figures are frequent, there may be a few large aberrant cells and some small areas of keratinization (squamous metaplasia) especially at the surface. Invasion of lymphatics, veins and extra-vesical tissues tends to occur late.

Grade IV. Epidermoid carcinoma (i.e., poorly differentiated transitional cell carcinoma). Again the superficial areas may be papillomatous so the diagnosis depends on demonstrating invasion of the bladder wall. The tumour is un-differentiated and the cells are arranged in irregular solid masses, often re-sembling the pieces of a jig-saw puzzle, and in invasive narrow cords, which may branch and intersect. The cells are large and spheroidal or polygonal, the nuclei are hyperchromatic and show very frequent mitoses. Aberrant cells with multiple or polymorphic nuclei are common.

Occasionally the greater part of a tumour is well differentiated, resembling a grade III, but there are areas of keratinization (squamous metaplasia) or of pseudoalveolar pattern (glandular metaplasia). Such changes indicate increased activity and raise the grading of what would otherwise be a grade III transitional cell carcinoma.

Squamous carcinoma. Only if a tumour shows widespread keratinization with "cell nests" in almost every cell mass should it be classified as squamous. Thus true squamous carcinoma of the bladder is an uncommon tumour.

Adenocarcinoma. Similarly, an adenocarcinoma has an alveolar structure throughout. The alveoli may be lined by single or multiple layers of cells which may be cubical or columnar. Often the cells in adenocarcinoma contain mucin and rarely the tumour may be a mucoid or colloid adenocarcinoma which is indistinguishable from a carcinoma of large bowel.

In all the subdividions of Grade IV carcinoma, invasion of lymphatics, extra-vesical tissues and of veins tends to occur early.

Grade V. Anaplastic carcinoma. This tumour resembles a round cell or spindle cell sarcoma, in fact, many "sarcomas" of bladder probably belong to this group of carcinoma. The bladder wall is diffusely infiltrated by large and small cells without architectural arrangement or grouping. Many multinucleated and polymorphic cells occur along with bizarre cells and nuclei. It is not possible to give a confident diagnosis of carcinoma in this type of undifferentiated neoplasm but the majority are not sarcoma.

This type of tumour is often highly radiosensitive and, following treatment, some have been known to remain tumour free for long periods (years).

In this classification the grades from I to V indicate increasing activity and thus a progressively poorer prognosis. There is a very sudden increase in activity and in mortality between grade II and grade III yet both may present as "papillomas" on cystoscopic and on bimanual examination.

In all reports both the numerical grading and the descriptive name of the neoplasm is given.

Furthermore, if biopsy samples are taken only from the superficial papilloma the two grades cannot be distinguished histologically. Thus it is imperative to state, when reporting such biopsies, whether or not bladder wall is included, so that a more adequate biopsy may be taken at a later date. Only when the base of the tumour has been adequately examined is one justified in concluding that a tumour is non-invasive. It is to emphasize this danger that the poorly differentiated papilloma is described as "papillary transitional cell *carcinoma*".

j) Comment on classifications

Although only six classifications of epithelial tumours of bladder have been considered, many others exist (ASH 1940, DART 1946, COLBY & SNIFFEN 1946). Nevertheless, they add little to those under discussion. The major issue is whether or not there is need to classify bladder tumours and if so which one should be used. It is recognized that the different histological patterns displayed by transitional cell carcinoma are but variants of the growth, and so many pathologists are opposed to a pigeonhole histological classification. Nevertheless, if a proper assessment of prognosis is to be obtained, some method of standard reporting is essential. If this is associated with a full clinical, bimanual and cystoscopic examination, an attempt at classification is justified. While some classifications are based on the cellular appearance of the growth and stage of invasion (KRETSCH-MER et al. 1934, DUKES & MASINA 1949, British Institute of Urology 1955, MELI-COW 1955, CRAIK 1957), others are based mainly on the extent of invasion (JEWETT 1952, MARSHALL 1952), although the latter does combine, secondarily, a cellular classification.

There appears to be some uniformity of opinion on the appearance of adenocarcinoma and squamous epithelioma, and not unnaturally, in view of the diverse histological patterns seen with transitional cell carcinoma, the main problem of histological classification occurs in this group. Since transitional cell carcinoma constitutes most of the cases of primary epithelial carcinoma, it becomes clear why the subject appears confused. However, when four of the classifications are reviewed (Table 1), a singularly common pattern emerges. All agree that the lesions may be papillary, or solid, although in one group (KRETSCH-

Table 1. *Correlation of grade and stage in epithelial tumour of bladder*
Transitional cell carcinoma
1. Grade

Author	Histological appearance		Cell differentiation	Other features
KRETSCHMER et al. (1934) . .	Papillary	Infiltrative carcinoma	Broder I—IV	—
DUKES and MASINA (1949) . .	Papillary	Solid	Low Average High	Squamous and glandular metaplasia
DUKES (1955)	Papillary	Solid	Differentiated anaplastic	—
MELICOW (1955).	Papillary	Solid	Broder I—IV	Metaplasia glandular and squamous

2. Stage of invasion

MELICOW (1955).	0	A	B	C	D		
DUKES and MASINA (1949) . .	I—IV						
MARSHALL (1952)							
(JEWETT and STRONG 1946)	0	A	B$_1$	B$_2$	C	D$_1$	D$_2$

MER et al.), the term infiltrating carcinoma is used. Each group used a different method to determine degree of cell differentiation. KRETSCHMER et al. and MELICOW apply BRODER's Grades I—IV, while DUKES and MASINA describe the growth as being of low, average or high malignancy. DUKES simply uses the term differentiated or anaplastic. As with the histological appearance, so common ground exists with assessment of the degree of cellular differentiation. Thus a papillary transitional cell carcinoma, BRODER's Grade I (KRETSCHMER et al., MELICOW) would be designated as a differentiated papillary growth (DUKES) or a transitional cell carcinoma of low grade 2b (i low) by DUKES and MASINA.

When, in addition to grading, the stage of invasion is used a truer picture of the lesion is formed. Some difference again exists in nomenclature. DUKES and MASINA and DUKES stage the growth 1 when it is localized to mucosa and 1a when the submucosa is involved. MELICOW uses the term BOWEN's or carcinoma *in situ* for the solid type of carcinoma localized to mucosa, and stages the tumour only when it involves the submucosa (Stage A). DUKES and his group stage the growth 2 when it is localized to bladder, whether or not it extends through or is superficial to the muscle coat. MELICOW applies stage B to a growth in or superficial to the muscle and C to a growth which has extended to the perivesical tissue. The main distinction between the two stagings lies in what significance should be attached to extent of tumour spread in the bladder. In this respect MARSHALL and JEWETT believe in the importance of the muscularis, as the dividing line between superficial and deep invasion, and the relationship to prognosis.

Nevertheless, it appears that a good relationship does exist between the macroscopic and microscopic appearance of the tumour and its extension in or through the bladder wall, and when these observations are correlated with clinical information (cystoscopic appearance, bimanual examination) a good appreciation of the growth can result. This can only be done by close cooperation between pathologist and clinician. Any of the combined grade and stage classifications are satisfactory and once one has been agreed on, it is imperative that both surgeon and pathologist should adhere closely to it, for only in this way will they understand a common problem. Although there is a close similarity in most classifications, as already shown, prejudices still remain to offset the universal establishment of one of them. This is unfortunate, for only then will pathologist and urologist be in a position to compare prognosis and operative technique. Nevertheless, some degree of uniformity exists both in America and Britain. Most American pathologists use BRODER'S grading and some uniform type of staging; in Britain, mainly due to the energy and activity of CUTHBERT DUKES, one or other of his classifications is almost universally used and has been adopted by the Institute of Urology.

The classification used by CRAIK and his group is not widely known, yet it does much to overcome some of the outstanding problems of the others. It emphasizes the importance of muscle invasion and sets a standard for classification which is less ambiguous than the others. It has the added advantage of bringing together grade and stage, which can be expressed as a single unit. This obviates one of the main disadvantages of all other methods. Using this classification, the first point is to determine the degree of invasion of bladder wall. If muscle is involved the grade will be 3 or 4, the final figure depending on whether or not squamous metaplasia or pseudoglandular changes are present. If the growth extends only to submucosa and *not* to muscle the tumour is in Grade II. In the classification numerical grading from I to V indicates increasing activity and a corresponding poorer prognosis, a feature not so easily appreciated in the other classifications. As CRAIK points out "there is a very sudden increase in activity and in mortality between Grade II and III, yet both may present as papillomas on cystoscopic and on bimanual examination."

The only real objection to this classification is the conflicting nomenclature used. Thus Grade II is referred to as "papillary transitional cell carcinoma" and Grade III "transitional cell carcinoma", yet both tumours have a papillary arrangement. Since the author describes in detail what he means by each grade, it would appear that the confusing nomenclature could be dropped without any loss of merit, and Grades I to V retained.

Certain lesions have not yet been incorporated in the scheme mainly because of lack of material. Such conditions include "intra-epithelial carcinoma" and "invaginated papilloma" a tumour with the histological structure of a papilloma, but which presents as a solid tumour clinically because the epithelial folding occurs not on fronds but in the thickness of the submucosa. Both conditions are usually associated with an established "invasive carcinoma" (CRAIK).

2. Adenocarcinoma

Although this type of growth occurs only in 0.5 to 2 per cent of recorded cases, it has aroused considerable interest and is the subject of some excellent papers (BEGG 1931, PATCH & RHEA 1935, PATCH & PRITCHARD 1946, HOWARD & BERGMAN 1948, KRETSCHMER 1949, WHEELER & HILL 1954). Most of them are reports of one or two cases (CAWKER 1947, LANE 1948, COPPRIDGE, ROBERTS & CULP 1951). Recorded cases have been reviewed fully by WHEELER and HILL who divide adenocarcinoma of bladder into primary and metastatic. 78 primary growths were found to occur in a normally placed exstrophied bladder or had their origin in urachal elements.

a) Primary adenocarcinoma

a) In a normally placed bladder. 14 cases were found in this group. Their age varied from 24 to 76 years with an average of 59.8 years, and there was a

slight preponderance of females. The growth was located usually at the base
or on the lateral walls of the bladder.

b) 12 cases occurred when the growth was found in an *exstrophied bladder*.
The age of the patients varied from 21 to 62 with an average of 46 years. There
was a preponderance of males in the ratio of 3:1.

Primary growths in both normally placed and exstrophied bladders have
certain features in common. Transition from non-neoplastic bladder epithelium
to adenocarcinoma can be demonstrated, as well as the coexistence of cystitis
cystica and cystitis glandularis.

b) Primary adenocarcinoma of urachal origin

Such tumours are located in the dome of the bladder and in their series,
WHEELER and HILL have considered all adenocarcinoma of this site to arise
from epithelium of urachal canal. They report 52 cases from the literature
(occurring in the dome and presumably urachal in origin). The ages vary from
23 to 83 with an average of 49 years and the sex ratio is approximately 3 males
to 1 female.

For full details of the anatomy and embryology of the urachal canal and
its relationship to adenocarcinoma reference should be made to the papers by
BEGG (1930, 1931, 1936) who draws attention to the importance of that portion
of the urachus which penetrates the muscle wall. It is usually at the upper
part of this or the lower end of the extravesical section that new growths
originate. Sometimes, where the canal communicates with the bladder, the
opening is at the top of a papilla and the tumour appears as a flat ulcer. Predominantly, the tumour involves the muscularis rather than the submucosa
and this point is of importance in deciding a urachal origin for adenocarcinoma.
In addition, urachal adenocarcinomas do not show any evidence of cystitis
cystica or glandularis, but urachal remnants connected with the neoplasm may
be demonstrated. Such a growth usually presents as a suprapubic mass.

Aetiology of adenocarcinoma of bladder. While epithelial elements
in the urachal canal give rise to adenocarcinoma in the dome of the bladder,
other factors play a part in the cause of primary growths in the normally placed
and exstrophied bladder.

1. Arising from glands normally present in the bladder. In the trigone and
suburethral regions of the bladder, tubular structures (ALBARRAN'S glands) are
sometimes found. Such structures were believed by some (VIRCHOW 1863,
KAUFMAN 1884) to be aberrant prostatic glands and by others to arise from the
posterior urethra (ASCHOFF 1894). SINDONI (1933) from an examination of the
bladders of 11 children (newborn to 10 years) and 9 adults, could find no evidence
of true glands. Nevertheless, it is believed by some that adenocarcinoma may
arise from paraprostatic (ASH 1939) or from mucus-secreting, subtrigonal or
subcervical glands (HOWARD & BERGMAN 1948). It would appear that examination of a bigger series of bladders of different age groups is required to determine
the incidence of these mucus-secreting glands before it can be decided that they
are the source of malignant epithelial tumours.

2. Origin from inclusions of germinal cells of the lower intestinal tract is believed
by some to explain adenocarcinoma in normal and exstrophied bladders. BAR-
RINGER (1920) considered an origin from remnants of allantois or cloaca while
LECENE and HOVELACQUE (1912) and DUPONT (1922) believed that islands of
mucus-secreting epithelium and glands form in exstrophied vesical mucosa and
submucosa due to developmental inclusions from intestine. Since the kidney

pelvis, ureter and bladder have a different embryological origin, the occurrence of gland structures of a hyperplastic or proliferative nature in the kidney pelvis and ureters alone or in combination with similar lesions in bladder (ABESHOUSE 1943) refutes the embryological theory of origin of adenocarcinoma. Again, it has been found that no glandular elements are present in the exstrophied bladder of the newborn child, but such elements occur in the adult exstrophied bladder. This suggested some other factor (metaplasia) as the cause of their presence.

3. *Metaplasia of bladder epithelium.* Most accounts of the origin of adeno-carcinoma in the normal placed or exstrophied bladder attribute it to a meta-plastic change in the bladder epithelium (PATCH & RHEA 1935, LOWREY 1939, SAPHIR & KURLAND 1939, EMMETT & MCDONALD 1942, PATCH & PRITCHARD 1946, KRETSCHMER 1949, WHEELER & HILL 1954). The early lesion is the appearance of Brunn's glands with subsequent development of cystitis cystica and glandularis and eventually adenocarcinoma. All three stages may be seen and serve to differentiate the growth from one of urachal origin. The cause of metaplasia in these cases is obscure although infection and urinary stones are believed to be contributory factors.

c) Secondary or metaplastic adenocarcinoma

The histological appearance of primary and secondary adenocarcinoma of bladder may be identical and so before the diagnosis of primary growth is made, the possibility of a metastasizing tumour must be eliminated. There are three probable sites of origin, large intestine, prostate and female genital tract. Adeno-carcinoma arising in large intestine is usually indistinguishable from primary growths in bladder, but the usual signs and symptoms of the large bowel lesion are generally evident by the time the bladder is involved. Similarly *endometrial adenocarcinoma* usually produces late invasion of bladder and the vaginal bleeding, invariably present, is a pointer to curettage. Invasive *adenocarcinoma of the ovary* with its variable histological patterns may present difficulties and should always be considered when the bladder tumour presents with atypical appear-ances. When the primary arises in *prostate* the glandular elements are much smaller, while intracytoplasmic mucin occurs much more frequently in primary adenocarcinoma of bladder.

3. Squamous epithelioma

True squamous carcinoma of bladder is an uncommon condition and should be distinguished from transitional cell carcinoma showing squamous metaplasia. In the past, the two conditions have been frequently confused, mainly due to a failure to define clearly the criteria on which the diagnosis of squamous epi-thelioma of bladder should be made.

A squamous epithelioma should consist entirely of typical squamous epi-thelial elements, and show widespread keratinization with "cell nests" in almost every group of cells (Fig. 38). No neoplastic transitional cell elements should be seen. Unlike squamous metaplasia, the two cases we have seen of squamous epithelioma were associated with leukoplakia of bladder and different gradations between the two conditions were evident. In transitional cell carcinoma with squamous metaplasia, transitional cell elements usually predominate and are easily recognized, although occasionally they show such extreme degrees of anaplasia as to resemble sarcoma. In one case in our series, while part of the tumour was composed of anaplastic transitional cells, which had infiltrated the muscle coat, an adjacent portion of the growth showed well marked cell nest formation with keratinization. Such a growth was a transitional cell carcinoma with squamous metaplasia and not a squamous epithelioma. When a squamous epithelioma dedifferentiates, as seen in cervix and antrum, the whole growth shows the anaplastic change.

If these criteria are applied it is found that a large number of cases reported as squamous epithelioma of bladder are no longer acceptable. In reviewing our own series of 30 cases reported as squamous epithelioma only two of them fulfil the above requirements; the remainder are undoubtedly transitional cell carcinoma with squamous metaplasia. Again the statement (CONNERY 1953) that "all squamous cell tumours are infiltrating and as vicious in character as can be found; with the exception of adenocarcinoma they have no equal", applies only to transitional cell carcinoma with squamous metaplasia. Obviously the prognosis in squamous epithelioma will depend on the degree of infiltration at the time of operation (McDONALD & THOMPSON 1948), and it is not justifiable to regard all examples of squamous epithelioma, especially the well-differentiated Grade I Broder Group, in the same light as one would the transitional cell carcinoma showing squamous metaplasia.

From the two cases studied personally, the naked eye appearance is not characteristic. Flat, raised whitish growths have been described by ABESHOUSE and TANKIN (1956). Such tumours appear to occur in any site in the bladder.

Aetiology. Most authors agree that although leukoplakia itself cannot be considered as a precursor to squamous epithelioma, similar factors are probably at work in both conditions. MONTGOMERY (1943) and CONNERY (1953) found that 20 per cent of all cases of leukoplakia go malignant.

4. Carcinoma complicating vesical diverticulum

While carcinoma and diverticulum of the bladder are very common, carcinoma in a diverticulum, like carcinoma in the dome of the bladder, is a comparatively rare lesion. Most reports are of single cases (EWELL 1930, BRIGGS 1930, HICKS 1930, LE COMTE 1932, STEWART & MUELLERSCHOEN 1932). MAYER and Moore (1954) note 150 cases in the literature to that date and report 8 cases "to emphasize the importance of the closest scrutiny of all vesical diverticula so that this serious and often obscure complication may not be overlooked." Since bladder diverticula occur predominantly in males, it follows that carcinoma complicating diverticulum will be more common in males. SMITH and SUDER (1950) found the ratio of 30 males to one female. The growth occurs predominantly in the 5th to 6th decade.

The growth may vary considerably in appearance just as the diverticulum can vary greatly in size. The diverticulum arises mainly from the right or left lateral wall of bladder, sometimes from the base.

The tumour may be a small papillary, pedunculated, ulcerated or rounded cauliflower-like mass which projects into or may fill the lumen of the diverticulum. A certain percentage of these tumours will be missed on cystoscopic examination because of their small size and location within the diverticulum.

Again, the growth can be found projecting from the diverticular orifice, as a reddish-white papillary mass.

Microscopically, the lesion may appear as a simple papilloma, or as any of the different grades of transitional cell carcinoma (Grades I—IV BRODER). Squamous epithelioma has been described in a few cases (TARGETT 1896, Case 4;

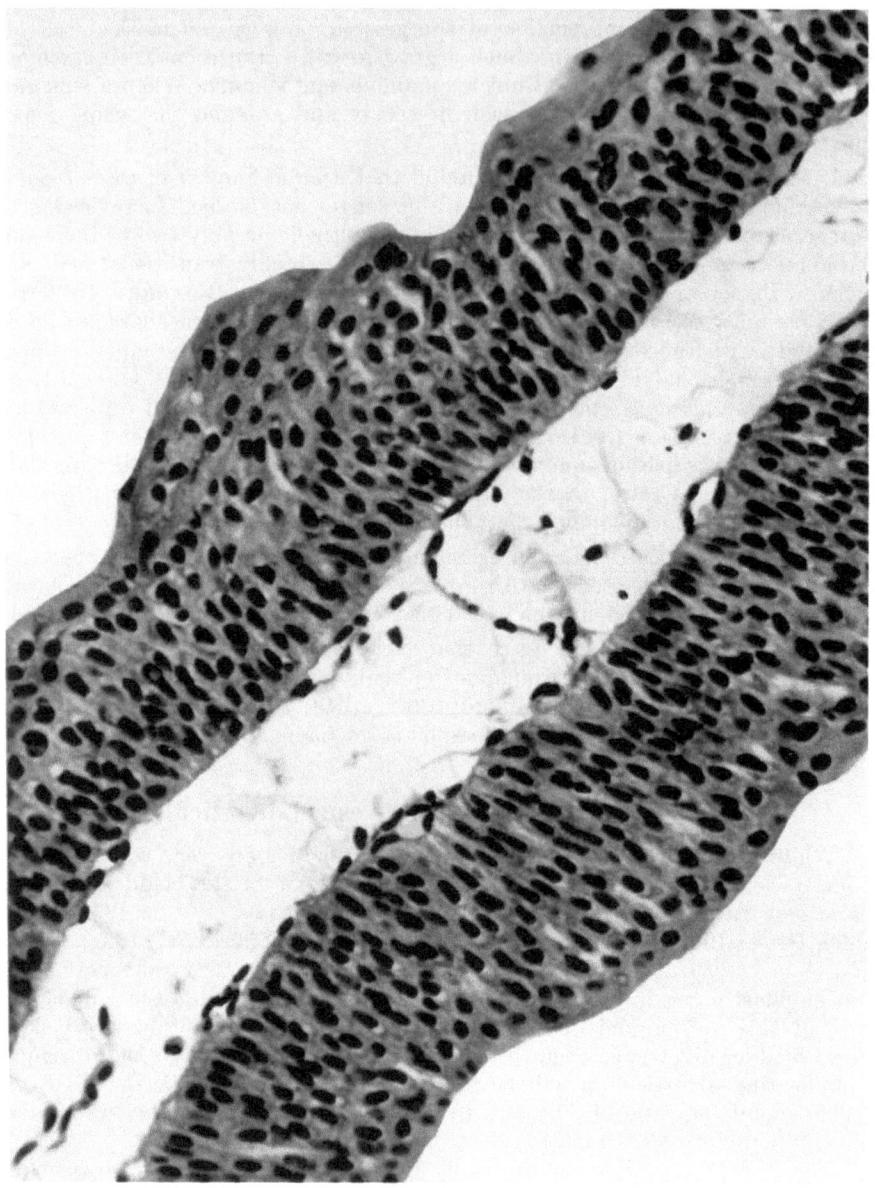

Fig. 44. A typical simple transitional cell papilloma. Haematoxylin and Eosin. × 500

SCHWARTZ 1923; HUNT 1929, Cases 2 & 4; MÜLLER 1954). Other microscopic appearances recorded are cavernous haemangioma and sarcoma (BLUM 1923), but in many of them the criteria for making this diagnosis are insufficient and it would appear that most of them may be transitional cell carcinoma with squamous metaplasia. True squamous epithelioma in a diverticulum, like that in bladder, is an uncommon finding.

The growth may be found localized to the diverticulum or may show evidence of invasion of the wall, yet the adjacent bladder may remain uninvolved as in the case reported by STEWART and MUELLERSCHOEN (1932). Nevertheless, distant metastases to lung, liver, spleen, left adrenal, kidney, mesentery, ribs and femur had occurred.

5. Simple epithelial tumours

Although the different types of primary epithelial carcinoma described are encountered in cystectomy (partial and complete) and biopsy specimens, it is only from biopsy material that one expects to find *simple transitional cell papilloma* of bladder (Fig. 44). This is a pedunculated growth covered by normal transitional epithelium. The cells are spindle-shaped, show no pleomorphism and few mitotic figures and are arranged at right angles to the fibro-vascular core. The growth is usually single and does not recur after extirpation.

Classification of biopsy material and application of histology grading

Since all methods of classification include staging of the growth, it follows that none of them can be satisfactorily applied to biopsy material unless part of the bladder wall or at least the base of the tumour is included in the specimen sent for examination. MILNER (1949) found a close relationship in grade from carefully and deeply made biopsies with resectoscope with the grade derived from partial and total cystectomies. DEAN (1948) on the other hand found the histological grade in biopsy specimens lower in approximately half the cases when compared with subsequent total cystectomy. Nevertheless, he points out that careful attempts to obtain extensive and deep biopsies were not always made.

CRAIK stresses in his classification that there is a very sudden increase in activity and mortality between Grade II and Grade III, yet both tumours may present as papillomas on cystoscopic and bimanual examination. Furthermore, if biopsy samples are taken only from the superficial papilloma the two grades cannot be distinguished histologically. When reporting such biopsies it is imperative to state whether or not bladder wall is included so that a more adequate biopsy may be taken at a later date. Only when the base of the tumour has been examined adequately is one justified in concluding that a tumour is non-invasive.

The position of the biopsy specimen is aptly summed up by MARSHALL, "Really satisfactory biopsies are necessary as no pathologist can do better than the quality of the material with which he is supplied. No tumour is better than its worst part."

II. Primary mesenchymal tumours of bladder

If the classification of primary epithelial tumours of the bladder is considered a problem, one cannot but admit that the position of mesenchymal tumours is no less complex. These tumours are rare, especially the benign growths. They will, therefore, be dealt with briefly before considering their more important malignant counterparts.

1. Benign mesenchymal tumours

The most common is the *leiomyofibroma* (Fig. 45) (KLEITSCH 1951) and the literature contains reports of not more than 50 such tumours (SEXTON 1952). Three types are described: a) Submucous. b) Interstitial. c) Peripheral. Depending on their position they may or may not have a pedicle. They are smooth, encapsulated, firm or soft tumours, their consistency varying with the presence

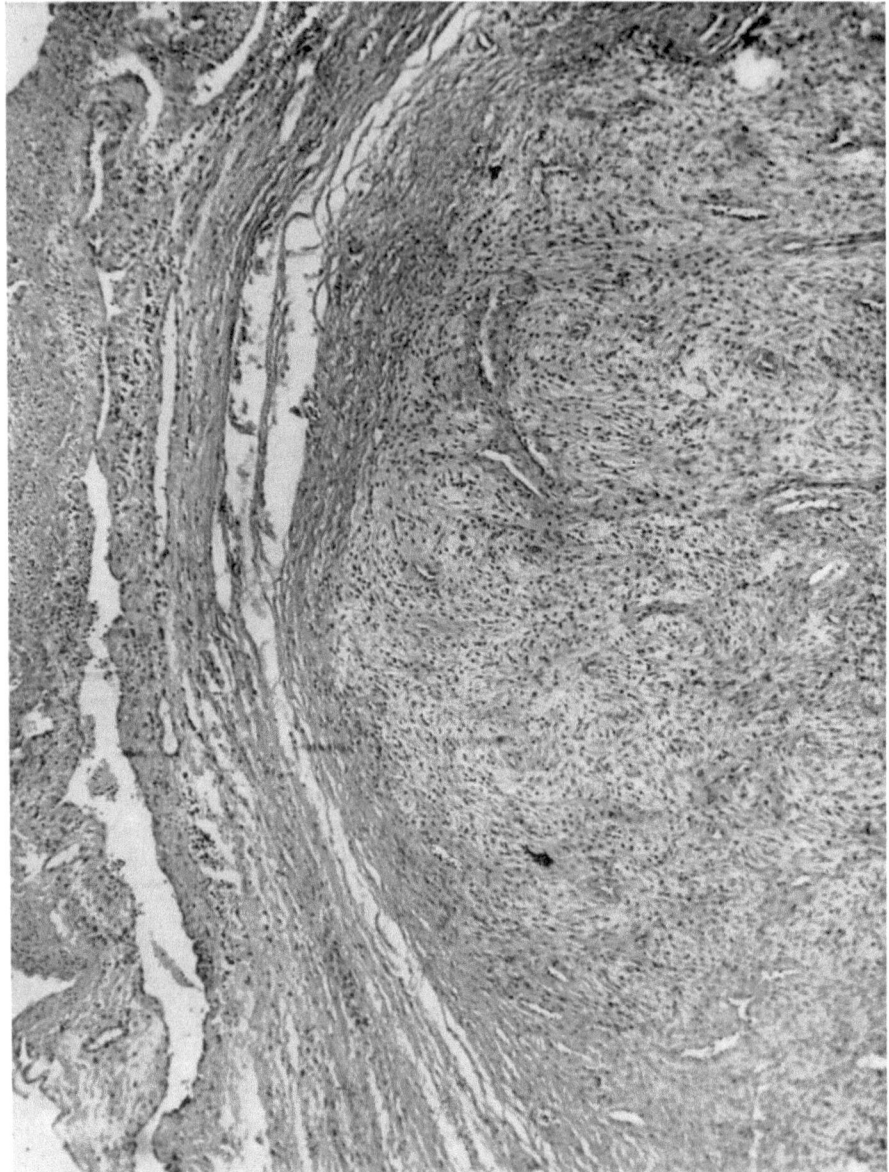

Fig. 45. This is a simple leiomyofibroma which was situated in the neck of the bladder. Haematoxylin and
Eosin. × 100

or absence of calcification. hyaline degeneration or necrosis. The size varies
within wide limits and the prognosis after operation is generally good (KRETSCH-
MER 1931). Malignant change is uncommon. Our own case, a small leiomyofibroma, was found on routine examination of tissue from the bladder neck
in a prostatectomy specimen. *Fibroma* behaves in the same fashion, but HAN-
BURY (1952) believes that all *rhabdomyoma* should be regarded as malignant.

    Angioma of the bladder is also worthy of mention. In a recent review of the
literature GRAHAM and BULKLEY (1955) found records of 45 cases and added

a further one. Most angiomas arise from the fundus of the bladder. They are generally bluish-black in colour, soft, irregularly shaped sessile tumours. The size varies and large masses which penetrate the paravesical tissues have been reported. Angiomas may be present elsewhere, usually in the skin. GRAHAM and BULKLEY stress the importance of differentiating these tumours from actual varicosities at the base of the bladder. However, the usual site of an angioma is the dome. In one of our cases a small, very vascular, simple epithelial papilloma was removed from near the external meatus. At first glance the extreme vascularity of the tumour suggested an angiomatous pattern, but the epithelial nature of the tumour was never in doubt.

2. Malignant mesenchymal tumours

Though the first report of sarcoma of urinary bladder was published as early as 1853 by GUERSANT, CECIL (1926) could find only 229 cases in the world literature. MUNWES (1910) and SMITH (1925) considered that sarcoma formed only 4.5 per cent of all malignant bladder tumours; CAULK (1926), PACK and LE FEVRE (1930) and RATLIFF and VALK (1939) put the figure even lower, possibly in the region of 0.3 per cent.

The rarity of these tumours alone makes their assessment doubly difficult. The pathologist can seldom draw sufficient cases from his own experience and must, of necessity, make recourse to old material which is frequently badly preserved and unsuitable for the application of more modern staining methods. Yet, despite these technical difficulties, such an investigation is necessary for it has now been shown that the original diagnosis cannot be accepted in many instances (SAPHIR & VASS 1938; HANBURY 1952, MOSTOFI & MORSE 1952). Indeed many so called cases of sarcoma listed in the past have proved to be anaplastic carcinoma (CRANE & TREMBLAY 1943, DUKES & MASINA 1949). In one of our cases the tumour at post-mortem appeared to be a sarcoma (Figs. 41 & 42), but after examination of a biopsy specimen, removed previously in another hospital, it was proved to be an anaplastic carcinoma. One could only conclude that deep X-ray therapy given in the interim period may have considerably altered the picture in this case. Indeed SCHOURUP (1955) mentions this effect of deep X-ray therapy and he also believes that surgical interference and chronic inflammatory change may alter the appearance of tumours.

The early literature contains many surveys on bladder sarcoma. One of the most comprehensive was that of CECIL (1926). He enumerated 14 different types of sarcoma, but the majority of these were grouped according to their microscopic appearance (e.g., round, spindle, alveolar, giant or mixed cell types) without reference to histogenesis. More than 50 per cent of the cases reviewed were included in these groups and in another 16 per cent the type of sarcoma was not stated. Of the remainder fibrosarcoma and myxosarcoma accounted for 11 per cent.

In a later review of 151 cases CRANE and TREMBLAY (1943) agreed that most round and spindle cell sarcoma reported in the past were in actual fact anaplastic carcinoma, and they considered that most of those recorded were cases of fibrosarcoma. Recently, however, few if any reports on fibrosarcoma can be found in the literature. It may be that these cases are not being recorded, since they have been reported in the past. On the other hand, we have not had one case in this hospital in the last 15 years, though anaplastic carcinoma occurs fairly frequently and the occasional leiomyosarcoma and rhabdomyosarcoma are seen. No one will deny the difficulty which exists in distinguishing the various tumours, especially the leiomyosarcoma and the fibrosarcoma, and one is not surprised that confusion should have existed in the past in their differentiation. More recently, however, with special staining techniques it may have been possible to recognize the smooth muscle origin in a large number of cases and this may in part have been responsible for the apparent decrease in the number of fibrosarcoma being reported.

In the present survey it is proposed to review sarcoma of the bladder in seven groups as follows, depending on the histogenesis of the tumours in each case:

a) Fibrosarcoma (including fibromyxosarcoma).

b) Leiomyosarcoma.

c) Rhabdomyosarcoma.

d) Primary lymphosarcoma.

e) Osteogenic and chondrosarcoma.

f) Angiosarcoma.

g) Miscellaneous.

a) Fibrosarcoma (including fibromyxosarcoma)

In reviewing the subject of fibrosarcoma of bladder, it is well to remember what has been said previously (p. 136) and to realize that the present review of this condition has of necessity been based on material from old literature.

This tumour occurs four times more frequently in males than in females (GABE 1932), especially in the first decade and after the fourth and fifth. According to MUNWES (1910) 15 per cent of the cases develop in children under ten. Initially the tumour is single, but frequently multiple growths are present by the time the case comes under observation. POZNANSKI (1914) described three types of growth.

1. Pedunculated. These are usually thicker than the ordinary epithelial papilloma. They are frequently soft and friable, sometimes lobulated and generally smooth, due to their covering of normal mucosa. Ulceration is rare (MUNWES 1910) except as a late feature. Their size varies and they are generally very red or blue in colour on account of their rich blood supply.

2. Polypoid with broad base; otherwise similar to above.

3. Infiltrating. Instead of growing into the lumen of the bladder these tumours infiltrate early. According to CECIL (1926) the commonest site is the base or trigone, but this is not accepted by all. ALBARRAN (1892) claimed they were frequently found on the anterior wall and MUNWES (1910) reported a higher incidence of tumours on the posterior and lateral walls. However, no matter the site, they should all be regarded as rapidly growing and highly malignant. Microscopically, a variety of patterns can be seen depending on the degree of differentiation of the tumours. Myxomatous change may be present. Mitotic figures are generally numerous especially in the more anaplastic type and tumour cells can be seen in close relationship to blood vessels, which are generally thin-walled and numerous. Tumour cells may also invade the epithelial "cap" but, though epithelium is lost in places, ulceration does not readily occur.

The prognosis for fibrosarcoma of the bladder is poor; local recurrence after operation, infiltration of the bladder wall and extension to neighbouring organs are frequent. The question of distant metastases, however, is controversial. Most authors agree that they occur late, but EWING (1922) believed that secondaries appeared earlier than in the epithelial tumours. On the other hand, MIXTER denied their occurrence at all, but most authors report an incidence between 2 and 20 per cent (ALBARRAN 1892, CONCETTI 1896, CECIL 1926).

b) Leiomyosarcoma

The first report to appear in the literature on myosarcoma of the bladder was that of GUSSENBAUER (1875), and later the smooth muscle nature of the tumour was stressed by RÖDER (1904). It was some time before the first comprehensive survey was made (KRETSCHMER & DOERHING 1939). Since then further cases have been added to the literature (SHIVERS & HENDERSON 1939), CRANE & TREMBLAY 1943, LEV & BELL 1947, MACKLES et al. 1948, JONES et al. 1950, KATZEN 1952, CECIL 1953, SILBAR &SILBAR 1955) bringing the total to approximately thirty.

Like fibrosarcoma it shows a predilection for men but it can occur at any age and any part of the bladder wall may be involved though usually not the trigone (LEV & BELL 1947, SILBAR & SILBAR 1955). Various theories have been put forward to explain its occurrence: partially atrophied Müllerian body (VERHOOGEN), urachus (PELLICT & DUPREY), supernumerary ureters (REIGAL), hydatid of Morgagni and lower layer of utricle (BLUM) have all been inculpated; and chronic inflammation and hypertrophy of the vesical musculature have not been overlooked as contributory factors.

The tumours are usually pedunculated and lobulated, and begin as greyish-white, smooth swellings. Ulceration is frequent and eventually they may become encrusted with lime salts. Microscopically, three types are recognized by CECIL (1953).

a) Fairly well differentiated tumour, resembling leiomyoma. Cohen, however, states that the tumour cells are shorther and rounder than those of the benign tumour and the nuclei are larger and hyperchromatic. No capsule can be seen and the stroma is scantier. Few mitotic figures are present.

b) Composed of short spindle cells with oval nuclei.

c) Most malignant with considerable variation in the morphological character of the cells. However, some fusiform cells may show a resemblance to smooth muscle and may be arranged in bundles (Fig. 46). These may give a clue as to the exact nature of the tumour. Their cytoplasm is generally coarsely fibrillar and, when stained with Mallory's method, intracellular fibrils may be seen concentrated round the nucleus (CECIL 1953) which usually contains small nucleoli. Hyaline and myxomatous degeneration of the stroma may be present and there may be perivascular lymphoid infiltration in these areas (SILBAR & SILBAR 1955). The vessels are usually prominent. No capsule can be identified and in some areas the tumour growth can be seen in close proximity to the normal bladder musculature.

In their case LEV and BELL (1947) reported the presence of large polygonal clear cells resembling xanthoma cells, with copious clear cytoplasm and relatively small nuclei. CAYLOR and WALTERS (1930) and POWEL (1932) describe multinucleated giant cells but in the opinion of CAPPELL and MONTGOMERY (1937) the former was actually a myoblastoma.

Until recently the prognosis in leiomyosarcoma of the bladder was considered to be exceptionally poor, death occurring within a year. In their more recent review SILBAR and SILBAR (1955) were not so pessimistic. They analysed the results in the 30 reported cases and found that 16 of the patients were dead within a year of operation, 6 were living over one year, 2 over three years, 1 over four years and 1 over ten years. In the remaining 4 no follow-up report was given. Deaths in the majority of cases were due to local recurrence, ascending infection and obstruction. Only two cases of distant metastases are reported in the world literature (WEYERBACHER & BALCH 1937, MUNGER 1939). These were

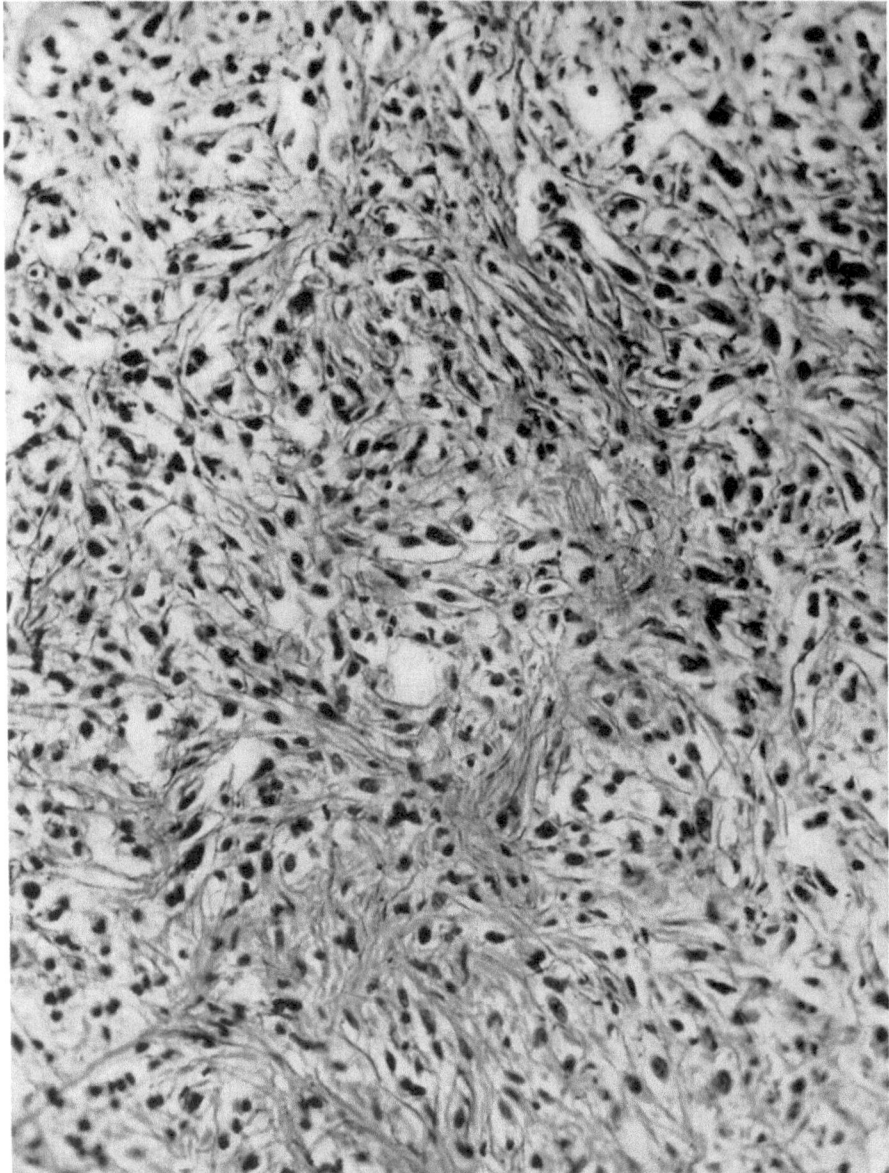

Fig. 46. A rather poorly differentiated leiomyosarcoma. Fusiform cells arranged in bundles can be seen.
Haematoxylin and Eosin.   × 320

in the liver and lung respectively. It may be that the relative infrequency of
distant metastases, coupled with better operative techniques and more thorough
removal of the tumours, is responsible for the slight improvement in prognosis.

c) Rhabdomyosarcoma

Though rare these tumours are probably next in order of frequency. In
recent years they have attracted considerable attention and a total of 36 cases
has now been reported in the literature (SLOTKIN & DAVIS 1954). The first

tumour of this type was described by CATTANNIO 1907 in 1884 in a young boy of twelve. KHOURY and SPEER (1944) reported a case and reviewed the literature at time. HANBURY (1952) included KHOURY and SPEER's series of 17 cases and brought the review up to date with a total of 28 cases. More recent cases have been added by HIGGINS (1952), MORSE and JARMAN (1953), RIDDELL and KUDISH (1953), and SLOTKIN and DAVIS (1954).

One of the striking points HANBURY (1952) made with regard to these tumours was the high incidence in children under two years. In the 26 cases (including his own) in which the age was given, twelve occurred in children under two, and another six in the two to ten year old group. The remainder were scattered fairly evenly in the different decades. The tumours were also commoner in males in the preponderance of 19 to 9. In his review of the literature HANBURY also found general agreement with regard to the site of the tumour. The majority arise from the trigone or the lower part of the bladder. Some project into the first part of the urethra, causing obstruction and in these cases it is often difficult to rule out a primary prostatic origin. No agreement can be reached, however, with regard to the actual histogenesis of the tumour and various theories have been put forward:

1. HOUETTE and WILMS believe that the tumours arise from embryonal heterotopias, probably anomalies of development of the Wolffian duct.

2. SHATTOCK (1909) refers to HENLE's illustrations and stated that striated muscle fibres are found normally around the external sphincter, surrounding the prostate and extending upwards and backwards round the neck of the bladder. "Vagrant or displaced sarcoblasts" from these muscle elements are said to give rise to the tumour.

3. WILLIS believes that the tumour arises from either "immature prospective muscular tissue or indifferent mesenchymal tissue with the potency of aberrant differentiation into muscle fibres" and such embryonic mesenchyme is found around the upper part of the urogenital sinus.

The fact that the tumour occurs more commonly in infants suggests the existance of some embryological anomaly. Moreover WILMS, and SHATTOCK's hypotheses do not explain KRETSCHMER's case (1947) of rhabdomyosarcoma occurring in the dome of the bladder. HOUETTE's case (1929) was found in association with a congenital diverticulum in a male child of 13 months. This is the only case reported in association with a diverticulum.

Normally the tumours are pedunculated and arise from the trigone. CAPPELL (1948) recognizes two types:

1. Coarsely lobulated polypoidal growth with many bulbous processes, suspended by a common stalk. These tumours project into the lumen of the bladder. Myxomatous change is common and the processes may be shed off as grape-like cysts. These may fill the bladder as in RIDDELL and KUDISH's case (1953). In many respects this type of tumour resembles sarcoma botryoides.

2. Mushroom-shaped tumour with a broad stalk. Occasionally the tumours are sessile. They are frequently multiple and may spread to involve the whole bladder. The surface epithelium is smooth, ulceration being an uncommon feature. Extra-vesical growth may occur as in HANBURY's case (1952).

Microscopically there is a difference of opinion with regard to the classification of these tumours. Of the earliest works on striated muscle tumours probably that of ABRIKOSSOFF (1926) is best known. He described four different types under the heading of "myoblastoma", but as CAPPELL and MONTGOMERY (1937) pointed out "all types may be found in the one tumour", and they preferred a broader division into two groups:

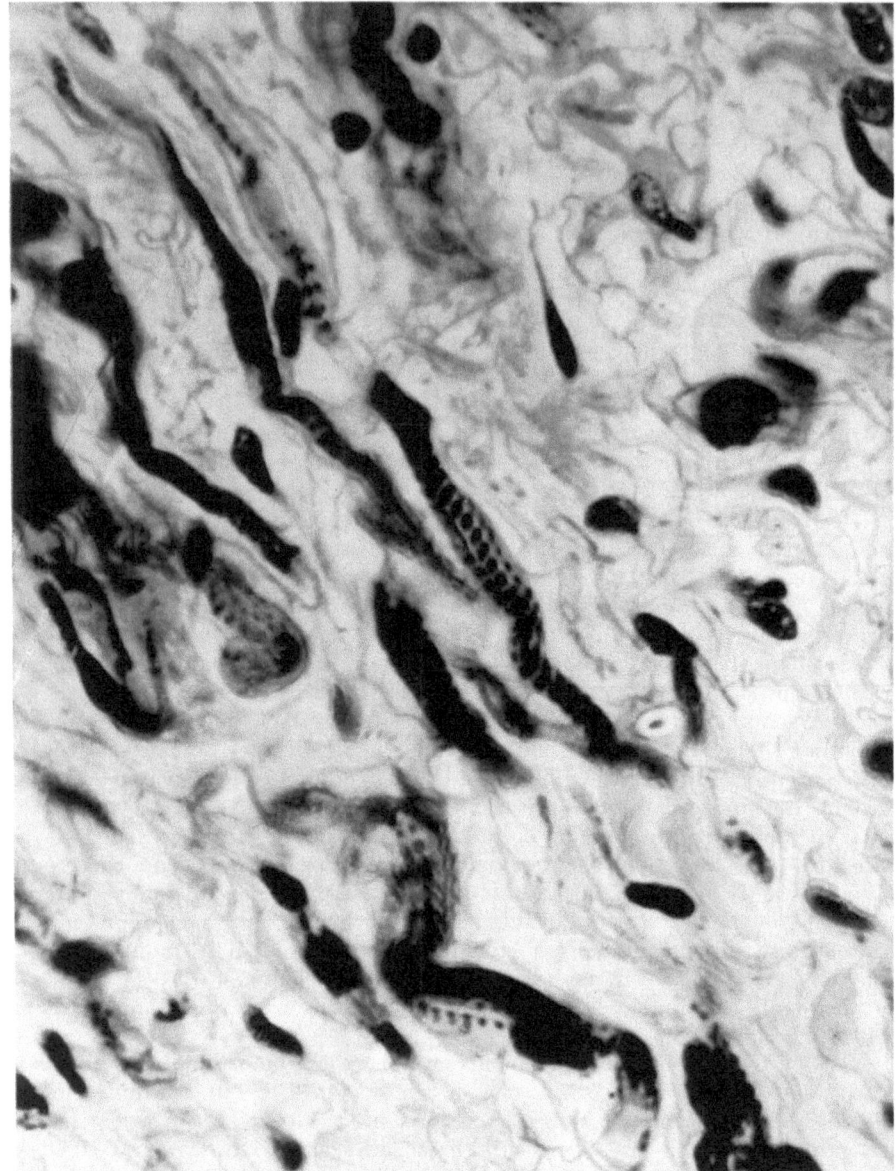

Fig. 47. Well developed cross striations in a rhabdomyosarcoma. Phosphotungstic acid haematoxylin. × 800

1. Those showing well defined cross striations. These could be sub-divided into a) simple, b) malignant.

2. Those without well defined cross striations, consisting of primitive, not very pleomorphic cells, round, oval or elongated, 20—25 μ in length, occurring in strands. These cells have a strongly acidophilic cytoplasm and correspond to ABRIKOSSOFF's myoblast. On this account they called this type of tumour a myoblastoma. Later, however, CAPPELL (1948) deprecated this term, as it suggested the existence of a highly malignant tumour arising from myoblasts,

whereas in actual fact the tumour appears to arise from adult striated fibres and is generally slow growing. It also carries a better prognosis and on this account alone CAPPELL (1948) feels it should still be recognized as a separate entity.

WILLIS (1948) disagrees with CAPPELL. He feels that the term leads to confusion by "perpetuating erroneous and dubious concepts". It is mentioned here, however, to stress the point that it carries a better prognosis.

The microscopic appearance of these tumours is well described by CAPPELL and MONTGOMERY (1937). The majority have an abundant poorly differentiated stroma, often myxomatous, in which may be scattered fairly well differentiated striated muscle fibres (Fig. 47) or their precursors. The striations may be obvious in haematoxylin and eosin stained sections, but special stains (phosphotungstic acid haematoxylin) may be necessary to make the diagnosis certain. However, should this still be in doubt CAPPELL and MONTGOMERY have drawn up certain criteria which they state are necessary for the recognition of a tumour of myoblast origin. These are:

a) an acidophil, granular or vacuolated cytoplasm (spider cells),

b) cells which are spindle-shaped or of strap-like or tadpole form,

c) frequent multinucleated cells with oval vesicular nuclei in which there are well marked chromatin knots and a single very large karyosome; clustering of nuclei or formation of cigar shaped rows may be seen,

d) precursors of striation. WÖLBACH (1907) described these as rows of darkly stained paired dots which he said were the earliest precursors of myofibrils.

The tumours are usually covered by the normal bladder epithelium. This may be absent in parts or occasionally squamous metaplasia may be seen. Ulceration is a late feature. No capsule can generally be recognized and infiltration of the bladder musculature is usually evident. Two of HANBURY'S cases (1952) were well differentiated and showed no evidence of invasion. He considered these to be rhabdomyomatous tumours of the bladder. Both had been diagnosed previously as "myxomatous polypi". HANBURY, however, states that it is probably advisable to regard all rhabdomyomatous tumours as malignant. Though metastases are unusual, local recurrence is frequent. The smooth appearance is deceptive because definite invasion of muscle is usually present. The prognosis is also bad because of the position of the tumour. Urinary obstruction and secondary infection are frequent sequelae and according to HIGGINS (1952) are responsible for most of the deaths. Only in the myoblastoma is the prognosis better and there are fewer recurrences after operation (CAPPELL & MONTGOMEY 1937).

Before leaving these tumours some mention should be made of the myxomatous change which is frequently seen. Tumour cells may be thinly spaced in the very oedematous stroma and the underlying lesion overlooked (HANBURY 1952, MOSTOFI & MORSE 1952). Hence, the diagnosis of myxomatous polypi or occasionally myxosarcoma may be wrongly made. Indeed LAZARUS and ROSENTHAL (1932) described a case of myxosarcoma of the bladder in a boy of two years, which might well have been rhabdomyomatous in nature. They reviewed the literature and found a further 9 cases, all occurring in children under 7 years, the majority being two or under. Except in one case where the whole bladder was involved (WEISS & DREYFUS 1928) the tumours had arisen from the base or trigone. One wonders, therefore, about the exact nature of these tumours. Certainly in making the diagnosis of fibromyxosarcoma the question of rhabdomyomatous origin should always be carefully considered.

10*

d) Primary lymphosarcoma

This is a rare condition. In a recent review JACOBS and SYMINGTON (1953) found only twelve cases (including their own) recorded in the literature. Unlike other bladder sarcoma the majority of the patients were females, the ratio being 3 to 1. The average age was over 40 years, but the range was wide and CHAFFEY's case (1885) occurred in a young boy of 3 years.

According to WILLIS (1948) the incidence of lymphosarcoma in any particular site appears to depend on the amount of lymphoid tissue normally present. Though its presence in the normal bladder is controversial, there is no doubt that prominent lymph follicles can be seen in response to chronic infection (HINMAN & CORDONNIER 1935). It is not true to assume from this that all cases of primary lymphosarcoma of the bladder are the result of cystitis follicularis. Though some reports give a history of recurrent cystitis over many years (CHAF-FEY 1885, MOROGNA 1927, KREUTZMANN 1942, RATHBUN & WEHRBEIN 1944), it would appear that primary lymphosarcoma can be present in the bladder in the absence of any bladder signs and symptoms (MOLONEY 1947).

The tumours may be single or multiple, firm in consistence and white in colour. They present as smooth round swellings projecting into the lumen of the bladder or as a diffuse infiltration of the bladder wall. The cells are mainly lymphoblasts and lymphocytes and lymph follicles with prominent germ centres are frequently seen. There is usually infiltration of the muscle coat, but the epithelium generally remains intact though in the later stages ulceration may occur. Mitotic figures vary depending on the malignancy. Apart from MO-LONEY's case the majority were considered to be relatively benign. Hence the prognosis after operation and deep X-ray therapy is generally good. Our own patient is still alive and well 9 years after operation but it is essential before committing oneself to ensure that the bladder lesion is in actual fact primary and not a secondary deposit and that the blood is normal. From existing reports it seems that bladder involvement in leukaemia is a rarity (KIRSHBAUM & PREUSS 1943, WATSON et al. 1949).

e) Primary osteogenic sarcoma and chondrosarcoma

Like lymphosarcoma these tumours are rare. CRANE and TREMBLAY (1943) found only 8 cases in the literature, their own making the ninth. Six of these they regarded as primary. More recently, however, JONES et al. (1950) reported a case, but in their opinion the osteogenic change had actually occurred in a leiomyosarcoma and they could still detect the latter elements in the tumour. These were not present in CRANE and TREMBLAY's case nor was there any evidence of a fibroblastic origin. The authors, therefore, suggested a possible origin from embryologic rests, probably Wolffian body. It is well known that renal epithelium can stimulate bone growth experimentally (NEUHOF 1917, HUGGINS 1931), and CRANE and TREMBLAY suggested that the close proximity of this tissue, stimulated development towards osteoid tissue in these rests.

Since the number of reported tumours is small it would be rash to generalise on their naked-eye appearance. They tend to have a brown, irregular crusted surface, covered with urinary salts, and on section they are white, hard and gritty. CRANE and TREMBLAY describe the microscopic appearance as consisting of "irregular anastomosing bands of dense eosinophilic material between which are small groups of polyhedral to spindle shaped cells with round, vesicular but also hyperchromatic nuclei". At many points these cells were arranged along the margins of the intercellular substance like osteoblasts. Large osteoclastic giant cells and groups of cartilage cells were also present. Mitotic figures were numerous and tumour could be seen invading capillaries. Pulmonary metastases had occurred and had a similar appearance to the primary.

As an aid to diagnosis the range of phosphatase activity of the tumour was estimated and found to be maximum at pH 9.5. The tumour also contained large amounts of alkaline phosphatase histologically.

With so few cases, one cannot at this stage be dogmatic about prognosis. CRANE and TREMBLAY's case developed metastases but in JONES' case, where the osteoid change was a secondary feature in a leiomyosarcoma, the patient was alive 9 months after operation.

f) Angiosarcoma

Though CRANE and TREMBLAY placed these tumours in a separate group they were only able to find a report of one such case in the literature (GAROFALO 1927) and they also mentioned another case of Kaposi' disease (ELJASZ 1932). It seems that simple angioma of the bladder is more common and it has been mentioned previously.

g) Miscellaneous group

Only a few rarities remain to be considered. It would seem that the neurogenic sarcoma must be regarded as such, or the other possibility is that it has been overlooked and passed as a fibrosarcoma. CRANE and TREMBLAY certainly only mention one case in their review.

Of the teratoma, the dermoid cyst is the only one recorded in the literature as involving the bladder. In 1895 CLADO collected 32 cases from the literature. Since then other reports have been added, but a primary bladder origin was not proved in all (CAUFFIELD 1956). The majority of cases appear to arise in females. Whether this sex preponderance is actual is still in doubt. The close proximity of the ovary, the relatively common occurrence of ovarian teratoma and the inability in a number of cases to define the primary site of the tumour, combine to render the picture obscure.

The classical symptom is pilimiction. Less often calculi and sebaceous material and rarely teeth have been passed in the urine. It is not surprising that many of the investigators in the past, faced by such dramatic symptomatology, have tended to overlook the morbid anatomy of these tumours apart from enumerating briefly the various tissues present (sebaceous glands, nerve tissue, squamous epithelium, hairs, cartilage, etc.). Ulceration appears to be a fairly frequent complication and extra-vesical growth is also common.

In dealing with biopsy specimens it should be remembered that squamous metaplasia, cystitis glandularis and follicularis are frequent concomitants of epithelial tumours. In one of our biopsy specimens the material consisted almost entirely of well marked mucus-secreting glands, some squamous epithelium, lymph follicles and a little transitional epithelium. If the epithelial origin of the tumour had not been recognized, one might have considered the possibility of a dermoid in this case.

Finally, mention should be made of a very rare tumour of which only two cases so far have appeared in the literature, namely *plasmacytoma*. The first case was recorded by MARION and LEROUX in 1924 and recently their compatriots (AUVIGNE et. al. 1956) reported a second. In both instances the tumours were smooth, soft, fleshy and red in colour. In the first case, the tumours were multiple. In the second a short pedicle was present. This patient was investigated more fully. X-ray examination of the bones and plasma electrophoresis revealed no abnormality. Surgical removal of the tumours was carried out in both instance and on section the material was found to consist of masses of typical plasma cells. The first patient was reported alive and well 6 months after operation. The second showed no evidence of recurrence four years later, but she died some months later at the age of 67 years; the cause of death never being fully determined.

References

A. Aetiology

BAKER, K.: The carcinogenic activity of dihydroxy benzidine. (3:3' dihydroxy 4:4' diamino diphenyl). Acta Un. int. Cancr. 7, 46 (1950). — BARSOTTI, M., and E. C. VIGLIANI: Bladder lesions from amines-statistical considerations and preventive measures. Congr. int. Mal. prof. 9, 484 (1949). — Bladder lesions from aromatic amines, statistical considerations and prevention. Arch. industr. Hyg. 5, 234 (1952). — BERENBLUM, I., and G. M. BONSER: Experimental investigation of aniline cancer. J. industr. Hyg. 19, 86 (1937). — BONSER, G. M., D. B. CLAYSON and J. W. JULL: An experimental inquiry into the cause of industrial bladder cancer. Lancet 1951 II, 286. — BOYLAND, E., D. M. WALLACE and D. C. WILLIAMS: Urinary enzymes in bladder cancer. Brit. J. Urol. 27, 11 (1955). — BOYLAND, E., and G. WATSON: 3-Hydroxyanthranilic acid, a carcinogen produced by endogenous metabolism. Nature (Lond.) 177, 837 (1956). — CASE, R. A. M.: Cohort analysis of cancer mortality in England and Wales, 1911—1954, by site and sex. Brit. J. prev. soc. Med. 10, 172 (1956). — CLEMMESEN, J., u. A. NIELSEN: Cancer incidence in Denmark 1943 to 1953. 1. Population at risk. 2. Tumours of urinary system and prostate. Dan. med. Bull. 3, 33 (1956). — DI MAIO, G.: Affections of the bladder due to amines. Congr. int. Mal. prof. 9, 476 (1949). — DOLL, R., and A. B. HILL: Lung cancer and other causes of death in relation to smoking. Brit. med. J. 1956 II, 1071. — GEHRMAN, G. H., J. H. FOULGER and A. J. FLEMING: Occupational diseases of the bladder. Congr. int. Mal. prof. 9, 472 (1949). — GOLDBLATT, M. M.: Vesical tumours induced by chemical compounds. Brit. J. industr. Med. 6, 65 (1949). — GROSS, E.: Das Carcinom vom Standpunkt der Gewerbetoxikologen. Angew. Chem. 53, 368 (1940). — HAMTOFT, H., M. LINDHARD: Tobacco consumption in Denmark. Dan. med. Bull. 3, 188 (1956). — KENNAWAY, E. L.: Further experiments on cancer-producing substances. Biochem. J. 24, 497 (1930). — LILIENFELD, A. M., M. L. LEVIN and G. E. MOORE: The association of smoking with cancer of the urinary bladder in humans. Arch. intern. Med. 98, 129 (1956). — MANSON, L. A., and L. YOUNG: Biochemical studies of toxic agents. 2. The metabolism of 2-naphthylamine and 2-acetamidonaphthalene. Biochem. J. 47, 170 (1950). — McDONALD, D. F., and R. R. LUND: The role of the urine in vesical neoplasm. 1. Experimental confirmation of the urogenous theory of pathogenesis. J. Urol. (Baltimore) 71, 560 (1954). — POOLE-WILSON, D. S.: Modern trends in urology (ed. E. W. Riches), 34, The effects of carcinogenic agents on the urinary tract, pp. 393 to 406. London: Butterworth 1953. — The treatment of malignant tumours of the bladder by irradiation therapy. Brit. J. Urol. 26, 326 (1954). — REHN, L.: Blasengeschwülste bei Fuchsin-Arbeitern. Langenbecks Arch. klin. Chir. 50, 588 (1895). — SCHAR, W.: Experimentelle Erzeugung von Blasentumoren. (Die Wirkung langdauernder Inhalation von aromatischen Amidoverbindungen.) Dtsch. Z. Chir. 226, 81 (1930). — SCOTT, T. S.: The incidence of bladder tumours in a dyestuffs factory. Brit. J. industr. Med. 9, 127 (1952). — WALLACE, D. M.: The natural history and possible cause of bladder tumours. Ann. roy. Coll. Surg. Engl. 18, 366 (1956). — WALPOLE, A. L., M. H. C. WILLIAMS and D. C. ROBERTS: Carcinogenic action of 4-aminodiphenyl and 3:2'-diamethyl-4 aminodiphenyl. Brit. J. industr. Med. 9, 255 (1952). — WILEY, F. H.: The metabolism of B. naphthylamine. J. biol. Chem. 124, 627 (1938).

B. and C. Symptomatology and Diagnosis

ASH, J. E.: Epithelial tumours of the bladder. J. Urol. (Baltimore) 44, 135 (1940). — CRABBE, J. G. S., W. C. CRESDEE, T. S. SCOTT and M. H. C. WILLIAMS: The cytological diagnosis of bladder tumours amongst deyestuff workers. Brit. J. industr. Med. 13, 270 (1956). — JEWETT, H. J.: Urology (ed. M. Campbell), Vol. 2. Tumours of the bladder, Chap. 3, 10, 44, 1094. Philadelphia and London: W. B. Saunders 1954. — MASINA, F.: Mucosal changes in relation to bladder tumours. Brit. J. Urol. 24, 344 (1952). PAPANICOLAOU, G. N.: Cytology of the urine sediment in neoplasms of the urinary tract. J. Urol. (Baltimore) 57, 375 (1947). — Atlas of exfoliative cytology. Cambridge, Mass.: (Commonwealth Fund) Harvard University Press 1954. — PAPANICOLAOU, G. N., and V. F. MARSHALL: Urine sediment smears as diagnostic procedure in cancers of urinary tract. Science 101, 519 (1945).

D. Treatment

BRICKER. E. M., H. BUTCHER and C. A. McAFEE: Late results of bladder substitution with isolated ileal segments. Surg. Gynec. Obstet. 99, 469 (1954). — CORDONNIER, J. J.: Urinary diversion utilizing isolated segment of ileum. J. Urol. (Baltimore) 74, 789 (1955). — ELLIS, F.. and R. OLIVER: Treatment of papilloma of bladder with radioactive colloidal gold AU 198. Brit. med. J. 1955 I, 136. — FRANKSSON, C.: Tumours of the urinary bladder; a pathological and clinical study of four hundred and thirty-four cases. Acta chir. scand. Suppl. 151, 1—203 (1950). — HIGGINS, C. C.: Cystectomy for carcinoma of the bladder; review of sixty-eight cases. Cong. Soc. Int. Urol. 9, Trans. 1, 129 (1952). — HUMPHREY, G. A.: Permanent cutaneous ureterostomy: a review of 174 cases. Cancer (Philad.) 9, 572 (1956). — IRVINE. W. T.. C. ALLAN and D. R. WEBSTER: Prevention of the late complications of ureterocolostomy by methods of faecal exclusion. Brit. J. Surg. 43, 650 (1956). — JACOBS. A.: Discussion on the treatment of carcinoma of the bladder. Proc. roy. Soc. Med. 45, 197 (1952). — Discussion on "The use of intestine in urology". Brit. J. Urol. 28, 406 (1956). — JACOBS. A.. and W. B. STIRLING: The late results of ureterocolic anastomosis. Brit. J. Urol. 24, 259 (1952). — JOHNSON. T. H.: Urinary diversion with voluntary control of faeces and urine; revised operative technique. J. Amer. Geriat. Soc. 4, 751 (1956). — LEADBETTER. W. F.. and J. F. COOPER: Regional gland dissection for carcinoma of the bladder; a technique for one-stage cystectomy. gland dissection. and bilateral ureteroenterostomy. J. Urol. (Baltimore) 63. 242 (1950). — LOWSLEY, O. S., and T. H. JOHNSON: A new operation for creation of an artificial bladder with voluntary control of urine and faeces. J. Urol. (Baltimore) 73. 83 (1955). — MASINA. F.: Segmental resection for tumours of the urinary bladder. Brit. J. Surg. 41. 494 (1954). — MILLEN. J. L. E.: Intracavitary Irradiation: Seventh Internat. Congr. of Radiology. Copenhagen 1933. — MILLIN. T.. and F. MASINA: Total cystectomy: a consideration of the technique. Brit. J. Urol. 21, 108 (1949). — MILNER. W. A.: The role of conservative surgery in the treatment of bladder tumours. Brit. J. Urol. 26. 375 (1954). — MULLER. J. H.: Radiotherapy of bladder cancer by means of rubber balloons filled in situ with solutions of a radioactive isotope (CO 60). Cancer (Philad.) 8. 1035 (1955). — POOLE-WILSON. D. S.: The treatment of malignant tumours of the bladder by irradiation therapy. Brit. J. Urol. 26, 326 (1954). — PYRAH. L. N., and F. P. RAPER: Some uses of an isolated loop of ileum in genito-urinary surgery. Brit. J. Surg. 42. 337 (1955). — RICHES. E. W.: The place of total cystectomy in the treatment of bladder growth. Ann. roy. Coll. Surg. Engl. 18. 178 (1956). — WALLACE. D. M.: Personal communications 1957. — WHITMORE. W. F.. and V. F. MARSHALL: Radical surgery for carcinoma of the urinary bladder; one hundred consecutive cases four years later. Cancer (Philad.) 9. 596 (1956). — WILDBOLZ. E.. and G. G. PORETTI: The treatment of cancer of the bladder by radioactive cobalt. J. Urol. (Baltimore) 74. 93 (1955). — YATES-BELL, J., and C. HENRIQUES: The use of radioactive gold grains in the treatment of bladder growths. Brit. J. Urol. 29. 97 (1957).

E. Discussion on methods of treatment

CONWAY. J. F.. and A. C. BRODERS: Submucous extension of squamous epithelioma of the urinary bladder. J. Urol. (Baltimore) 47. 461 (1942). — GIRONCOLI, F. DE: Cystectomy or not for cancer of the bladder. Urologia (Treviso) 22, 69 (1955). — JEWETT. H. J., and J. F. CASON: Infiltrating carcinoma of bladder; curability by segmental resection. S. med. J. (Bgham, Ala.) 41. 158 (1948). — MOSTOFI, F. K.: Study of 2678 patients with initial carcinoma of bladder. Survival rates in relation to therapy. J. Urol. (Baltimore) 75, 480 (1956). — NICHOLS, J. A., and V. F. MARSHALL: The treatment of bladder carcinoma by local excision and fulguration. Cancer (Philad.) 9. 559 (1956). — WHITMORE, W. F., and V. F. MARSHALL: Radical surgery for carcinoma of the urinary bladder; one hundred consecutive cases four years later. Cancer (Philad.) 9, 596 (1956).

F. Secondary tumours of the bladder

BRUNSCHWIG, A.: Complete excision of pelvic viscera for advanced carcinoma. a one-stage abdomino-perineal operation with end colostomy and bilateral ureteral implantation into the colon above the colostomy. Cancer (Philad.) 1, 177—183 (1948). — GRAVES, R. C., C. J. E. KICKHAM, KICKHAM and I. NATHANSON: The bladder complications of carcinoma of the cervix. Surg. Gynec. Obstet. 63, 785 (1936). — GUNEM, E., and J. BALAT: Secondary malignant tumours of the urinary bladder. Metastasis from primary foci in distant organs. J. Urol. (Baltimore) 75, 6 (1956). — HELLSTRÖM, J.: Discussion on papillomatosis of the urethra. (Ausworth, A.) Brit. J. Urol. 28, 1, 12 (1956). — HERBUT, P.: Urological pathology. Secondary tumours. Vol. 1, pp. 305—309. London: Henry Kempton 1952. — HERMANN, H.: Metastatic tumours of the urinary bladder originating from carcinomata of the gastro-intestinal tract. J. Urol. (Baltimore) 22, 257 (1929). — LAZARUS, J.: Primary malignant tumours of retrovesical region. J. Urol. (Baltimore) 55, 190 (1946). — OPPENHEIMER,

GORDON D.: Late invasion of the bladder and prostate in cancer of the rectum or recto-sigmoid following abdomino-perineal resection. Ann. Surg. 117, 456 (1943). — PYRAH, L.: Retrovesical tumours: a report of three cases. Brit. J. Urol. 26, 1, 75 (1954). — SCOTT, W.: Complete cessation of persistent bladder papillomatosis following nephroureterectomy and segmental cystectomy for papillary carcinoma of a ureter. J. Urol. (Baltimore) 65, 235—240 (1951). — WALLACE, D. M.: The natural history and possible cause of bladder tumours. Ann. roy. Coll. Surg. Engl. 18, 366 (1956).

3. Metastasis

FRANKSSON, C.: Tumours of the urinary bladder; a pathological and clinical study of four hundred and thirty four cases. Acta chir. scand. Suppl. 151, 1—203 (1950). — JEWETT, A. J., and G. H. STRONG: Infiltrating carcinoma of the bladder. Relation of depth of penetration of the bladder wall to incidence of local extension and metastases. J. Urol. (Baltimore) 55, 366—372 (1946). — LEADBETTER, W. F., and J. A. COLSTON: Brain metastasis in carcinoma of the bladder. J. Urol. (Baltimore) 38, 267—277 (1937). — PAQUIN JR., A. J., and S. I. ROLAND: Secondary carcinoma of the penis. A review of the literature and a report of nine new cases. Cancer (Philad.) 9, 626—632 (1956).

G. Non-epithelial tumours of the bladder

FEGGETTER, GEORGY Y.: Sarcoma of the bladder. Brit. J. Surg. 25, 382 (1937).

H. Pathology

ABESHOUSE, B. S.: Exstrophy of bladder complicated by adenocarcinoma of bladder and renal calculi. J. Urol. (Baltimore) 49, 259 (1943). — ABESHOUSE, B. S., and L. H. TANKIN: Leukoplakia of renal pelvis and bladder. J. Urol. (Baltimore) 76, 330 (1956). — ABRIKOSSOFF, A. I.: Über Myome ausgehend von der quergestreiften willkürlichen Muskulatur. Virchows Arch. path. Anat. 260, 215 (1926). — ALBARRAN, J.: Les tumeurs de la vessie, pp. 96—111. Paris 1892. Quoted by J. Gabe 1932. — ASCHOFF, L.: Ein Beitrag zur normalen und pathologischen Anatomie der Schleimhaut der Harnwege und ihrer drüsigen Anhänge. Virchows Arch. path. Anat. 138, 119 (1894). — ASH, J. E.: Comments on pathology of bladder epithelial tumours. Urol. cutan. Rev. 43, 705 (1939). — Epithelial tumour of the bladder. J. Urol. (Baltimore) 44, 135 (1940). — AUVIGNE, R., J. AUVIGNE et J. KERNEIS: Un cas de plasmocytome de la vessie. J. d' Urol. 62, 85 (1956). — BARRINGER, B. S.: Colloid adenocarcinoma of the bladder. Surg. Gynec. Obstet. 30, 86 (1920). — BEGG, R. C.: Urachus; its anatomy, histology and development. J. Anat. (Lond.) 64, 170 (1930). — Colloid adenocarcinoma of bladder vault arising from the epithelium of the urachal canal: with a critical survey of the tumours of the urachus. Brit. J. Surg. 18, 422 (1931). — Colloid tumour of urachus invading the bladder. Brit. J. Surg. 23, 769 (1936). — BLUM, V.: Harnblasendivertikel. Neue Erfahrungen und kritische Literaturstudie. Z. urol. Chir. 12, 290 (1923). — Quoted by H. L. KRETSCHMER and P. DOERHING 1939. — BRIGGS, W. T.: Carcinoma in diverticulum of bladder. J. Urol. (Baltimore) 24, 517 (1930). — CAPPELL, D. F.: Tumours of striated muscle. Ann. roy. Coll. Surg. Engl. 2, 80 (1948). — CAPPELL, D. F., and G. L. MONTGOMERY: On rhabdomyoma and myoblastoma. J. Path. Bact. 44, 517 (1937). — CATTANIO, Quoted by J. G. MONCKEBERG, Über heterotope mesodermale Geschwülste am unteren Ende des Urogenitalapparates. Virchows Arch. path. Anat. 187, 471 (1907). — CAUFFIELD, E. W.: Dermoid cysts of bladder. J. Urol. (Baltimore) 75, 801 (1956). — CAULK, J. R.: Sarcoma of the bladder. J. Urol. (Baltimore) 16, 211 (1926). — CAWKER, C. A.: Case report: mucinous adenocarcinoma of urachus invading urinary bladder. Canad. med. Ass. J. 57, 58 (1947). — CAYLOR, H. D., and W. WALTERS: Leiomyosarcoma of urinary bladder. J. Urol. (Baltimore) 24, 303 (1930). — CECIL, A. B.: Leiomyosarcoma of the urinary bladder. J. Urol. (Baltimore) 70, 257 (1953). — CECIL, H. L.: Sarcoma of the bladder: report of a case upon whom a total cystectomy was done. J. Urol. (Baltimore) 16, 471 (1926). — CHAFFEY, W. C.: Lympho-sarcoma of the bladder. Trans, path. Soc. Lond. 36, 287 (1885). — CLADO, S. 1895: Quoted by E. W. CAUFFIELD 1956. — COHEN, J. S.: Quoted by A. B. CECIL 1953. — COLBY, F. H., and R. C. SNIFFEN: Carcinoma of the bladder: a classification of epithelial tumours and a study of the effects of external radiation. Trans. Amer. Ass. gen.-urin. Surg. 38, 221 (1946). — CONCETTI, L. 1896: Quoted by R. F. O'NEIL, Bilateral ureterostomy for palliation in a case of tumour of the bladder. Boston med. surg. J. 172, 677 (1915). — CONNERY, D. B.: Leukoplakia of urinary bladder and its association with carcinoma. J. Urol. (Baltimore) 69, 121 (1953). — COPPRIDGE, W. M., L. C. ROBERTS and D. A. CULP: Glandular tumours of bladder. J. Urol. (Baltimore) 65, 540 (1951). — CRAIK, J. 1957: Personal communication. — CRANE, A. R., and R. G. TREMBLAY: Primary osteogenic sarcoma of the bladder: complete review of sarcomata of bladder. Ann. Surg. 118, 887 (1943). — DART, R. O.: Grading of epithelial tumours of urinary bladder. J. Urol. (Baltimore) **36**, 651 (1946). — DEAN, A. L.: Comparison of malignancy of bladder tumours as shown by cystoscopic biopsy and subsequent examination of entire excised organ. J. Urol. (Baltimore) **59**, 193 (1948). — DUKES, C.: Institute of Urology (Univ. of Lond.), Classification of tumours of bladder. (Histological grouping) I.O.U. Broadsheet No 1. 1955. — DUKES, C., and F. MASINA: Classification of epithelial tumours of bladder. Brit. J. Urol. **21**, 273 (1949). — DUPONT, R. 1922: Quoted by J. E. ASH 1939. — ELJASZ, A. 1932: Quoted by A. R. CRANE and R. G. TREMBLAY 1943. — EMMETT, J. L., and J. R. McDONALD: Proliferation of glands of urinary bladder simulating malignant neoplasm. J. Urol. (Baltimore) **48**, 257 (1942). — EWELL, G. H.: Primary carcinoma in diverticulum of bladder. J. Urol. (Baltimore) **24**, 649 (1930). — EWING, J. 1922: Quoted by J. GABE 1932. — GABE, J.: Case of sarco-carcinoma of urinary bladder with brief review of the literature of sarcoma of bladder. Brit. J. Urol. **4**, 145 (1932). — GAROFALO, F. 1927: Quoted by A. R. CRANE and R. G. TREMBLAY 1943. — GRAHAM, J. B., and G. J. BULKLEY: Angioma of bladder. J. Urol. (Baltimore) **74**, 777 (1955). — GUERSANT, 1853: Quoted by J. GABE 1932. — GUSSENBAUER, C.: Exstirpation eines Harnblasenmyoms nach vorausgehendem tiefen und hohen Blasenschnitt, Heilung. Langenbecks Arch. klin. Chir. **18**, 411 (1875). — HANBURY, W. J.: Rhabdomyomatous tumours of urinary bladder and prostate. J. Path. Bact. **64**, 763 (1952). — HICKS, J. B.: Carcinoma in diverticulum of bladder. J. Urol. (Baltimore) **24**, 205 (1930). — HIGGINS, T. T. Rhabdomyosarcoma of the bladder. Brit. J. Urol. **24**, 158 (1952). — HINMAN, F., and J. CORDONNIER: Cystitis follicularis. J. Urol. Baltimore) **34**, 302 (1935). — HOUETTE, C.: Rhabdomyome diverticulaire congénital de la vessie. Anal. anat. path. **6**, 267 (1929). — Quoted by E. N. KHOURY and D. D. SPEER 1944. — HOWARD, L. H., and R. T. BERGMAN: Mucous adenocarcinoma of urinary bladder. J. Urol. (Baltimore) **50**, 455 (1948). — HUGGINS, C. B.: The formation of bone under the influence of epithelium of the urinary tract. Arch. Surg. (Chicago) **22**, 377 (1931). — HUNT, V. C.: Malignant disease in diverticula of the bladder. J. Urol. (Baltimore) **21**, 1 (1929). — JACOBS, A., and T. SYMINGTON: Primary lymphosarcoma of urinary bladder. Brit. J. Urol. **25**, 119 (1953). — JEWETT, H. J.: Carcinoma of bladder: influence of depth of infiltration on 5 year results following complete extirpation of primary growth. J. Urol. (Baltimore) **67**, 672 (1952). — JEWETT, H. J., and G. H. STRONG: Infiltrating carcinoma of bladder. Relation of depth of penetration of bladder wall to incidence of local extension and metastasis. J. Urol. (Baltimore) **55**, 366 (1946). — JONES, H.. M. C. MORUS and C. J. ROSS: Osteogenic leiomyosarcoma of the bladder. Brit. J. Surg. **38**, 242 (1950). — KATZEN, P.: Leiomyosarcoma of bladder: report of case. J. Urol. (Baltimore) **67**, 518 (1952). — KAUFMAN, E.: Über Enkatarrhaphie von Epithel. Virchows Arch. path. Anat. **97**, 236 (1884). — KHOURY, E. N., and F. D. SPEER: Rhabdomyosarcoma of urinary bladder: clinico-pathological case report with review of literature including tabulation of rhabdomyosarcoma of prostate. J. Urol. (Baltimore) **51**, 505 (1944). — KIRSHBAUM, J. D., and F. S. PREUSS: Leukaemia; clinical and pathologic study of 123 fatal cases in series of 14,400 necropsies. Arch. intern. Med. **71**, 777 (1943). — KLEITSCH, W. P.: Parasitic fibromyoma of the bladder. J. Urol. (Baltimore) **65**, 60 (1951). — KRETSCHMER, H. L. 1931: Quoted by E. Sexton 1952. — Rhabdomyosarcoma of bladder; report of case and review of literature. Arch. Path. (Chicago) **44**, 350 (1947). — Primary mucus-secreting adenocarcinoma of bladder. J. Urol. (Baltimore) **61**, 754 (1949). — KRETSCHMER, H. L., B. S. BARRINGER, W. F. BRAASCH, A. L. DEAN, R. S. FERGUSON, E. L. KEYES and G. G. SMITH: Cancer of the bladder: a study based on 902 epithelial tumors of the bladder in the carcinoma registry of the American Urological Association. J. Urol. (Baltimore) **31**, 423 (1934). — KRETSCHMER, H. L., and P. DOERHING: Leiomyosarcoma of urinary bladder. Arch. Surg. (Chicago) **38**, 274 (1939). — KREUTZMANN, H. A. R.: Primary lymphosarcoma of bladder. J. Urol. (Baltimore) **48**, 147 (1942). — LANE, T. J. D.: Uncommon bladder condition simulating carcinoma; glandular proliferation in epithelium of urinary tract with special reference to cystitis cystica and cystitis glandularis. Brit. J. Urol. **20**, 175 (1948). — LAZARUS, J. A., and A. A. ROSENTHAL: Myxosarcoma of bladder: case report of child 2 years of age. J. Urol. (Baltimore) **27**, 695 (1932). — LECENE, P., et A. HOVELACQUE: Les cancers développés sur la vessie extrophiée. J. Urol. méd. chir. **1**, 493 (1912). — LE COMTE, R. M.: Neoplasms primary in bladder diverticula. J. Urol. (Baltimore) **27**, 667 (1932). — LEV, M., and W. E. BELL: Leiomyosarcoma of urinary bladder. J. Urol. (Baltimore) **57**, 251 (1947). — LOWREY, S. R.: Adenocarcinoma of bladder. J. Urol. (Baltimore) **42**, 118 (1939). — MACKLES, A., S. IMMERGUT, D. M. GRAYZEL and Z. R. COTTLER: Carcinoma and sarcoma of bladder: report of unusual simultaneous occurrence of both tumors. J. Urol. (Baltimore) **59**, 1121 (1948). — MARION, G., et LEROUX: Plasmocytome vésical. J. d'Urol. **18**, 121 (1924). — MARSHALL, V. F.: Relation of pre-operative estimate to pathologic demonstration of extent of vesicle neoplasms. J. Urol. (Baltimore) **68**, 714 (1952). — MAYER, R. H., and T. D. MOORE: Carcinoma complicating vesical diverticulum. J. Urol. (Baltimore) **71**, 305 (1954). — McDONALD, J. R., and G. J. THOMPSON: Carcinoma of urinary bladder: pathological study with special reference to invasiveness and vascular invasion. J. Urol. (Baltimore) **61**, 435 (1948). — MELICOW, M. M.: Tumors of the urinary bladder: a clinico-pathological analysis of over 2500 specimens and biopsies. J. Urol. (Baltimore) **74**, 498 (1955). — MILNER, W. A. Transurethral biopsy: an accurate method of determining true malignancy of bladder carcinoma. J. Urol. (Baltimore) **61**, 917 (1949). — MIXTER, C. G.: Quoted by J. GABE 1932. — MOLONEY, G. E.: Lymphosarcoma of bladder: report of case. Brit. J. Surg. **35**, 91 (1947). — MONTGOMERY, O.: Disease of the skin. Philadelphia: Lea & Febiger 1943. — MOROGNA, P. 1927: Quoted by A. JACOBS and T. SYMINGTON 1953. — MORSE, W. H. and W. D. JARMAN 1953: Quoted by E. A. SLOTKIN and R. D. DAVIS 1954. — MOSTOFI, F. K., and W. H. MORSE: Polypoidal rhabdomyosarcoma (sarcoma botryoides) of bladder in children. J. Urol. (Baltimore) **67**, 681 (1952). — MÜLLER, G.: Harnblasendivertikel und Karzinom. Z. Urol. **47**, 230 (1954). — MUNGER, A. D.: Primary mesothelial tumour (leiomyoma malignum) of bladder with secondary involvement of abdominal cavity. J. Urol. (Baltimore) **42**, 229 (1939). — MUNWES, C.: Zur Statistik und Kasuistik der Blasensarkome. Z. Urol. **4**, 837 (1910). — NEUHOF. H.: Fascia transplantation into visceral defects: an experimental and clinical study. Surg. Gynec. Obstet. **24**, 383 (1917). — PACK, G. T., and R. G. LE FEVRE: Age and sex distribution and incidence of neoplastic disease at Memorial Hospital, New York City, with comments on "cancer ages". J. Cancer Res. **14**, 167 (1930). — PATCH, F. S., and J. E. PRITCHARD: Mucinous adenocarcinoma of bladder. Trans. Amer. Ass. gen.-urin. Surg. **38**, 213 (1946). — PATCH, F. S., and L. J. RHEA: Genesis and development on BRUNN's nests and their relation to cystitis cystica, cystitis glandularis and primary adenocarcinoma of bladder. Canad. med. Ass. J. **33**, 597 (1935). — PILLIET, and DUPUY: Quoted by H. L. KRETSCHMER and P. DOERHING 1939. — POWELL. B. F.: A case of leiomyosarcoma of the bladder. Brit. J. Urol. **4**, 259 (1932). — POZNANSKI, 1914: Quoted by J. GABE 1932. — RATHBUN, N. P., and H. L. WEHRBEIN: Lymphosarcoma of urinary bladder. Trans. Amer. Ass. gen.-urin. Surg. **36**, 95 (1944). — RATLIFF, R. K., and W. L. VALK: Sarcoma of bladder: report of three cases. J. Urol. (Baltimore) **42**, 559 (1939). — RIDDELL, H. I., and H. G. KUDISH: Rhabdomyosarcoma of bladder: review and report of a case. J. Urol. (Baltimore) **70**, 472 (1953). — RIEGAL: Quoted by H. L. KRETSCHMER and P. DOERHING 1939. — RÖDER: Dtsch. med. Wschr. **30**, 485 (1904). Quoted by M. LEV and W. E. BELL 1947. — SAPHIR, O., and S. K. KURLAND: Adenocarcinoma of urinary bladder. Urol. cutan. Rev. **43**, 709 (1939). — SAPHIR, O., and A. VASS 1938: Quoted by K. SCHOURUP 1955. — SCHOURUP, K.: Rare malignant tumours of urinary bladder. Acta path. microbiol. scand. Suppl. **105**, 145 (1955). — SCHWARTZ, O. A.: Über Carcinom in Divertikeln der Harnblase. Z. urol. Chir. **13**, 47 (1923). — SEXTON, E.: Fibroma of bladder. J. Urol. (Baltimore) **67**, 309 (1952). — SHATTOCK, S. G.: Rhabdomyoma of the urinary bladder. Proc. roy. Soc. Med. Sect. Path. **3**, 31 (1909/10). — SHIVERS, C. H. DE T., and K. P. HENDERSON: Tumours of bladder: review of 101 cases. J. Urol. (Baltimore) **42**, 761 (1939). — SILBAR, J. D., and S. J. SILBAR: Leiomyosarcoma of bladder: 3 case reports and review. J. Urol. (Baltimore) **73**, 103 (1955). — SINDONI, M.: Glandula vescicali e tumori a struttura glandulare della vescica. Arch. ital. Urol. **10**, 309 (1933). — SLOTKIN, E. A., and R. D. DAVIS: Rhabdomyosarcoma of the bladder. N.Y. St. J. Med. **54**, 2837 (1954). — SMITH, E. C.: Primary sarcoma of bladder. Canad. med. Ass. J. **15**, 628 (1925). — SMITH, P. G., and G. L. SUDER: Primary carcinoma in bladder diverticula. Urol. cutan. Rev. **54**, 321 (1950). — STEWART, H. L., and G. J. MUELLER-SCHOEN: Malignant tumour of a diverticulum of urinary bladder. J. Urol. (Baltimore) **27**, 685 (1932). — TARGETT, J. H.: Diverticula and new growths of bladder. Trans. path. Soc. Lond. **47**, 155 (1896). — VERHOOGEN: Quoted by H. L. KRETSCHMER and P. DOERHING 1939. — VIRCHOW, R. 1863: Quoted by H. L. KRETSCHMER 1949. — WATSON, E. M., H. R. SAUER and M. G. SADUGOR: Manifestations of lymphoblastomas in genito-urinary tract. J. Urol. (Baltimore) **61**, 626 (1949). — WEISS, A. G., et Mlle DREYFUS: Sarcome de la vessie chez un nourrisson. Rev. franç. Pédiat. **4**, 801 (1928). — WEYERBACHER, A. F., and J. F. BALCH: Leiomyosarcoma of bladder, with report of case and review of literature. J. Urol. (Baltimore) **38**, 278 (1937). — WHEELER, J. D., and W. T. HILL: Adenocarcinoma involving urinary bladder. Cancer (Philad.) **7**, 119 (1954). — WILLIS, R. A.: Pathology of tumours. London: Butterworth & Co. 1948. — WÖLBACH, S. B.: Congenital rhabdomyoma of the heart: report of a case associated with multiple nests of neuroglia tissue in the meninges of the spinal cord. J. med. Res. **16**, 495 (1907).

Various Organic Diseases

Leslie N. Pyrah

With 26 Figures

A. Infarction of the kidney

I. Renal infarction from arterial occlusion

A renal infarct may result from embolus or thrombosis of the renal artery or one of its branches, or from prolonged arterial spasm, resulting for instance from severe shock. Gradually increasing constriction of the lumen of one of the arteries from any cause may produce a similar result more gradually.

1. Pathology

An arterial infarct is usually ischaemic, due to the complete cutting off of the blood supply to a sector of the kidney, most of the branches of the renal arteries being end-arteries. If there is an associated thrombosis of the corresponding vein, the infarct will be a combined arterial and venous infarction; alternatively a combined infarct may be at first an anaemic infarct, but becomes a combined infarct following subsequent infiltration of blood across the walls of capillaries or small vessels in the infarcted zone, together with secondary venous obstruction.

A recent renal infarct resulting from arterial occlusion, is wedge-shaped, greyish-yellow in colour with a haemorrhagic border, and the base of the wedge is beneath the renal capsule; renal infarcts may be single or multiple in the same kidney and. depending on the size of the branch of the renal artery which is occluded. may be limited to the cortical part of the kidney or may involve the cortex and medulla. If the corresponding vein is not blocked, the infarcted zone is rapidly drained of blood and the infarcted tissue dies and becomes gradually transformed into fibrous tissue. Total infarction of the kidney, which is rare, would lead to total necrosis of the kidney. Gradually the affected part of the kidney becomes reduced in size and the tissue becomes firm and dry, and infiltrated at its edge with leucocytes which ultimately disappear. Later, granulation tissue grows into the infarct, connective tissue cells helping in the process of organisation. The young connective tissue gradually undergoes contraction producing a puckered scar, and if there are multiple infarcts some lobulation and local shrinkage of the kidney takes place (ALLEN). Microscopically, rudimentary renal tubules may be found in the scar tissue at the edge of the infarcted area and occasionally within it, the epithelial cells, however, having changed their character and sometimes showing attempts at regeneration of the epithelial elements. At the apex of the scarred zone, the closed or partially recanalised artery to the area may be detected (KAUFMANN). The end-stage is a puckered scar on the convex border of the kidney, often disposed transversely; the renal capsule becomes intimately fused with the cortical portion of the scar. Calcific deposits are occasionally seen in the scar of an old infarct.

An example of renal infarction produced by a mechanism differing from that of simple arterial embolism or thrombosis, namely, by the slowly increasing narrowing of a main branch of the renal artery, may be seen in tuberculous kidneys as the lesion progresses, and the changes have been described by LIE-

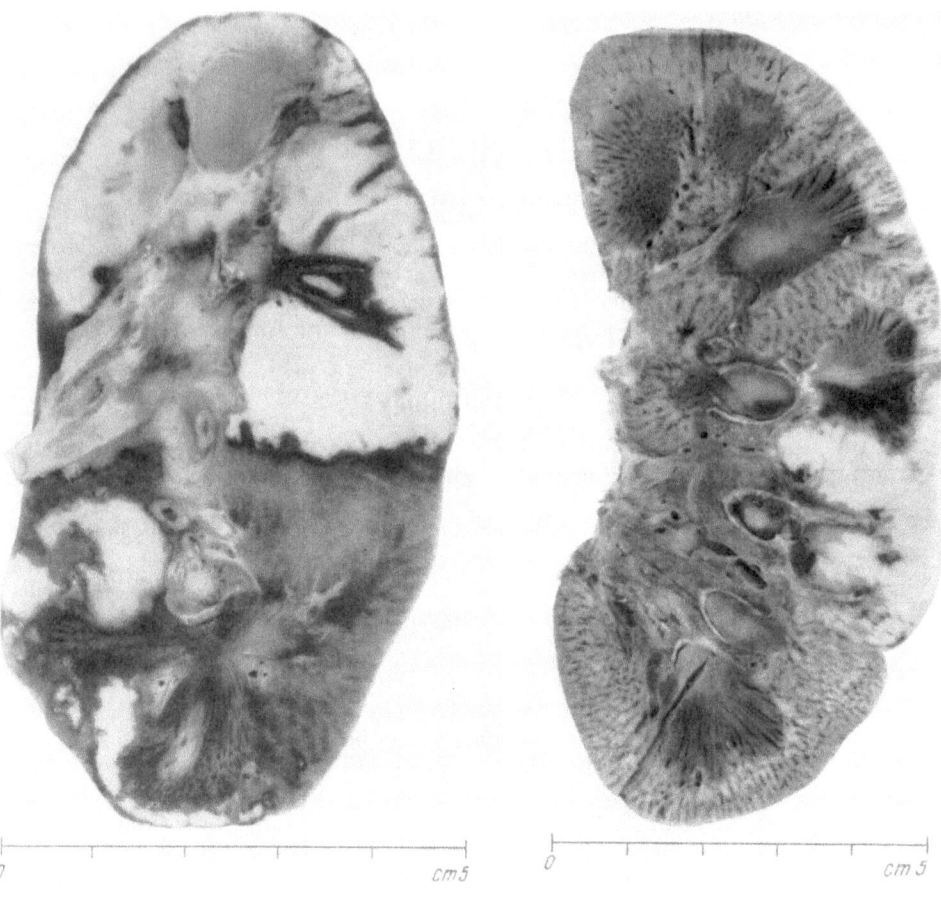

Fig. 1   Fig. 2

Fig. 1. Pale infarct of kidney. Specimen is a kidney from a hypertensive man of 69, showing a big, thrombotic infarct involving a broad stretch of the superficial cortex and parts of the adjoining interpyramidal cortex and medulla. The firm, opaque, cream-coloured necrotic tissue of the infarct is separated from the congested renal substance around it by a continuous zone of greyish-white granulation tissue, showing that the lesion is of some weeks' standing. The deep grooves at the margin of the infarct are probably the remains of old foetal fissures, although they may have been accentuated by contraction of the newly-formed fibrous tissue surrounding the infarct. The irregular pits on the outer surface of the infarct are due to coarse arteriosclerotic scarring, and their presence supports the view that vascular thrombosis, superimposed upon old-standing atheroma, was the cause
of the lesion

Fig. 2. Thrombotic infarction of kidney. Specimen shows extensive infarction in the left kidney of a woman of 57, who had subacute bacterial endocarditis superimposed upon mitral stenosis. The infarction involves almost the whole of the upper part of the kidney and a number of discrete irregular patches of cortex in the lower half, and it has apparently been caused by occlusion of a large number of small peripheral vessels, because most of the major arteries are patent, although one or two are occluded. The infarcted areas have the typical opaque greyish-white colour of necrotic tissue, and their capsular surfaces project above the level of the adjoining tissue, showing that they are distinctly swollen. They are demarcated from the healthy parenchyma by sinuous zones of intense hyperaemia, but there is no evidence of a reparative reaction around them, showing that they are all fairly recent
lesions

BERTHALL. The tuberculous lesion may produce a thickening of the wall of one of the arteries within the kidney and this is followed by a progressive narrowing of the lumen and a consequent retardation of the local circulation and ultimately a local anaemia or ischaemia of a wedge-shaped zone. Hyalinisation of the glomeruli and tubular atrophy are followed eventually by fibrosis and contracture. An alternative method of vascular blocking is that in which the tuberculous process, after spreading into the wall of the artery, produces a secondary endarteritis which results in a roughening of the intima, which in turn may induce local arterial thrombosis, and thus infarction and subsequent necrosis in the area of tissue supplied by the blood vessel. Usually the infarcted area is gradually infiltrated by the tuberculous disease, so that the early stages in the development of this type of infarct may be obscured.

A septic infarction of the kidney, such as is found in cases of bacterial endocarditis, produces initially the same changes as an aseptic infarct, though later infection may spread into it and into the neighbouring kidney substance from the occluding infected embolus: on the surface of such an infarct the renal capsule is inflamed and beneath it beads of pus may be seen.

Histological and histo-chemical changes occur in the infarcted tissue very rapidly and have been studied in the experimental animal. GOEBEL et al. and HALPERT et al. showed that when the renal arteries are ligated in the rat, cytoplasmic changes are present in the kidney in the first hour. Histo-chemical changes were first seen after two hours and consisted in a slight decrease in both 5-nucleotidase activity and PAS-positive material. After four hours there was a decrease in local alkaline phosphatase activity. Fatty degeneration of the tubular epithelium was not observed until enzymatic activity had decreased markedly. PETERS and FABER showed that after a branch of the renal artery in the rat had been occluded for three hours there was a complete cessation of succinic dehydrogenase activity in the infarcted area. GAVAN and KAUFMANN also showed in the rat, demonstrable decrease of succinic dehydrogenase and cytochrome oxidase activity four hours after the ligation of one renal artery; after thirty-six hours both enzymes were completely inactive except in a narrow subcapsular zone.

2. Aetiological groups

a) Cardiac disease

Cardiac disease is the commonest cause of renal infarct resulting from arterial occlusion. The great majority are symptomless and are discovered as an incidental finding at autopsy. A few give rise to clinical symptoms which may lead to diagnosis.

Clots moving from the left auricular appendix, in patients suffering from auricular fibrillation, or from the wall of the heart as a complication of infarction of the heart muscle, are the most common cause of arterial infarction of the kidney. SPERLING described infarcts of the kidney in 76 per cent of 300 autopsies of patients dying from endocarditis. Conversely, BARNEY and MINTZ found a valvular or myocardial lesion in 95 per cent of 136 autopsies in which renal infarcts were noted. HOXIE and COGGIN reviewing 14,411 autopsies performed at the Los Angeles County Hospital during nine years, found renal infarction in 205 patients (1.4 per cent), the presence of infarction having been diagnosed clinically, however, in only 2 patients. The ages of the patients varied between four months and eighty-eight years; 5 were infants of less than one year; one-eighth were under thirty years of age and slightly more than half were over fifty. The incidence in males and females did not differ significantly. In 76.6 per cent of the 205 patients, cardiac lesions were present. The chief causes of the embolism or thrombosis of the renal vessels were: sub-acute bacterial endo-carditis, 59; coronary occlusion with mural thrombi, 29; sclerosis of the aorta or the renal arteries, 24; other causes were rheumatic heart disease with or without auricular fibrillation, hypertension, syphilis of the heart, aortic aneurysm, venous thrombosis and peri-arteritis nodosa. Bilateral infarction was found in 102 patients. The infarcts were usually small, being from 1 to 2 cm. in diameter. In only 3 cases was the infarct equal to one-third the size of a kidney, and equal to one-half the kidney in one patient; the whole kidney was infarcted in each of 9 cases. Infarcts were commonly found in other organs also.

b) Arterial disease

Atheromatous plaques from the aorta may become detached and give rise to embolism and subsequent thrombosis of one of the branches of the renal artery. Primary thrombosis of the renal artery may result from peri-arteritis nodosa or may be secondary to arterio-sclerotic narrowing of the renal artery or one of its branches. Rarely thrombo-angeitis obliterans and syphilitic end-arteritis may give rise to thrombosis of the renal artery (ALLEN). Arterial spasm following severe and prolonged shock, may produce a local anaemia resulting in an infarct. According to ALLEN bilateral cortical necrosis of the kidneys may be caused by prolonged spasm of the smaller cortical vessels with conse-quent diffuse infarction.

c) Septic infarcts

The embolus may be infected if it has arisen from a breaking-down valve or a detached vegetation in a case of infective endocarditis and the infarction is then a septic infarct; typical cases are described by SCHWARTZ. A septic infarct of the kidney may arise as a complication of pyaemia resulting, for example, from acute osteomyelitis or puerperal infection. Infected renal infarcts have been recorded in infancy and childhood as a complication of tonsillitis, diphtheria and entero-colitis, and in adults suffering from pulmonary tuber-culosis and acute and chronic pyelonephritis.

d) Infarction of the kidney following trauma

Severe trauma to the loin is an occasional cause of renal infarction, and this is of practical surgical importance because it may be possible to give effective treatment. HIRSCHBERG and SOLL (1942) surveyed 376 cases of renal infarction of which only 2 were the result of trauma. Cases following trauma were reported by BARNEY and MINTZ (1933) and REXFORD and CONNOLLY (1945). In some cases following severe injury renal infarction may be accompanied also by fat embolism. In CARVER'S (1951) first case, a male, aged 23, there was a traumatic renal infarction associated with massive fat embolism of the kidney. The man had extensive bodily injuries necessitating an operation for fractures in the leg, following which death occurred eight days later: at autopsy, there was fat embolism in the glomerular and peritubular renal veins, together with a cortical renal infarct and also fat embolism of the lungs. A similar combination of widespread fat emboli in various organs including the kidney, together with renal infarction, was found in his second case, a man aged 68, who sustained multiple fractures of the skull and long bones.

3. Clinical picture, diagnosis and treatment

In the majority of cases infarction of the kidney gives rise to no symptoms and since it often occurs during a terminal illness such as cardiac failure, its clinical importance in such cases is not great. On the other hand, a minority of patients present with clinical symptoms which may cause difficulty in diagnosis, and a few patients present important symptoms such as hypertension and anuria which can sometimes be treated successfully if they are recognised. An excellent summary of many of the reported cases in the literature is given by REGAN and CRABTREE.

Although infarction of the renal artery or one of its branches has been commonly reported in series of autopsy findings, in patients with cardiac disease, according to TEPLICK and YARROW who described two cases, not more than 100 cases have been reported clinically up to 1955 and very few of these were diagnosed in life. Fifty cases were collected by REGAN and CRABTREE in 1948; 30 had auricular fibrillation and 21 had other conditions which might have been associated with embolism.

a) Usual clinical picture

A patient who develops symptoms is usually known to be suffering from cardiac or arterial disease. At the onset he experiences a severe continuous pain in the upper abdomen or in one loin, often associated with vomiting; the pain becomes localised in the loin after a few hours and is severe for two or three days, after which time its intensity is gradually reduced. A rise in temperature is common on the second or third day. The kidney is not palpably enlarged. Albuminuria and red cells are often found in the urine. The pain and the temperature may gradually subside and the patient's condition shows considerable clinical improvement within a few days.

If it is convenient for an intravenous pyelogram to be done, it is found that a total arterial infarction results in a nonfunctioning kidney. On cystoscopy there is no efflux from the ureter of the affected kidney but the retrograde pyelogram reveals a normal outline of the pelvis and calyces. Successive intravenous pyelograms during succeeding weeks may continue to show a nonfunctioning kidney. Aortography, if done, will show non-filling of the renal vessels on the affected side (WEISS, ADLER-RACZ, FOLEY, REXFORD and CONNOLLY, MUNGER, TEPLICK and YARROW). Gradually the infarcted kidney will undergo atrophy. The patient may remain well or may die in a few weeks or months from cerebral embolism or cardiac failure.

If the infarct affects one main branch of the renal artery, pyelograms may suggest a renal tumour. FRAENKEL reported the case of a man of 54 who had renal pain, in whom both the intravenous pyelogram and the aortogram suggested the presence of a renal tumour; exploration showed the presence of an anaemic infarct (Fig. 2A).

A biggish septic infarct of the kidney has to be diagnosed from carbuncle of the kidney which is a similar infected localised parenchymal lesion, usually the result of a staphylococcal bloodborne infection. Both conditions may give rise to aching pain in the affected loin and there is fever; the urine is usually sterile unless the lesion communicates with a calyx and the retrograde pyelograms are usually normal, though with a large carbuncle, distortion of one or more calyces and of the ureter may be seen. In infarctions resulting from infective endocarditis a positive blood culture may be ultimately obtained though repeated examinations may be necessary.

Many renal infarcts complicating pyaemia are quite unsuspected, though renal pain in the course of such an illness may suggest their presence.

The principal features in the diagnosis of total occlusion of the renal artery, therefore, are the sudden onset of severe abdominal pain which settles in the loin after an interval, occurring in a patient who is suffering from cardiac disease, a non-functioning kidney of normal size, and a normal retrograde pyelogram. At the onset the condition may be mistaken for renal colic due to the passage of stone, debris or blood-clot, for gallstones, or indeed for an acute abdominal catastrophe. Such a syndrome occurring in a patient with cardiac disease should, however, always raise the question of infarction of the renal

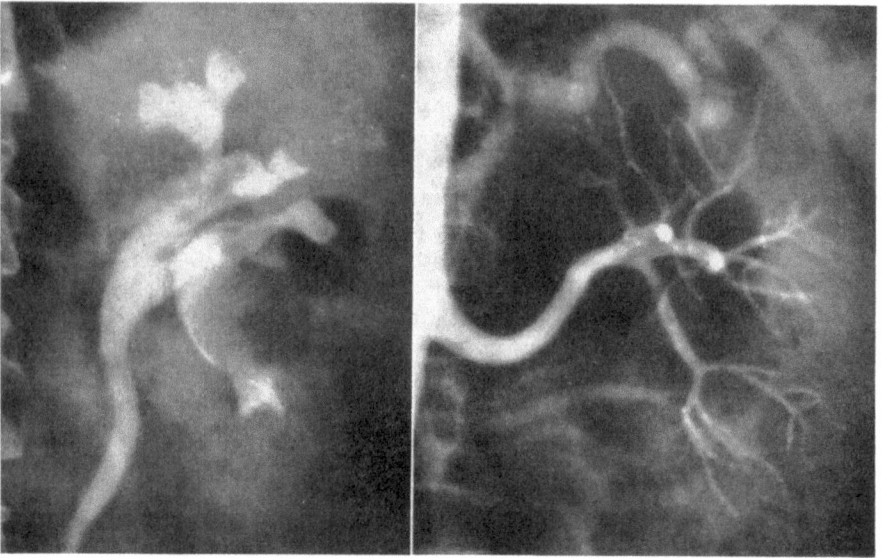

Fig. 2 A. Radiographs of a case of Mr. G. J. FRAENKEL, of infarction of left kidney proved by operation: 1. Intravenous pyelogram of left kidney showing a spheroidal space-occupying lesion in the lower half, with deformity of the calyces. 2. Angiogram of the left kidney. There is a curved displacement of the arteries around the space-occupying lesion. This was thought to be compatible with a diagnosis of cyst

artery. In the very much rarer bilateral occlusion of the renal arteries, severe pain is accompanied by shock and later anuria, and death in uraemia. Such a sequence will occur if one kidney is infarcted and the remaining kidney has been previously badly damaged by disease.

Generally speaking, surgical removal of a totally infarcted kidney is not strictly necessary if the embolus is sterile and the patient's condition seems to be improving. The affected kidney gradually undergoes contraction and the renal elements become gradually replaced by fibrous tissue; such a kidney, examined in later years, would have the appearance of a small, hypoplastic kidney. Probably many small kidneys that have been previously considered to be the result of chronic atrophic pyelonephritis have resulted from arterial infarction.

Occasionally, in survivors, in a kidney which has appeared to be the seat of a total infarction, recovery of function has been recorded after a few months; in such cases one of the main branches rather than the main trunk of the renal artery, must have been occluded (BEN-ASHER, SHEA et al). Kidneys examined radiographically months or years after an unsuspected arterial infarction, may give the picture of an atrophic pyelonephritis; moreover the scarring of the kidney and even the cystic degeneration of an infarction may cause such distortion of an affected calyx as to simulate a renal tumour (TEPLICK and YARROW).

b) Renal injury and infarction

Total renal infarction may be a serious complication of a renal injury or of a surgical operation, for example upon the abdominal aorta, and may not be readily recognised clinically.

The clinical picture is that of a severe injury involving the renal area, followed by pain, shock, haematuria and the development of a tender swelling in the loin; the syndrome is indistinguishable from that of ruptured kidney. The intravenous pyelogram shows absent or diminished function of the affected kidney. In some cases gross injury of the renal artery has been found, while in others it has not been possible to demonstrate such a condition. VON RECKLINGHAUSEN (quoted by ASCHNER 1922) reported the autopsy of an eight-year old boy who died eight days after a severe fall, in whom there was a circular injury of the wall of the left renal artery, distal to which there was extensive thrombosis of all branches of the artery and complete necrosis of the kidney. In a case recorded by REXFORD and CONNOLLY, the patient, a man of 27, had a severe crush injury followed by pains mostly in the upper abdomen, thorax and shoulders; there was no demonstrable fracture. Investigation revealed a few red cells in the urine and later albuminuria. Following persistent pain in the loin, it was found necessary to remove the left kidney, which, when sectioned, showed three bluish-gray, triangular, soft infarcts, the largest being 3 cm. across, more than 50 per cent of the kidney tissue having undergone necrosis: the patient made a good recovery. A further case of total renal infarction following injury is recorded by HIRSCHBERG and SOLL.

An interesting case of traumatic infarction of the kidneys following an operation for resection of the abdominal aorta for vascular obstruction, the operation being complicated by anuria, was recorded in a case record of the Massachusetts General Hospital (1955). The patient was a man aged 59, who had part of the abdominal aorta, including the bifurcation, resected and replaced by an homologous arterial graft. Some hours after operation the peripheral pulse disappeared, necessitating a further operation during which there was a period of severe hypotension. Following anuria for three days the patient died. At autopsy both renal arteries were found to be the seat of extensive atheroma, especially marked at their junction with the aorta. It was shown that many of the arcuate and smaller interlobular arteries of both kidneys were completely occluded by atheromatous material of the same type as that found loosely adherent to the wall of the renal arteries; extensive focal infarction had resulted from these multiple renal emboli. It was thought that the necessary manipulation of the aorta during the operation, had loosened arterio-sclerotic debris from the wall of the renal arteries close to their junction with the aorta, resulting in multiple emboli of both kidneys with consequent anuria.

c) Hypertension following renal infarction

A renal infarction may be followed by the development of hypertension and thus constitute one example of unilateral renal disease which may be followed by hypertensive changes; it is, however, a comparatively unusual cause. In the series of 46 cases of hypertension resulting from unilateral renal disease collected by LANGLEY and PLATT infarction of the renal artery occurred in only one instance.

Fatal cases in which renal infarction, proved at autopsy, had been followed by hypertension were recorded by FISHBERG (3 cases), and by YUILE (2 cases). WAINWRIGHT reported a case of heart failure in a man of 47 who when first seen was not hypertensive. Later, following severe pain in the right loin, the blood pressure rose to 210/100 and he died several days later. At autopsy the right renal artery was completely occluded by a dark red thrombus extending from its aortic orifice to its bifurcation and the kidney showed a large irregular pale infarct.

Bilateral renal infarction may be followed firstly by hypertension and later by fatal anuria. PRINZMETAL, HIATT and TRAGERMAN reported a patient suffering from chronic rheumatic heart disease who became hypertensive a few days after an attack of severe abdominal pain; in the succeeding days there developed gradually increasing suppression of urine, nitrogen retention, and death from uraemia. At autopsy both renal arteries were found to be occluded by thrombi, probably of embolic origin, resulting in almost complete infarction of both kidneys.

Transient hypertension may follow embolism of the renal artery, as in a case reported by FISHBERG. In a case suffering from mitral stenosis with auricular fibrillation, recorded by BEN-ASHER, following symptoms referred to one kidney, which was shown on intravenous pyelography to be non-functioning, the blood pressure became elevated; after three weeks the function of the affected kidney was shown on further pyelography to have returned to normal, and the patient became normotensive. WOLFE referred to similar cases.

In a few cases it has been possible successfully to remove a kidney, the seat of extensive infarction, which has been the cause of hypertension, though the patient may have endured a severe illness. Such cases were recorded respectively by BOYD and LEWIS, by SHEA, SCHWARTZ and KOBILAK, and by HUNTER and MCELMOYLE. BOURNE described the case of a girl of 17 who during a severe cardiac illness developed pain and tenderness in the left loin and hypochondrium: three weeks later she developed a mild right hemiplegia and passed into a state of semi-coma from which she gradually emerged. Subsequent investigation showed a non-functioning left kidney, and nephrectomy was performed for infarction of the renal artery, which resulted in a return of the blood pressure to normal; the patient was well three years later.

The time relationship between renal embolism and the development of hypertension appears to vary; in a case reported by HOWARD et al. the interval was 17 days, while in PERERA'S case hypertension developed within one year, the blood-pressure returning to normal following nephrectomy. PONTASSE reported a case in which hypertension developed within five months of the development of embolism of the renal artery, the blood-pressure returning to normal after nephrectomy; HALLER et al. reported a similar case in which hypertension was observed 95 days after embolism and again the blood-pressure returned to normal following nephrectomy.

d) Anuria

In a few reported cases, unilateral total or subtotal infarction of the kidney (sometimes associated with lower nephron nephrosis of the opposite kidney), or bilateral renal infarction, have been a cause of anuria and sometimes the diagnosis may be difficult or impossible. This syndrome is of current interest since cases of anuria now reach certain centres for treatment on the artificial kidney in the hope that the lesion responsible for the anuria is recoverable, as it may be if the infarction is unilateral. A case in which renal infarction resulting from periarteritis nodosa was complicated by anuria, and which illustrates the complexity of the diagnosis was recorded as a case record of the Massachusetts General Hospital (1946).

II. Thrombosis of the renal vein with haemorrhagic infarction

Thrombosis of the renal vein occurs occasionally in the neo-natal period, in infancy, sometimes in later childhood and also rarely in adults. It is most commonly found complicating infantile diarrhoea though it may follow other infections such as pneumonia, puerperal infection, pyaemia or pyelonephritis.

Venous infarction of the kidney is more serious than arterial infarction since the latter is usually aseptic. The thrombosis is apt to be bilateral either at first or by subsequent extension from a unilateral thrombosis.

The first full description of renal vein thrombosis in infants suffering from ileo-colitis was given by RAYER: v. BECKMAN reported 10 further cases and POLLAK observed 12 cases. BARENBERG et al. reported 5 cases of renal vein thrombosis out of 20 in which autopsy had been performed for infantile diarrhoea; in these 5 cases the age of the child at the onset of the diarrhoea was from 3 to 13 days, and the children lived for from 4 to 11 days after the beginning of the illness. Cases have been recorded by NORDWALL, HEPLER (who collected 40 cases from the literature), MARSHALL and WHAPHAM, BARENBERG et al., CAMPBELL (who referred to 40 cases in infants and children), and McCLELLAND and HUGHES. MILBURN found that only 258 cases had been reported up to 1952: half the cases were unilateral and the incidence was equal in the sexes.

1. Pathology

The precise mechanism of the production of the thrombosis is not known, but infection or toxins resulting from the primary condition, together with associated dehydration or electrolyte imbalance with consequent haemoconcentration are probable contributory causes. Neither is it understood why the newborn should be especially predisposed to this condition. The venous pressure in newborn children is said to be rather low; polycythaemia is physiological during the neo-natal period but this can hardly be of aetiological importance. The predominant status of bed-rest in infants may cause a greater tendency to blood-clot formation owing to increasing amounts of thromboplastin in the circulation (MILBURN). OSLER and McCRAE suggested on anatomical grounds that because of its greater length and because of the pressure exerted on its walls by the aorta, the left renal vein is more prone to thrombosis than the right.

At autopsy the affected kidneys are enlarged, often to twice the size of a normal kidney, are deep red or purple in colour and show severe haemorrhagic changes resulting from diapedesis into the interstitial tissues, glomeruli and tubules, which often destroys these and renders the line of demarcation between the cortex and the medulla almost indistinguishable. Sometimes the haemorrhagic change is most marked in the medullary zone. Usually the renal vein and its larger tributaries are plugged with blood-clot.

Thrombosis of the renal vein in infants does not, however, always cause complete obstruction so that it is not, apparently, always associated with renal infarction. Thus BEHR found 37 examples of thrombosis of one or both renal veins at autopsy but in only 3 had there been a resulting haemorrhagic renal infarction. Similarly infarction of the smaller tributaries of the renal vein may occur without massive infarction of the entire kidney and without massive thrombosis of the renal vein. Morison and also Kobernick et al. found that the arcuate and the interlobular veins, sometimes over wide sectors of the kidney, may be the seat of the initial thrombosis, possibly spreading later to the renal veins, but resulting sometimes in generalised infarction of the kidney without a thrombosis of the renal vein itself. Milburn reported the case of a child eight days old from whom an enlarged infarcted kidney was removed in which there was no thrombosis of the renal vein; the greater part of the renal tissue was infarcted, the cellular elements being degenerate or destroyed, with changes in different parts of the kidney varying from interstitial haemorrhage to mild necrosis and total destruction; nevertheless thrombi were not demonstrated in the lumen of the interlobar, arcuate or interlobular vessels, indeed no vascular pathology could be demonstrated. Zuelzer et al. are of the opinion that there is a common factor which produces either, or both, haemorrhagic infarction and renal vein thrombosis, and suggests that this common factor is stasis and that it is the degree of completeness and suddenness of this which determines one or the other, or both; perhaps when stasis is gradual, haemorrhagic infarction without thrombosis occurs, due to anoxia of the capillary walls, with diapedesis of cells, followed by frank interstitial haemorrhage.

In unilateral cases with haemorrhagic infarction with or without renal vein thrombosis, there is evidence that the opposite kidney may show the changes of lower nephron nephrosis with resulting fatal anuria. One of the fatal cases reported by McClelland and Hughes showed this double syndrome. It is not surprising that this lesion should occur in such infants, as the two basic causes of lower nephron nephrosis, namely shock and increased blood pigment excretion, are present. Death may also occur following the extension of clot across the inferior vena cava to the opposite renal vein.

It would appear, therefore, that venous changes in the kidney may occur in the following pathological forms (Milburn):

1. Primary thrombosis of the renal vein, with infarction or without infarction.

2. Primary haemorrhagic infarction of the kidney without renal vein thrombosis.

3. Haemorrhagic renal infarction secondary to renal inflammation with or without venous thrombosis.

The clot is often infected so that secondary changes in the kidney secondary to this may be found at autopsy or operation, and this tendency to infection enhances the gravity of the condition.

2. Symptoms and diagnosis

a) Infants

The children affected by this condition are usually neonates or infants; thus Regan and Crabtree reported 6 cases of haemorrhagic infarction of the kidney in infants of the ages 9 days, 13 days and 29 days, 2 months, 3 months and 2 years respectively. The thrombosis was found in association with gastro-enteritis in 4 cases, pneumonia in one and cellulitis of the groin in one.

The clinical picture in infants suffering from this condition is variable and the diagnosis consequently difficult. The infant is unable to complain of pain, and severe diarrhoea with prostration resulting from the consequent dehydration, is the usual clinical sequence. At some stage in the illness one or both kidneys may be palpably enlarged, and oedema, haematuria, albuminuria, oliguria and terminal anuria commonly follow in fatal cases. In some fatal cases there has been nothing to draw attention to the kidney. 95 per cent of the infants die unless the affected kidney is removed surgically, and bilateral cases are uniformly fatal following a progressive and septic course.

If intravenous pyelography is practicable, which it often is not, a non-functioning kidney may be found. Retrograde pyelography usually shows a filling defect or no filling of the renal pelvis, or extravasation of the medium into the kidney substance (CAMPBELL and MATTHEWS). The diagnosis must be made from Wilm's tumour, cystic kidney, perinephric abscess and pyonephrosis and renal carbuncle.

b) Older children and adults

In older children and in adults pain is the rule though in some reported cases it is not emphasized (WHITE and PORTER); sometimes the pain radiates to the genitalia simulating renal colic (ASCHNER, FALCI, LOEB). Typically there is a sudden onset with pain, accompanied by tenderness in the affected loin. Haematuria or red cells in the urine are almost uniformly present. The kidney is palpably enlarged, sometimes forming a considerable mass in the loin. An intravenous pyelogram shows a non-functioning kidney.

In adults, thrombosis of the renal vein may accompany perirenal suppuration, from rupture of a perinephric or cortical renal abscess or from pyelonephritis, and HYMAN reported one case complicating an adrenal or a renal tumour.

Cases also occur resulting from puerperal infection. REGAN and CRABTREE reported 2 cases in women aged 20 and 33 years, who developed the infarction 5 and 65 days respectively after childbirth. BIERMATH reported a case of renal thrombosis following a septic abortion; the patient died after attempted nephrectomy. MELICK and VITT reported a case of thrombosis of the right renal vein with total renal infarction, which followed childbirth; following fever and generalised pains, especially in the leg, the patient was found to have a staphylococcal septicaemia and later a non-functioning right kidney; ten weeks after the onset nephrectomy was successfully performed.

ASCHNER reported cases of renal vein thrombosis in a man of 67 who had prostatic obstruction which was complicated by gastric cancer, a second case following an abscess in the cortex of the kidney resulting from a blood-borne septicaemia, and a third case in a man of 50 with a bacillus coli septicaemia following instrumentation of the urethra.

According to ALLEN, thrombosis of the main renal vein may produce the clinical picture of the nephrotic syndrome. Thus, there is loss of protein in the urine because of increased glomerular permeability from the effects of the elevated venous pressure. There follows the reversal of the albumin-globulin ratio and clinical oedema. This mechanism, of course, is only one of the modes of production of the clinical nephrotic syndrome. The kidney itself is moderately enlarged and shows tubular atrophy and interstitial oedema. DEROW, SCHLESINGER and SAVITZ found in the case of a 15 year old boy who had the nephrotic syndrome, that at autopsy there were enlarged, swollen kidneys and there had been earlier thrombosis of the inferior vena cava, the portal, the splenic and the renal veins, all of which had been recanalised but new fresh fibrin thrombi were present. The nephrosis was presumably secondary to the venous thrombosis.

3. Treatment

If the diagnosis of total venous infarction of the kidney can be made in life, nephrectomy offers a chance of survival and a few successfully treated cases have now been recorded: it is necessary to be satisfied that the opposite kidney is healthy.

Until June 1950 the kidney had been successfully removed for this condition in infants in only 3 cases, 2 being reported by CAMPBELL and MATTHEWS and one by SANDBLOM. MILBURN described a fourth successful case in 1952, a male infant 8 days old, who was admitted to hospital critically ill with fever, vomiting, dehydration and acidosis, abdominal distension and oliguria, and a palpable swelling in the right loin. After preliminary correction of the acidosis, the right kidney, the seat of a haemorrhagic infarction due to renal vein thrombosis, was successfully removed on the ninth day of life.

Nephrectomy has been performed successfully in a number of cases in adults. MARION successfully performed nephrectomy on a 17 year old woman who developed renal vein thrombosis three days after curettage for puerperal sepsis.

If renal vein thrombosis, believed to be bilateral, has resulted in anuria, treatment on the artificial kidney would be worth while in the hope that the thrombosis of at least one of the renal veins is only partial and capable of canalisation in three weeks or so, the patient being sustained by the artificial kidney meanwhile.

III. Thrombosis of the renal vein associated with thrombosis of the inferior vena cava

Thrombosis of one or both renal veins rarely complicates thrombosis of the inferior vena cava. The condition may occur in infants or children with diarrhoea; alternatively it may, at any age, complicate pelvic sepsis and thrombophlebitis, trauma, aneurysm of the aorta and extensive pelvic malignant disease (WEBER et al).

A typical case is recorded by DEROW et al. A male child, four months old, was admitted to hospital on account of vomiting and diarrhoea. On the fourth day after admission he developed oedema of the scrotum and haematuria, and a large, tender, palpable mass was found in each loin. Oedema gradually developed in the lower extremities and a period of anuria preceded a fatal issue sixteen days after admission. Autopsy revealed bilateral thrombosis of the renal veins and of the inferior vena cava with haemorrhagic infarction of both kidneys, both of which were enlarged to approximately four times their normal dimensions. STEVENS and TOMSYKOWSKI reported a similar case in a 19 month old child who developed vomiting, diarrhoea and haematuria; and FERIOZI, RICE BURDICK and TROENDLE reported a further case in a 4 month old boy.

BARNEY and MINTZ described a case of thrombosis of the renal vein resulting from thrombosis of the veins of the legs, the pelvis and the inferior vena cava, complicating a perinephric abscess in a woman aged 31.

As in other forms of renal infarction, hypertension may occur as a late complication in survivors, and the anatomical changes in such cases may be bizarre. PERRY and TAYLOR reported a case of bilateral renal vein thrombosis in a boy who later developed a fatal hypertension. Following an attack of whooping cough at the age of 9, the feet and legs became swollen during convalescence, resulting, as subsequently appeared, from a thrombosis of the inferior vena cava. He subsequently developed hypertension from which he died at the age of 12. Autopsy showed long-standing thrombosis of the inferior vena cava which was partly recanalised. The left renal vein, which, had been similarly thrombosed, was reduced to the size of a fibrous cord and the kidney to complete atrophy. The right renal vein had been similarly thrombosed, but on this side a biggish aberrant vein had provided a channel for the return of some of the venous blood, enabling the kidney to survive and to function.

B. Perirenal haematoma

By the term perirenal (or circumrenal) haematoma is meant a massive haemorrhage around the kidney. In most reported cases, the condition has followed some intrinsic disease of the kidney or of its vascular supply, less commonly the cause of the bleeding has been traced to pathological conditions of organs near to the kidney, of which the suprarenal gland is the most common, and in a few cases, autopsy has failed to reveal a precise cause.

BONET (1700) described the first case of spontaneous perirenal haematoma in a child suffering from an exanthem. The early surgical literature contains many references to the condition from such varied causes as aneurysms, diseases of the suprarenal glands, chronic nephritis and haemorrhagic cysts of unknown aetiology. RAYER (1839) described a perirenal haematoma originating from the suprarenal gland and named the syndrome "apoplexia renum". WUNDERLICH (1856) gave the first full description of the disease, together with a classification, and the condition was called WUNDERLICH's disease by COENEN, though several earlier examples of the condition resulting from various causes had been recorded by other observers. COENEN (1910) presented 13 cases; he believed that nephritis predisposed to subcapsular haemorrhage because of the increased fragility of the renal blood-vessels in that condition and also because of raised intrarenal pressure. POLKEY and VYNALECK (1933) reviewed the literature and collected 178 cases. Many single cases or short series have since been reported, such as those by HERITAGE, COPPRIDGE, CARVER, MILLER and CORDONNIER, KEEFER, ROLNICK and DAVIDSOHN, and FORT.

I. Aetiology

In the series of 178 cases collected by POLKEY and VYNALECK the ages of the patients ranged from 6 days to 89 years; 70 per cent of the cases were between 20 and 60 years, and in 40 per cent from 30 to 50 years. Males outnumbered females in the proportion of 3 to 2.

The following list enumerates the primary diseases reported in the literature, of which perirenal haematoma has been a major and sometimes fatal complication; in considering perirenal haematoma therefore, the underlying pathology must always be considered.

1. Traumatic haematoma

Ruptured kidney; crush injuries involving damage to soft, visceral or osseous structures in or near the loin.

2. Spontaneous haematoma

a) Diseases of the kidneys

Nephritis; tumours (including hypernephroma, carcinoma, adenoma and tumours of the renal pelvis); hydronephrosis; infections; tuberculosis; calculous disease; congenital cystic disease; simple cysts.

b) Diseases of the adrenal glands

Tumours (including phaeochromocytoma and carcinoma).

c) Diseases of the blood vessels

Arteriosclerosis (with associated hypertension); aneurysms of the abdominal aorta, the renal artery or the ovarian arteries; periarteritis nodosa; infarction and thrombosis of the renal artery or vein.

d) Diseases of the blood or blood-forming organs

Haemophilia; polycythaemia; leukaemia; scurvy; purpura; Hodgkin's disease

e) Infections

Renal infections (including tuberculosis); appendicitis; malaria; typhoid; syphilis; pyaemia; the exanthems.

f) Diseases of the retroperitoneal organs and tissues

Perinephritis; acute haemorrhagic pancreatitis; retroperitoneal tumours.

g) Idiopathic, no cause being found

The series of 178 cases of spontaneous perirenal haematomas collected by POLKEY and VYNALECK illustrates the relative frequencies in the different aetiological groups: Diseases of kidneys, 112 cases (63 per cent); diseases of blood vessels, 29.2 per cent; idiopathic, 26 cases (14.6 per cent); infections, 10 per cent; diseases of the adrenal glands, 11 cases (6 per cent); diseases of the blood or blood-forming organs, 4.5 per cent.

II. Pathology

HÜBNER showed that temporary clamping of the renal pedicle often resulted in subcapsular haemorrhage. POLKEY and VYNALECK ligated the renal vein in 22 dogs. Perirenal haematomas were found in 2 of 5 animals which died post-operatively and in 14 of 17 which were killed at various periods of time afterwards. Some of the haematomas were small, being usually located near the lower pole of the kidney; others were large, sometimes completely surrounding the kidney; in one-third of the cases they were subcapsular. They considered that the haematomas were the result of venous congestion.

In the great majority of the recorded cases, the haematoma is spontaneous and there has not been any history of injury, cases of obvious ruptured kidney being, of course, excluded. Rarely, however, in a kidney which is the seat of pathological change, slight local trauma or an obvious blow to the loin or a sudden twist of the body involving the lumbar region, may determine the onset of the bleeding.

If the cause of the bleeding arises from a focus within the kidney, some parenchymal renal bleeding as well as a perforation of the cortical renal tissue or a ruptured renal vessel, are found in association with the circumrenal haematoma. The bleeding may remain subcapsular, in which case a large part or almost the whole of the renal capsule is stripped by massive blood-clot; cases in which the renal capsule had not been torn by a perirenal haematoma were reported by HAEBLER, CATHELIN, LENK, RICKER, v. BECK and LÄWEN. When the bleeding is more abundant, and probably when a massive arterial haemorrhage occurs from a renal focus, the accumulating blood ruptures through the renal capsule into the perinephric space, until the haemorrhage is ultimately arrested by the formation of clot. At first the blood remains within the fatty capsule of the kidney, though later it extends downwards to the iliac fossa, upwards towards the diaphragm, and inwards to the vertebral column, where its spread is limited by the attachment of Gerota's fascia to the vertebral bodies; the blood may be further diverted along the course of the large abdominal vessels and that of their main branches into the common mesentery. The haematoma occasionally ruptures into the peritoneal cavity. Usually the parietal muscles and the perinephric fat prevent the extension of the bleeding into the sub-cutaneous tissues but occasionally this may occur; moreover the overlying sub-cutaneous tissues may become oedematous from an outpouring of tissue fluids.

An extensive haematoma may cause death from shock; such a fatal ending is especially liable to occur in cases which rupture into the peritoneal cavity. Most recorded cases, however, have come to operation.

There is a group of cases in which subacute or even chronic haemorrhage may occur and in which the clinical course is less urgent. In a long-standing haematoma of limited size, which has not been operated on, the clot may gradually organise into a mass of fibrous tissue. Alternatively a cyst may slowly develop from the absorption of the central part of the clot and the formation of a thick wall of fibrous tissue; the wall of such a cyst may undergo varying degrees of calcification. Subcapsular haematomas occasionally become converted into cysts. Occasionally the haematomas become infected.

Bilateral perirenal haematoma is a rare occurrence, there being only 5 examples in the 178 cases collected by POLKEY and VYNALECK; 2 of these resulted from nephritis and one each from aneurysm of the renal vessels, periarteritis nodosa, and suprarenal haemorrhage respectively. LINK reported a further bilateral case resulting from periarteritis nodosa in a 36-year old man who suddenly developed a haematoma round the right kidney which was treated by removal of some of the clots and by drainage, and following his recovery, a similar condition developed ten weeks later round the left kidney for which a left nephrectomy was performed.

III. Clinical picture and diagnosis

The clinical picture of traumatic rupture of the kidney and its associated perirenal haematoma is described in another chapter.

Sometimes it is known that the patient suffers from some renal or other pathological condition, though often the haematoma is the first symptom. In most cases the symptoms of perirenal haematoma are of sudden onset and necessitate urgent surgical intervention because of acutely developing symptoms. A minority run a subacute course, with recurring phases of bleeding. A few cases are insidious, the haematoma developing slowly over weeks or months.

1. Acute cases

In cases of spontaneous haematoma with sudden and severe bleeding, the onset is sudden, and the patient experiences severe pain in the corresponding loin and iliac fossa. In the early stages of the haemorrhage the pulse rate is raised, and there is perspiration and pallor; if the haemorrhage becomes severe, there are the general signs of loss of blood, with faintness or collapse, shock, sub-normal temperature and a low red cell count. In other cases there is an intermittent elevation of temperature which leads to the suspicion of perirenal

inflammation. Gradually, a tender mass, not moving on respiration and sometimes increasing rapidly in size, can be felt in the affected loin and there is
rigidity of the overlying muscles. Bruising of the skin is rare though oedema
may be noted. Occasionally, in addition to there being a large, tender mass
filling the loin, the umbilicus and the surrounding tissues for a distance of one
half-inch or more are infiltrated with blood, as in HERMAN's case of perirenal
haematoma resulting from a large renal lipoma. There may be marked abdominal
distension resulting from ileus and the patient may then complain of nausea,
vomiting and constipation. If the haematoma is associated with intra-renal
disease, there may be haematuria.

The combination of sudden renal pain, with a mass in the loin and signs of
internal haemorrhage should suggest the diagnosis. In some such cases the
clinical course may be rapid and severe and a fatal outcome may result unless
surgical treatment is promptly put in hand. The cause of death may be shock
from massive haemorrhage, but paralytic ileus, uraemia and pneumonia may
accelerate death. The immediate prognosis depends upon a correct diagnosis
followed by prompt surgical intervention as well as upon the application of
resuscitative measures before and after operative intervention; the remote
prognosis depends upon the cause of the haemorrhage.

In the most acute cases little investigation is possible and the surgeon must
rely on physical signs alone in deciding the appropriate treatment. In cases
which are less acute radiography may be helpful. The shadow of a large mass
in the loin may be seen on the plain radiograph and such a shadow may be
diffuse if the renal capsule has been ruptured, or circumscribed if the haematoma
is subcapsular. The line of the psoas major muscle may be obscured or completely
absent. The intravenous and retrograde pyelograms may be normal, though
the shadow of the ureter may be deviated towards the mid-line. There may also
be the following radiographic abnormalities: A lumbar scoliosis with the con-
cavity towards the affected side; elevation of the corresponding side of the
diaphragm; distension of the adjacent part of the colon with air; displacement
of the kidney laterally in the presence of a large haematoma (EKMAN). A barium
enema may show the colon to be displaced to the outer side of a large mass
(ELMER and WYNGARDEN).

In some cases the patient may be a known sufferer from some intra-renal
disease, for instance calculous disease or tuberculosis. General examination of
the patient may reveal some associated disease of the vascular system, the blood
or the blood-forming organs which may lead to a diagnosis.

In the absence of injury, the diagnosis may be difficult and the condition
may be mistaken for an acute abdominal condition such as appendix abscess,
acute cholecystitis, a sub-acute perforation of a gastric or duodenal ulcer or
even intestinal obstruction. The observation of rapidly developing shock with
a lump in the loin may suggest the nature of the condition, though in many
cases the correct diagnosis is only reached following an exploratory operation.
In women the sudden collapse with marked pallor may suggest an ectopic
pregnancy.

2. Subacute and chronic cases

In cases in which the haematoma has developed slowly over weeks or even
months (ROLNICK, ELMER and WYNGARDEN), a palpable mass in the loin is
sooner or later discovered. There is usually anaemia. In such cases the extravasation of blood has taken place more slowly, and the formation of a mass in
the loin is the principal symptom, with pain either minimal or absent. Occa-

sionally the first symptom is sudden weakness followed after an interval by the discovery of a lump in the loin and later by pain; COPPRIDGE described such a case, in which at operation a large subcapsular renal haematoma was found. The parenchyma showed the microscopic changes of chronic glomerulo-nephritis, with many small haemorrhages into the kidney substance; a renal vessel had ruptured, resulting in a tear in the renal cortex followed by a haematoma.

In cases in this group elevation of temperature, with leukocytosis, is commonly present. Haematuria may or may not have been observed. Such cases are liable to be mistaken for renal tumours, and even perinephric abscess and pyelo-nephritis, though the pyelogram may be normal. Obstructive anuria may result from pressure of the haematoma upon the ureter as in the case reported by ASCHNER and KLINGER in which an infected subcapsular perirenal haematoma caused sudden obstruction of a solitary kidney, resulting in anuria. Lumbar ecchymosis resulting from blood beneath the skin, which may gradually have tracked downwards to the corresponding side of the scrotum, may enable the correct diagnosis to be made pre-operatively as in 3 cases described by CIBERT, VACHER and CAVAILER, all of which were of the insidious type presenting with a lump in the loin. Late haemorrhagic cysts formed as a result of a haematoma may show calcification of the wall, as in the case of MILLER and CORDONNIER. Occasionally a perirenal haematoma totally unsuspected may be found at autopsy (UGELLI).

3. Important causes of perirenal haematoma

Since this rather rare condition is a complication of so many important pathological conditions of several different organs, all of which usually manifest themselves clinically in other ways, it is desirable to give a short summary of the mode in which it may present itself in relation to the basic pathological cause. In the succeeding paragraphs, examples of the varied causes of this rare syndrome, the unusual mode of presentation of some of the cases and the difficulties with which the surgeon is faced in making an accurate pre-operative diagnosis and prognosis, are referred to.

a) Renal tumours

A hypernephroma or carcinoma of the cortical part of the kidney may rupture spontaneously or following a trivial injury, with the production of a haematoma in the perinephric tissues before it has given rise to other symptoms. Such cases have been reported by EKMAN and NYSTROM. Similarly in a case of the writer's, the patient, a man of 43, was admitted to hospital with pain in the right iliac fossa, an elevated temperature and a tender lump in the region of the appendix; a pre-operative diagnosis of appendix abscess was made but operation revealed a hypernephroma of the lower pole of the kidney with a large recent haematoma.

Two cases of perirenal haematoma complicating renal tumours were reported by STEARNS and HERSHMAN. In one, a man of 63, the condition resulted from a papillary carcinoma of the kidney; the main complaint was vomiting followed later by upper abdominal pain; a diagnosis of acute cholecystitis was made and a laparotomy revealed a mass in the region of the right kidney; a second successful operation through the loin revealed a large haematoma arising from a papillary carcinoma in the upper pole of the kidney for which nephrectomy was done. Their second case was that of a man aged 42 who was admitted to hospital in extremis and who was found at autopsy to have a large perirenal haematoma associated with a kidney which was the seat of chronic atrophic pyelonephritis and incidentally a small papillary adenoma at its lower pole; the adrenal gland also showed evidence of haemorrhagic necrosis.

Renal and perirenal lipomata with circumrenal haematoma were recorded by Scriver, Herman and Jentzer. In Scriver's case the haematoma which complicated pregnancy, had ruptured into the peritoneal cavity. Intrarenal lipoma, of which approximately 16 are on record (Herman) are usually large, and give rise to renal pain and a palpable mass; in Herman's case, in which consideration of the pyelogram had led to the diagnosis of renal tumour, and also in 3 of the 16 cases to which he referred (namely those reported by Robertson and Hand, Hinz, and Nicholson and Gillespie), renal pain was the result of haemorrhage around the lipomata.

A case of perirenal haematoma caused by a renal angioma in a woman aged 29 was reported by Price, one resulting from a renal sarcoma was recorded by Powell and Clark, and one resulting from a sarcoma of the ureter, which was also associated with haematonephrosis, by Petkovic.

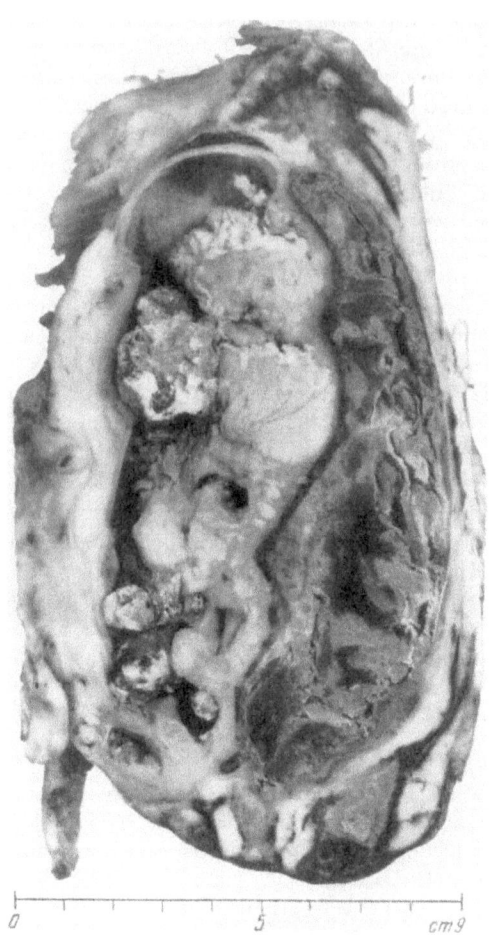

Fig. 3. Moderate-sized perirenal haematoma around a kidney which was the seat of marked hydronephrotic atrophy, a large stag-horn calculus and a transitional cell carcinoma of the renal pelvis. The patient was a woman of 67, who was known to have had stones for some years before operation. She had considerable pain in the kidney, evidently due to the haematoma, for three weeks before the kidney was successfully removed

b) Hydronephrosis

Perirenal haematoma resulting from hydronephrosis has been reported by Beatty, Carver, Counseller and Emmett, and Keefer. In Carver's case a large subcapsular haematoma resulting from hydronephrosis was successfully treated by nephrectomy; the renal capsule was intact. Nystrom recorded a case of a man of 23 who developed a perirenal haematoma following the rupture of a hydronephrotic kidney during a pole-jump.

c) Renal infection

A case of perirenal haematoma arising spontaneously in a patient with chronic pyelonephritis was reported by Irwin, and Moore recorded a similar case in a man with an atrophic kidney. A case following renal carbuncle was recorded by SAGARRA.

d) Nephritis

MARTIN recorded a case of a boy of 18 who developed sudden pain in the right loin which led to a diagnosis of retrocaecal appendicitis, and in whom at operation a large retroperitoneal perirenal haematoma necessitating nephrectomy was found; the kidney was hydronephrotic and was also the seat of widespread interstitial nephritis with areas of small, round-cell infiltration throughout the renal tissue; the patient made a good recovery and was fit three years later.

e) Arteriosclerosis and hypertension

KEEFER recorded the case of a man of 35 suffering from hypertension who complained of pain in the left flank of several weeks' duration, and who developed a temperature, anaemia and a palpable swelling in the loin believed to be a perinephric abscess or a renal tumour. At operation a perirenal haematoma was found, which was believed to have arisen from a spontaneous rupture of vessels in the renal cortex. MILLER and CORDONNIER reported a case in a man of 34 who suffered from hypertension and who developed discomfort in the right lumbar region. On examination, a tender, cystic mass was found in the right loin which was shown on radiography to be calcified. A pyelogram showed distortion of the outline of the renal pelvis and at operation a partly calcified, subcapsular renal haematoma was discovered. The kidney with the cyst was removed. The hypertension appeared to have been successfully relieved.

f) Periarteritis nodosa

Periarteritis nodosa is a necrotising inflammatory arteritis affecting the media of the medium-sized and small arteries and arterioles of almost any organ. The kidney, the heart and the liver are most commonly involved in that order, and the lesion produces weakness in the wall of the artery with the formation of secondary aneurysms which may rupture; alternatively thromboses and infarctions may occur. The rather confusing clinical picture may include polyneuritis and polymyositis with fever, abdominal manifestations, anaemia and general weakness, and some degree of nephritis resulting in albuminuria. Hypertension is common. Renal as well as perirenal haematoma may occur in such cases and may be associated with haematuria. The diagnosis may be difficult since the lesions of the kidney may also result in anuria or uraemia. When perirenal haematoma complicates periarteritis nodosa, whether acutely or insidiously, the patient has, therefore, frequently had an illness of longer or shorter duration presenting atypical symptoms. Cases have been described by WEVER and PERRY, KEEFER, ROLNICK and DAVIDSOHN, FORT, and LINK.

ROLNICK and DAVIDSOHN reported the case of a man of 54 who had urinary retention believed to be of prostatic origin. For ten days after the retention he had an elevated temperature and a dull pain over the right lower abdomen and right flank, and intravenous pyelograms showed deformity and compression of the calyces of the right kidney. A palpable mass developed in the loin, and the right kidney with a large perirenal haematoma was successfully removed, the kidney having the typical microscopic appearances of periarteritis nodosa.

FORT'S case was a man, aged 39, who was admitted to hospital with respiratory symptoms, loss of weight and aching in the muscles; a fortnight after admission he developed sudden pain and tenderness in the right side of the abdomen and a palpable mass developed in the loin, retrograde pyelograms demonstrating that the right kidney and ureter were displaced anteriorly. At operation a biggish haematoma was found around the kidney which was removed; the changes of periarteritis nodosa were found and the patient died of uraemia several months later. RALSTON and KVALE reported that, in a study of 30 cases of periarteritis nodosa, renal insufficiency was the immediate cause of death in one-third of the cases and in one of the cases there had been a large perirenal haematoma arising from a haemorrhagic infarct of the kidney.

g) Adrenal causes

A perirenal haematoma arising from a spontaneous rupture of a phaeochromocytoma of an adrenal gland in a woman of 54 was reported by McFARLAND and BLISS. A similar case, arising from a carcinoma of the adrenal gland, was reported by HERITAGE. CARVER reported a case of a huge, fatal perirenal haematoma in a male patient aged 47 who had complained for some months of aching in the left loin and also haematuria. The patient died following a very severe attack of pain in the loin accompanied by circulatory collapse. Autopsy showed a large perirenal haematoma resulting from spontaneous haemorrhage from the left suprarenal gland.

h) Diseases of the blood or blood-forming organs

Some of the diseases referred to above may give rise to haematomata in various parts of the body including the perinephric space. A case of a perirenal haematoma in a patient suffering from chronic myeloid leukaemia was reported by EKMAN.

i) Acute pancreatitis

CACHIN, TANRET, MONSAINGEON and LEVILLAIN reported 2 cases of acute haemorrhagic pancreatitis associated with perirenal haematoma and with anuria.

j) Idiopathic causes

Many cases have been described in which at operation or autopsy no cause for the bleeding has been found. In the case recorded by ELMER and WYNGARDEN a large subscapsular perirenal haematoma was removed together with the kidney but no intrarenal microscopic changes were found; some of the perirenal vessels showed arteriosclerosis. MARTIN recorded a fatal case of perirenal haematoma in a man of 35, in which both kidneys and suprarenals were found to be normal at autopsy. DARO and TODD described a case of spontaneous perirenal haematoma of unknown cause which followed an eight-month pregnancy in a woman aged 24; three days after delivery a swelling was noticed in the right upper quadrant of the abdomen, which later became painful and much worse after a fall. A large perirenal haematoma, together with the kidney, was removed but no pathological changes were found in the kidney itself.

In a fully recorded clinical discussion at the Massachusetts General Hospital (1949) a case was described of a perirenal haematoma in a woman of 65 who was admitted complaining of a sudden severe pain in the left loin, and who had a tender lump in the flank not moving on respiration; following operation, the kidney was shown to be normal and the cause of the haematoma was believed to be the spontaneous rupture of a perirenal artery. The differential diagnosis was fully discussed before operation and the most likely diagnosis was thought to be a left retroperitoneal extrarenal tumour, either an embryonal-cell carcinoma or neurogenic sarcoma or fibrosarcoma.

IV. Treatment

In 47 of the 178 cases reported by POLKEY and VYNALECK treatment was expectant and the mortality was 100 per cent. Generally speaking the perirenal haematoma which is associated with a ruptured kidney should be treated conservatively; it is only rarely that operation is necessary since the haematoma usually becomes self-limiting, shrinks and is eventually absorbed.

All other cases in which a perirenal haematoma is suspected should probably be submitted to an exploratory operation. Simple evacuation of the haematoma with drainage has given an occasional good result in recent years but, since the bleeding usually arises from the kidney, nephrectomy is generally necessary and has given the best results. Blood transfusions and other restorative measures, and antibiotics if there is an associated infection should be administered before operation. ASCHNER and KLINGER reported a case which was successfully treated by conservative measures, and LINK recorded a further case, resulting from periarteritis nodosa, which was similarly treated. 62 of 178 cases of POLKEY and VYNALECK were submitted to nephrectomy, with a mortality of 24 per cent. A high mortality usually approaching 50 per cent has been recorded in most published series.

C. Movable kidney

The normal kidney moves with respiration within its fatty coverings (though the perinephric fat is stated to take some part in the respiratory movement), and it probably also moves with changes in the position of the body. Since the renal artery and vein are attached to the great vessels, which are themselves fixed in position and have no respiratory excursion, the path followed by the kidney during movement is along the arc of a circle, an excursion which cannot usually be detected during clinical examination unless the degree of movement considerably exceeds that of normal. The renal pelvis and the upper part of the ureter move with the kidney, the ureter adapting itself to the renal movements; sometimes, for example, an excretory radiograph taken in full inspiration shows a kinked or slightly tortuous upper ureter, which becomes linear in a radiograph taken in full expiration. According to KELLY and LEWIS the extent of the movement of the kidney is 1.5 to 5.0 cm. in women and about half that distance in men. The adrenal glands and their fatty coverings, being secured, especially on the right side, to the great vessels by their own vascular pedicles, do not descend with the kidney. It is often possible to palpate the lower pole of a normal right kidney and less often that of the left, in individuals in whom the abdominal wall is thin.

According to BURFORD, the condition of movable kidney was first described by Franciscus de Pedemontanus in the fourteenth century and was later described as a clinical entity by RAYER in 1841.

I. Degrees of mobility of the kidney

HINMAN describes three degrees of mobility when estimated by clinical examination. In the first degree, the kidney can be grasped as it descends during full inspiration; in the second degree, the palpating fingers of the two hands can be made to meet above it during inspiration, and to hold it down; in the third degree the kidney moves to a level as low as the iliac fossa. The term "floating kidney" is reserved for a kidney which has a peritoneal covering which is more or less complete on its anterior and posterior surfaces and which, therefore, may prolapse or can be pushed by the palpating hand, into the abdominal cavity proper, not only downwards but towards the middle line.

BRAASCH et al. have divided renal ptosis into four grades based on the degree of descent of the kidney as determined by pyelography. The normal position of the renal pelves is usually opposite the first or second lumbar vertebra, the right kidney being at a slightly lower level than the left. In the first grade of renal ptosis, the renal pelvis is situated opposite the third lumbar vertebra, in the second grade the pelvis is opposite the fourth lumbar vertebra, and in the third grade it is opposite the fifth lumbar vertebra; cases in which the renal pelvis is situated at a lower level than the fifth lumbar vertebra are placed in the fourth grade. In their series of 230 cases, which included 73 bilateral cases, 31 were placed in the first grade, 85 in the second grade, 92 in the third grade, and only 12 in the fourth grade. The bilateral cases were mostly in the second and third grades, and ptosis involving the right kidney alone was present in two-thirds of all the cases.

Excessive mobility is a mechanical abnormality and not a disease. Probably in only a small proportion of all cases can symptoms properly be ascribed to it, and in a still smaller proportion do pathological changes, such as, for example, an obstructive factor at the pelvi-ureteric junction, complicate the condition.

PRATHER described 2 cases of medial ptosis of the kidney. During investigation for the cause of renal pain, retrograde pyelograms taken in the supine position were normal; radiographs taken with the patient in the lateral position, the painful kidney being uppermost, showed marked displacement of the kidney shadow towards the vertebral column; normal controls showed no such movement. Nephropexy resulted in relief of symptoms in both cases.

II. Aetiology

1. Incidence

KELLY analysed a large series and concluded that the incidence of movable kidney was 22.8 per cent in women and 2.1 per cent in men. The right kidney was affected more often than the left and both kidneys more often than the left kidney alone. Most cases were detected between the ages of 20 and 50. In BELL's series there were 300 cases in women and 10 in men. KIDD gives the incidence as about 18.4 per cent in women and 1 per cent in men. HINMAN states that movable kidney occurs in about 20 per cent of all women, in 2 per cent of men and rarely in childhood.

2. Causative factors

a) Visceroptosis and posture

Movable kidney is very often part of a general visceroptosis, the stomach, the colon, the intestines and the liver also lying at a lower level than normal and possibly having a greater respiratory excursion than normal. In many cases, however, the movable kidney is not demonstrably accompanied by the displacement of any other abdominal organ; according to KELLY, 149 of 237 cases of movable kidney were uncomplicated. In cases of visceroptosis, weakness and loss of tone of the muscles of the abdominal wall play a part. People with a normal erect posture and with well-developed abdominal and trunk muscles are less likely to have a movable kidney than the individual with a slouching posture. The thin woman with a slight postural kyphosis and a protruding abdomen associated with lordosis, and diminished muscle tone, is the type who may be expected to have a movable kidney. In such patients there is often a sub-normal amount of fat around the kidney and in the body generally.

b) Anatomical factors

WALKOW and DELITZEN observed that the paravertebral fossae were shallower than normal and more widely open at their lower ends, in patients with abnormal mobility of the kidneys, and that this applied more to women than to men. The relaxed abdominal wall which sometimes follows repeated pregnancies may be an explanation in some women, though movable kidney may occasionally be found in young women whose abdominal muscles appear on palpation to have normal tone.

c) Injury

THOMSON-WALKER reported a history of a blow. muscular strain or other injury preceding the discovery of a movable kidney in 11.4 per cent of cases. It is difficult, however, to prove the aetiological association of the two conditions, since an abnormally movable kidney may have been present before the alleged injury. The use of hard corsets is said to induce a degree of muscular weakness in the abdominal wall and may be a predisposing factor.

d) Associated renal disease

Abnormal mobility of the kidney may occur in association with hydro-nephrosis, but care must be exercised before attributing that condition to movable kidney. A kidney which is the seat of a renal tumour may have an increased respiratory excursion. but this movement differs in its cause from that of a true movable kidney.

III. Pathological changes

1. The kidney and perinephric fat

In conditions of excessive mobility the perinephric fat is usually diminished in amount and is sometimes entirely absent. Tough strands of fibrous tissue may be found connecting the true capsule of the kidney with the perirenal fascia: these strands are really condensations of fibro-fatty tissue resulting from the diminished amount of perinephric fat. The renal vessels, especially the artery, may be longer than normal, having become elongated by the range of movement of the kidney. The intimate relations of the right kidney to the duodenum and the ascending colon and those of the left kidney to the pancreas and descending colon are usually disturbed; sometimes fibrous bands connect the kidney to those organs. The kidney has been found in a congenital lumbar hernia and in a diaphragmatic hernia (THOMSON-WALKER), though such hernial sacs tend to allow the prolapse of adjacent organs into them; these conditions differ from true movable kidney.

2. Abnormal movements

The vascular pedicle in a kidney with severe abnormal mobility may undergo torsion to such an extent as partly to obstruct the renal vein, leading to congestion of the kidney, to subcapsular haemorrhages and even to partial or complete suppression of the urine from the affected kidney. The urine from such a kidney may contain blood cells and casts. Such torsion is temporary, the kidney returning to normal (THOMSON-WALKER).

Temporary obstruction of the pelvi-ureteric junction may also result from rotation of the kidney producing the clinical syndrome known as Dietl's crises. Such rotation may occur on the transverse axis of the kidney. Alternatively, a simple kinking of the ureter at the pelvi-ureteric junction may occur because of the descent of the kidney, and this is particularly likely to occur in the presence of fibrous adhesions in that position; such an obstruction may result in acute distension of the renal pelvis, which though temporary, may lead eventually to some degree of hydronephrosis.

IV. Clinical picture
1. Symptoms

The great majority of patients with movable kidney do not have clinical symptoms which are referable to them. The condition is often found accidentally on routine examination of the abdomen, and even kidneys with the greatest degree of mobility may give rise to no symptoms. Such being the case, it is often difficult to correlate with certainty the symptoms of which the patient complains, with the existence of a movable kidney which may be discovered on examination. Some patients continue to complain of the same pain even when a movable kidney has been fixed at operation.

In a proportion of cases, however, symptoms are complained of which can properly be related to abnormal renal mobility. Whereas in earlier decades surgeons frequently ascribed many symptoms, including those in a wide range of apparently renal, gastro-intestinal and colonic manifestations, to the presence of an abnormally movable kidney, and ordered for its relief either long periods of rest in bed, or operative treatment, opinion has now moved very much against these earlier vogues; indeed the surgeon of to-day, while frequently recognising abnormal clinical degrees of renal mobility, is much less anxious to relate such a finding to the patient's symptoms, unless these are of a more limited and circumscribed character than was formerly thought to be appropriate.

a) Symptoms relating to the kidney

Sometimes the patient is aware of the movement of a lump in the abdomen, which takes place when she is in a certain position, or following a certain movement. On assuming the erect posture, or on turning to the opposite side in bed, the patient may herself be able to palpate a lump as the kidney moves across towards the middle line of the abdomen, or may be able to follow its return to the normal position in the loin. Such movement may be accompanied by a sensation of discomfort, which is either ill-localised in the loin or situated just below the tip of the ninth costal cartilage, or the patient may experience a slightly sickening sensation which occurs when the kidney returns to its normal position. The oft repeated recurrence of this movement with its accompanying symptoms, may induce persistent discomfort in the loin and may, indeed, lead to the development of neurasthenic symptoms. The surgeon may readily confirm the observation made by the patient herself, palpation of the moving organ reproducing exactly the sickening sensation described by the patient. In such cases, which are uncommon, the symptoms really do appear to relate to the abnormal movement of the kidney. Some cases in this group are true "floating kidneys" and others fall into HINMAN's third group.

Some patients complain of pain which approximates to the usual distribution of renal pain, namely in the back and radiating downwards into the iliac fossa: sometimes, however, the pain lacks the sharply defined character of an obvious renal pain, raising doubts in the surgeon's mind. There may be associated backache, nausea and vomiting. A few complain of urinary frequency, dysuria and nocturia, but again, care must be exercised before such symptoms are ascribed to a movable kidney.

Attacks of severe pain, mimicking the renal colic of calculous disease, first described by DIETL in 1864 and known as Dietl's crises, occur in some patients. In such cases the crises have often been preceded by an aching pain in the region of the kidney, radiating to the iliac fossa, which is characteristically relieved by rest and aggravated by exercise and movement, especially bending, and even by constipation. Dietl's crises constitute severe intermittent attacks of colicky pain of renal distribution, which may last from a few minutes to a few hours and are usually accompanied by vomiting, nausea, perspiration and elevation of the pulse rate. The symptoms are believed to be occasioned by kinking of the pelvi-ureteric junction sufficient to cause retention of urine within the renal pelvis; temporary polyuria may follow the relief of the attack. Examination during such an attack may reveal a tender and occasionally a slightly enlarged kidney.

THOMSON-WALKER referred to occasional haematuria which may follow muscular effort, and to the frequent appearance of albuminuria and even tube casts due to venous congestion; it would seem, however, that these findings can be ascribed to abnormal mobility only when they follow a recent temporary torsion of the renal pedicle.

It has been suggested by some workers that nephroptosis may be a direct cause of hypertension and that such hypertension may be relieved by nephropexy (McCANN and ROMANSKY). It seems more probable, however, that if hypertension is associated with nephroptosis, it is only coincidental, and that intrinsic renal disease is the cause and that any relief following nephropexy will only be temporary (BRAASCH and GOYANNA).

b) Symptoms referred to other organs

Earlier writers have made much of the gastric symptoms supposedly due to movable kidney. Epigastric pain related or unrelated to food, flatulence, a feeling of gastric distension, and intermittent vomiting, have been said to be frequent symptoms, caused especially by a mobile right kidney, due to a drag of adhesions on the duodenum. Jaundice resulting from dragging by the mobile kidney on the common bile duct or the cystic duct has been described; it seems doubtful how far or how often such symptoms are due to a movable kidney. Similarly constipation and colonic distension and pain have been attributed to a mobile left kidney.

In patients who have a movable kidney and who complain of such extra-renal symptoms, the surgeon must carefully consider the presence of intrinsic disease of the stomach, gall bladder or colon before attributing them to a movable kidney. The relation of many such symptoms to a movable kidney is often very doubtful.

Neurasthenic symptoms are not uncommonly present in a patient who is found on examination to have a movable kidney and such symptoms may be aggravated, though probably not caused, by a movable kidney.

2. Investigations

The movable kidney can be palpated in the loin, usually without difficulty since most of the patients are of the lean, asthenic variety, with atonic abdominal walls. In the worst cases the entire kidney can be held down, when the patient inspires, between the examining hands on the abdomen in front and the loin behind; when the kidney is thus held, it causes pain of a slightly sickening character, which disappears as the kidney is released, to move upwards to its normal position.

In order to detect the full range of movement, the patient should be examined first in the recumbent position, the extent of movement being determined during each inspiration. She should also be examined in the erect posture and if the abdomen can then be adequately relaxed, the extent of the movement can be more readily determined. A "floating kidney" has a wide range of movement in an arc of a circle the centre of which is the renal vascular pedicle; such a range of movement has been called the "cinder-sifting" movement and is best detected under general anaesthetic.

Intravenous pyelography may reveal a normal kidney. Often, however, there is a kink at the pelvi-ureteric junction, best demonstrated if the radiograph is taken after full inspiration, the ureter then joining the renal pelvis more or less at a right-angle. The radiographic diagnosis of mobility and ptosis is best made by retrograde pyelography, the films being taken with the patient in the erect position in full inspiration and expiration and the extent of the movement then compared with that of a normal individual. Retrograde pyelography may also serve the purpose of deciding whether the discomfort produced by injection of the opaque medium up the ureter catheter reproduces pain of the kind of which the patient complains.

In a few patients the pyelogram may show a mild hydronephrosis of the pelvi-ureteric type. It will be difficult to be certain whether such a hydronephrosis is due to a neuromuscular defect or to an aberrant renal artery, occurring in a movable kidney. The characteristic rounded lower end of the renal pelvis and the narrowed pelvi-ureteric junction found with an aberrant renal artery should serve to differentiate the two conditions.

Slight pyelectasis associated with abnormal mobility not caused by organic pelvi-ureteric obstruction, should show a normally patent exit from the renal pelvis into the ureter, though there may be some angulation at the pelvi-ureteric junction.

Caution should be exercised in attributing slight degrees of hydronephrosis to renal ptosis. When the two conditions are found to be combined in one kidney, the conclusion may be drawn that the abnormal descent interferes with normal drainage of urine from the kidney and that it results in pyelectasis, which is possibly related to the grade of movement. BRAASCH et al. showed in their cases that such an assumption was not justified; in 230 cases the incidence of pyelectasis and delayed emptying time was approximately equal in movable kidney in the first three grades of ptosis and less frequent in the fourth grade, in which ptosis was most marked. These workers found varying degrees of pyelectasis or delay in the emptying time of the medium in almost one-third of 230 cases. They point out that such pyelectasis may result from either dynamic or adynamic factors; such factors include: 1. Mechanical obstruction at or near the pelvi-ureteric junction; 2. Adynamic or atonic dilatation resulting, for example, from a previous renal infection; 3. Imbalance of the renal innervation with subsequent atony. The absence of mechanical obstruction in some of these cases may be shown by the absence of retained fluid in the renal pelvis, in a film taken fifteen to twenty minutes after the retrograde catheter has been withdrawn; moreover, surgical exploration may fail to reveal any mechanical obstruction. In other cases, operation determines the presence and precise site of a mechanical obstruction at the pelvi-ureteric junction.

V. Diagnosis

The diagnosis of movable kidney as a cause of symptoms used to be made very commonly, and many varied operations have been done in an attempt to relieve its supposed symptoms. Surgical opinion has, in recent years, swung in the opposite direction. While clinical recognition of the varying degrees of mobility of the kidney is a matter of routine examination, the surgeon is now very cautious before attributing to it the many symptoms formerly ascribed to it, and loth to advise an operation unless the clinical evidence points strongly to its being the cause of such symptoms. Opinion has, therefore, moved very far from that held a generation or more ago.

1. Other causes of pain

Aching pain in the loin approximately of renal distribution, starting behind and radiating to the front, is a common symptom in women, and may be muscular in origin, or be caused by a prolapsed intervertebral disc between the twelfth dorsal and first lumbar vertebrae, or by the pressure upon nerves from the osteophytes of spinal osteoarthritis.

Movable kidney must not be confused with ectopic kidney which usually lies at a much lower level than a normal kidney and cannot be reduced into the loin.

Dietl's crises must be distinguished from acute renal colic resulting from stone, from hydronephrosis caused by an organic obstruction at the pelvi-ureteric junction resulting from a neuromuscular defect, stricture, folds or valves at the junction, or an aberrant renal artery; such cases will be differentiated on radiographic and pyelographic investigation. The association of haematuria with the renal colic would suggest a stone or tumour of the renal pelvis or parenchyma.

Other conditions causing acute attacks of pain in the right side of the abdomen which may conceivably be mistaken for Dietl's crises are acute retrocaecal appendicitis and gall-stone colic.

2. Swellings in the loin

a) Mucocele of the gall bladder

A large mucocele of the gall bladder is a cystic swelling which comes down under the right costal margin. It lies in the anterior part of the abdomen, its outline can sometimes be seen on the abdominal wall of a thin patient who is fully relaxed and in a recumbent position and it has a very marked excursion on respiration. The swelling is dull on percussion and the dullness is continuous with that of the liver. The colon, which is resonant on percussion, never lies in front of it.

b) Riedel's lobe of the liver

An abnormal downward extension of the right lobe of the liver is not un-common and when it is large and partly separated from the rest of the liver by a groove, it may be difficult to distinguish from a movable kidney. On percussion, a Riedel's lobe is dull and the dullness is continuous with that of the liver. The lower border is sharp and not rounded and there is a very considerable respiratory excursion.

c) Enlarged spleen

A moderately enlarged spleen on the left side may possibly be mistaken for a movable kidney.

d) Malignant colonic tumours

A malignant growth of the ascending colon on the right side or of the descending colon below the splenic flexure on the left side, may mimic a movable kidney. Both may give rise to a swelling which can be reduced into the loin but the swelling is usually irregular, hard and nodular and not reniform. The clinical symptoms of intestinal obstruction should assist in the differentiation.

VI. Treatment

Patients in whom a movable kidney has been discovered should not have their attention drawn to it by the casual remarks of the surgeon nor should such a patient have access to the clinical notes. The majority of patients in whom a movable kidney is discovered do not need any form of supportive or surgical treatment. There will be, however, a small residuum of patients in whom the surgeon believes that the symptoms complained of are caused by the excessive mobility of the kidney and in these, at first non-operative treatment, and later, if necessary, operative treatment should be carried out.

1. Non-operative treatment

Abdominal and general bodily exercises should be given to improve general muscle tone and the muscles of the abdominal wall, and also to correct abnormalities of posture. The patient should be encouraged to increase her weight in order to increase the perirenal fat, and thus possibly assist in the reduction of the mobility. An abdominal belt or corset reinforced by a pad which will exert pressure in the loin, is most suitable; the belt should be placed in position with the patient recumbent. Long periods of rest in bed, which formerly had many advocates, have now no place in the treatment of movable kidney.

2. Operative treatment

According to DODSON, Hahn in 1881 performed the first operation of nephropexy, resecting part of the perinephric fat and suturing the remainder to the lumbar incision.

In the very few patients in whom the surgeon decides that the symptoms are caused by a movable kidney, operation should be done. Nephropexy is indicated in cases in which the patient herself is aware of the movement of the kidney, which she can palpate, and in which pain is experienced during such movement. It should be seriously considered in patients who experience pain having a characteristic renal distribution, especially when there are early hydronephrotic changes, when the injection of opaque medium into the renal pelvis along a ureteric catheter reproduces the pain of which the patient complains, and when there is delay in the drainage of injected opaque medium from the renal pelvis. Operation is contra-indicated in cases who have a wide variety of symptoms relating to organs other than the kidney, and also in neurasthenic patients.

The aim of the operation of nephropexy is to fix the kidney in such a way that it acquires firm fibrous attachments to the muscular parietes thereby limiting its mobility. Securing the kidney in the correct position in the loin should allow of free drainage of urine from the renal pelvis and also avoid the kinking which results in Dietl's crises. Many different methods of fixation have been used.

The kidney is exposed through an oblique incision in the loin through the bed of the resected twelfth rib, the patient being placed on her side in the normal kidney position. The kidney is freed from its fatty coverings posteriorly and examined to exclude intra-renal disease. The renal pelvis and upper part of the ureter are examined to exclude a hydronephrosis from an organic cause at the pelvi-ureteric junction; any fibrous bands or adhesions found there, should be divided. When the surgeon is satisfied that the kidney is anatomically normal he proceeds with the fixation operation. A semi-circular incision is made through the true renal capsule, skirting both poles and the convex border of the posterior surface of the kidney, and the capsule is stripped towards the hilum from the underlying renal parenchyma. Using chromic catgut a series of four interrupted sutures an inch or more apart, are passed through the renal parenchyma near to the convex border taking a moderate bite of renal tissue; the stitches are then passed fairly deeply through the muscles of the parietes, fixing the kidney in its normal position, the upper pole lying in front of the twelfth rib. The sutures are tied, but only sufficiently tightly as to secure apposition between the kidney and the muscle and not tightly enough to crush the renal tissue. The raw area of the kidney from which the capsule has been stripped should adhere firmly to the apposed muscles. The wound is closed.

Some surgeons insert the stitches so that the uppermost stitch passes through the intercostal muscle above the twelfth rib, the others being sutured to the muscles below the rib. The reflected renal capsule may itself be stitched to the muscles as the principal or as a secondary method of anchorage.

An alternative method of treatment is to secure the kidney in its normal position by making a hammock of perinephric fascia below the lower pole. The steps of this operation, as described by DEMING, are as follows. The perinephric region is exposed by an oblique incision. The perinephric fascia is incised longitudinally on the posterior surface of the kidney and as far towards the middle line as possible. It is then dissected away in one sheet and turned forward along with the perinephric fat, from the posterior surface, the convex border and the anterior surface of the kidney. The right lobe of the liver should be separated from the muscles of the posterior abdominal wall if it is attached to it. When this is done, the kidney can then be raised so that its lower half lies over the last rib and the parietal muscles. In the position in which it is finally fixed, the upper pole of the kidney should be carried a little towards the middle line and the lower pole slightly outwards so as to give adequate drainage to the lowest calyx of the kidney. The pelvi-ureteric junction is freed from any fibrous bands or adhesions. The fibro-fatty sling below the kidney is then made. A series of interrupted mattress stitches of catgut are placed through the perinephric fascia and fat, and also the peritoneum if readily available, to secure them to the quadratus lumborum muscle. Owing to the abundance of the perinephric fat there is no difficulty in securing apposition without tension. The first suture is placed about a centimetre from the ureter and as high as possible on the quadratus muscle posteriorly. A series of five or six or even more sutures at the same level secure the perinephric fascia and fat to the quadratus muscle to close the lower aperture through which the kidney can move downwards. A second row of mattress sutures can be usefully placed at a lower level than the first in order to provide further support. The wound is then closed with drainage.

A suspension procedure such as is described above can be combined with fixation of the kidney. Thus, two or three sutures may be placed through the capsule of the kidney and the quadratus lumborum muscle at the upper pole, the middle and the lower pole (DODSON).

Results of the operation of nephropexy

MATHÉ, using his own technique for the operation of nephropexy in 139 cases, claimed cure or relief in 98.7 per cent. Pyelography carried out after the operation in 71 cases revealed a successful fixation of the kidney in every case.

WOODRUFF and SCHERER, using their technique of nephropexy in 59 patients, claimed cure or relief in 98.3 per cent; follow-up pyelograms in 48 patients showed successful fixation of the kidney in every case.

DODSON reported the results in 18 patients operated on for movable kidney with relief in 16 cases. Most of the cases were dealt with by the use of a sling of perinephric fat and fascia placed beneath the kidney, but in some the sling procedure was combined with fixation of the kidney to the quadratus muscle by catgut sutures.

Of 21 patients who had nephropexy, reported by BRAASCH et al, 11 experienced more or less subjective relief from their symptoms but only 3 had complete and permanent relief. In others the original pain was largely relieved but other symptoms continued. Ten patients were not subjectively benefited by the operation.

The experiences of BRAASCH et al probably reflect the feelings of most surgeons in regard to the operation of nephropexy for movable kidney. In order to achieve a reasonable proportion of satisfactory results, the greatest care must be exercised in the selection of cases for operation.

D. Changes in size and position of the bladder

I. The normal bladder

The normal bladder in the male lies in the pelvis behind the pubic bones and in front of the rectum from which it is separated by the recto-vesical pouch. In the female, the bladder is similarly placed behind the pubic bones while posteriorly it is in relation to the lower part of the body and cervix of the uterus and to the anterior vaginal wall. The bladder is an extraperitoneal organ, though it is covered by peritoneum over most of the vault and part of the sides. It lies in the cave of Retzius, a potential cavity containing areolar tissue, which is so lax that the bladder readily becomes distended with urine. When empty, the bladder collapses downwards because the fixed lower part, the trigone, is attached to the prostate and the prostatic urethra, which themselves are fixed to the triangular ligament. During distension, the bladder enlarges upwards, backwards and outwards and becomes an abdominal organ.

In childhood the bladder lies at a higher level than in adult life, and is, in fact, an abdominal organ.

II. Alterations in the position of the bladder

1. Physiological

During pregnancy the enlarging uterus causes the urethra to be elongated, and the bladder to become an abdominal organ, being lifted almost completely out of the pelvis.

2. Pathological

a) Fibroids of the uterine cervix

Fibroids of the uterine cervix or the lower segment of the uterus may similarly alter the position of the bladder, causing it to be displaced upwards into the abdomen; if the tumour is asymmetrical with regard to the middle line, the bladder may be deviated to the right or to the left as well as upwards.

b) A swelling in the pouch of Douglas

A swelling in the pouch of Douglas of a few inches in diameter, caused by an ovarian cyst, pelvic abscesses from any cause, pelvic haematocele and retroverted gravid uterus may raise the level of the bladder from the pelvis into the abdomen. Pressure on the urethra by the swelling may result in urinary retention.

c) Uterine prolapse

A cystocele caused by a prolapse of the anterior vaginal wall is a downward change of position of the bladder.

In prolapse of the uterus, the uterus descends and in the second degree of prolapse the cervix protrudes wholly or partly from the vulva; there may be an associated cystocele. In the third degree of prolapse, or procidentia, the whole of the uterus is expelled through the vulva, and the vagina in consequence is almost completely inverted. As a result of this inversion, the bladder may prolapse with the vagina and the uterus. In less advanced cases, the base of the bladder only is everted, but in severe cases the whole of the bladder lies in the procidentia and therefore outside the body.

d) Tumours of the pelvis

Massive tumours of the pelvis such as chondroma or chondro-sarcoma, arising from the cartilage of the sacro-iliac synchondrosis or from the pubic symphysis, are often asymmetrical and may bring about bizarre changes in the position of the bladder. Thus the organ may be displaced backwards from tumours arising anteriorly, or be moved to one side or the other of the pelvis.

e) Hernia of the bladder

α) Inguinal and femoral herniae

A sector of the normal bladder, or a diverticulum of a bladder, the neck of which is partly obstructed, may prolapse into an inguinal or a femoral hernial sac. When the bladder is collapsed and empty, any part of it is less likely to descend into a hernial sac than when it is distended. Inguinal hernia of the bladder is more common in men than in women and is more commonly associated with direct than with oblique inguinal hernia, being, therefore, of greater frequency in older than in younger patients. Femoral hernia of the bladder occurs almost exclusively in women. Hernia of the bladder occasionally occurs in children (WAKELEY, WATSON).

A portion of a bladder which has a capacity greater than normal, as in some cases of prostatic or urethral obstruction, is more likely to enter a hernial sac than is a bladder of normal size. As the hernial sac increases in size, it may exert traction on the bladder to which it is closely related, drawing it downwards so that a part of the bladder wall constitutes part of the actual hernial sac. The fatty coverings of the bladder usually come down with the organ, forming a distinct layer between the musculature of the bladder and the peritoneum of the sac. At operation for radical cure of the hernia the bladder should be easily recognised as a greatly thickened part of the wall of the inner side of the sac; the sac should be dissected from the peri-vesical fat and muscle, allowing the latter to be reduced upwards before the sac is transfixed.

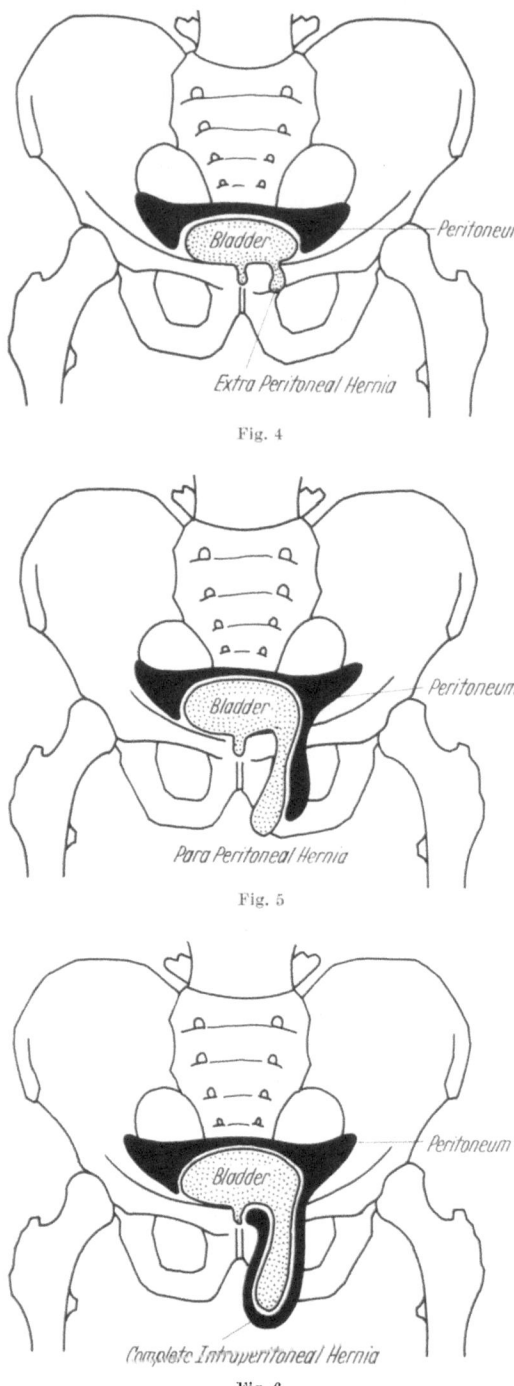

Extra Peritoneal Hernia

Fig. 4

Para Peritoneal Hernia

Fig. 5

Complete Intraperitoneal Hernia

Fig. 6

Figs. 4—6. Schematic drawings of three varieties of hernia of the bladder. 4 Extrapertoneal hernia of bladder, 5 Paraperitoneal hernia of bladder. 6 Intraperitoneal hernia of bladder. (By courtesy of R. A. MOGG, Esq. F.R.C.S., of Cardiff)

WAKELEY reported a series of 1,250 cases of inguinal hernia in which a hernia of the bladder was present in 27 (2.16 per cent). During the same period, in 196 operations for femoral hernia, a hernia of the bladder was present in 11 (5.6 per cent). WAKELEY recorded three anatomical varieties of hernia of the bladder, namely, extraperitoneal, paraperitoneal and intraperitoneal (Fig.s 4—6).

Extraperitoneal hernia of the bladder. In this type which is rare a small portion of the anterior or lateral extraperitoneal surface of the bladder wall descends into the inguinal or the crural canal, without, however, drawing a pouch of peritoneum downwards with it. Such herniae are of small size and in the case of inguinal herniae the direct variety is the commoner.

Paraperitoneal hernia of the bladder. The bladder comes to lie on the inner side of a direct or oblique inguinal hernial sac, usually the former, and is entirely outside the peritoneum. This variety is the most common and is easily recognised at operation by the increased thickness of the sac, in that part which is nearest to the abdominal cavity.

Intraperitoneal hernia of the bladder. This condition is unusual but is more common than the extraperitoneal variety. The hernia is nearly always a large inguinal hernia and the upper and posterior portion of the bladder descends into the sac, usually in association with loops of small or large intestine and of omentum.

A diverticulum of the bladder may descend into a hernial sac. POTT (quoted by

WAKELEY) recorded such a case in a child aged six. The swelling in the groin was believed to be a tumour but when an incision was made into it, a calculus from the bladder was extracted. Another case in a boy of eighteen months, was reported by CORNER and ROWNTREE.

A case of a large paraperitoneal hernia of the bladder, in which a large part of that organ was actually in the scrotum, was observed by MOGG. The patient was a man aged 70 who was operated on for a strangulated inguinal hernia, when the intestine was reduced, but because of his poor general condition, no attempt was made to repair the hernia, the wall of which contained a large portion of the bladder. Fig. 7 shows the hernia containing the bladder, in the right side of the scrotum and the right inguinal region when the bladder is distended; Fig. 8 shows the reduction in size in the hernial sac following micturition. Three years later the man was admitted with acute retention of urine; a cystogram (Fig. 9) then showed that the greater part of the bladder was in the scrotum, within the hernial sac.

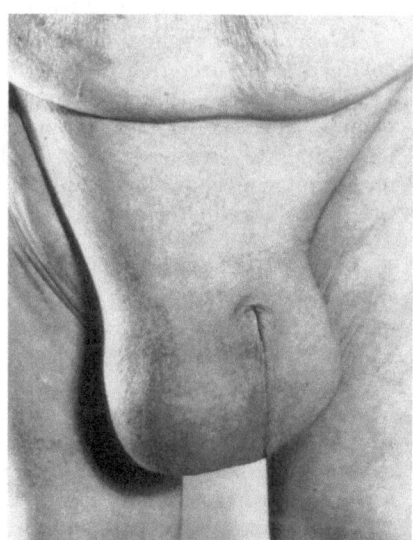

Fig. 7

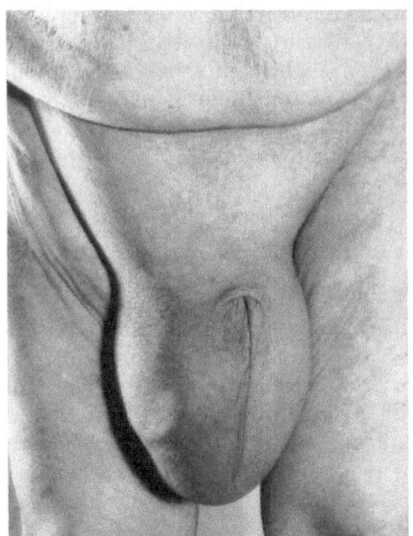

Fig. 8

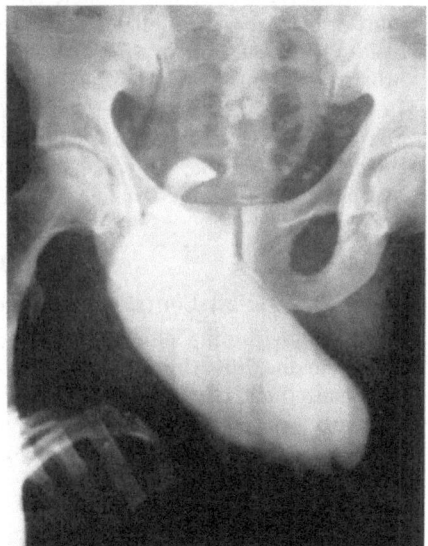

Fig. 9

Figs. 7—9. Mr. R. A. MOGG's case of a right inguino-scrotal hernia containing a large part of the bladder: 7 The hernia is in the scrotum and the bladder is moderately distended. 8 The hernia after micturition. 9 Cystogram showing the greater part of the bladder in the scrotum. (For description, see text. By courtesy of Mr. R. A. MOGG, F.R.C.S., of Cardiff)

β) Incisional hernia

The bladder may partly prolapse into a lower abdominal incisional hernial sac. The extent of the prolapse may be shown preoperatively by a lateral cystogram.

III. Changes in the size of the bladder

1. Enlarged bladder

a) Mega-ureter-megacystis syndrome

In some cases in children with bilateral mega-ureter, the bladder is considerably enlarged. The ureters are also grossly dilated; they are narrowed at the point at which they pass through the vesical detrusor muscle, but the ureteric orifices are gaping. The bladder wall is somewhat thickened and the capacity of the bladder is considerably reduced. This syndrome has been designated by WILLIAMS the mega-ureter-megacystis syndrome, to whose volume (XV) in the Encyclopedia the reader should refer for the full clinical picture and treatment.

b) Partial obstruction of the bladder neck and the urethra

In cases of prostatic enlargement, fibrous bladder neck, stricture of the urethra, congenital urethral valves, and in some cases of spina bifida, the bladder responds to prolonged obstruction by a gradual increase in its capacity; in the early stages of obstruction, muscular hypertrophy of the bladder wall occurs, but later compensation may fail, and the bladder enlarges and becomes atonic, when it is capable of holding from 20 to 30 oz. of urine. The atonic bladder may recover its tone and return to almost normal capacity following a period of catheter drainage and the removal of the obstruction.

In a few cases persistent atony of such large bladders in cases of prostatic obstruction, is responsible for failure to void even after the obstructing prostate has been removed; under such circumstances the treatment of the atony is a matter of some difficulty. The surgeon must ensure that no minor grade of residual obstruction of the bladder neck, by stricture or by tags of fibrous or prostatic tissue, is left behind which may even slightly interfere with voiding. In some cases a suprapubic cystotomy for three or four months may enable the bladder to recover its tone; voiding may then occur following the removal of the suprapubic tube. In some cases, removal of the dome of the bladder followed by repair, has enabled natural voiding to occur.

2. Contracted bladder

a) Contracted bladder following chronic cystitis

In long-standing chronic cystitis, which occurs usually in women who have had recurrent infection for many years, the bladder may be uniformly contracted as a consequence of fibrosis of the detrusor; the capacity of such bladders may vary from 3 to 6 oz. The patient complains of great frequency of micturition, particularly at night, rendering sleep difficult and sometimes gravely interfering with the general health. The operation of ileocystoplasty is of great value in such cases.

b) Tuberculous contracted bladder

In patients with tuberculous disease of the kidney, the bladder may be normal in size and shape if there is no tuberculous vesical ulceration; in more advanced cases, when tuberculous vesical ulceration is present, the bladder tends to become progressively smaller. At first, the diminution in size is largely due to spasm which is secondary to the ulceration; later, the reduction in size may be secondary to the scarring and fibrosis associated with deep tuberculous ulcers penetrating the bladder wall. When the bladder heals, as it usually does following nephro-ureterectomy of the diseased kidney and the prolonged administration of streptomycin, P.A.S. and isoniazid, it may resume its normal size if the ulceration has been limited in extent and shallow in its penetration; alternatively, if the tuberculous ulcers have been deep and extensive, the subsequent scarring may cause a generalised and permanent bladder contracture reducing the capacity of the organ to 2, 3 or 4 oz. Moreover, the shape of such bladders may be altered, since scarring may result in the formation of numerous septa which distort the bladder at the site of the previous tuberculous ulceration.

In the smallest, healed, tuberculous bladders ("thimble-bladder") in which the capacity may be less than 2 oz., there is often an associated dilatation of the prostatic urethra; the prostatic ducts may also dilate and give rise to pseudo-diverticula extending into the substance of the prostate (Figs. 10 and 11). Such

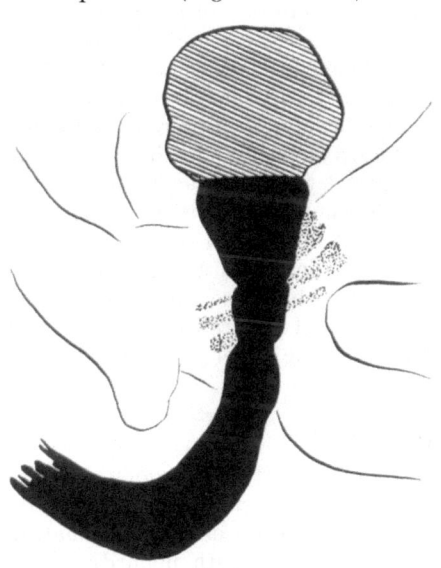
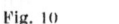

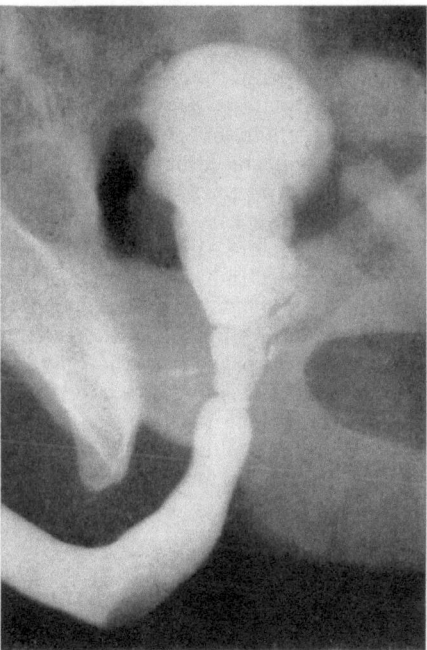

Fig. 10   Fig. 11

Fig. 10—11. Cysto-urethrogram of case of tuberculous contracted bladder. The bladder has a capacity of one to two ounces. The prostatic urethra (black in the left photograph) is greatly dilated and there are numerous pockets or diverticula leading from it into the substance of the prostate

contracted bladders are now usually treated, with considerable success, by the operation of ileocystoplasty (COUVELAIRE, CIBERT, PYRAH).

c) Contracted bladder following interstitial cystitis

In interstitial cystitis associated with Hunner's ulcer, which occurs usually in women around the menopause, the capacity of the bladder may be reduced to three or four ounces. Distension of the bladder is painful, necessitating frequent voiding. The operation of ileocystoplasty gives variable results in such cases (PYRAH).

E. Purpura of the bladder

Purpura occurs in the urinary tract as a local or symptomatic manifestation of one of the varieties of generalised purpura, namely thrombocytopenic purpura and anaphylactoid purpura (Schonlein-Henoch syndrome). There is some evidence that purpura may occur in the kidneys and the bladder as an independent local disease. The subject of purpura of the urinary tract was reviewed by KIDD and more recently by MARTIN.

I. Symptomatic purpura affecting the urinary tract

1. Thrombocytopenic purpura

Thrombocytopenic purpura is a disease of unknown etiology, believed by some to be due to an excessive destruction of blood platelets by the spleen and the reticulo-endothelial system, and by others to a primary defect in the capillary endothelium which results in platelets being used to plug breaches in the capillary walls, thus bringing about their effective disappearance from the circulating blood. Examination of the blood usually reveals a considerable reduction in the blood platelet count. Good accounts of this condition are given in articles by MARTIN and by ROBSON.

Although the skin and the mucous membranes are commonly affected by this type of purpura, it is rare for the urinary tract to be involved. A case was recorded by DUKES of a middle-aged woman who developed purpura accompanied by profuse haematuria from both kidneys ten days after an abdominal operation; the bleeding time was prolonged and the platelet count was reduced. The condition proved fatal and autopsy showed haemorrhages into the connective tissue and the mucosa of the renal pelvis with blood-clots in both ureters, and also purpuric patches in the bladder. ROSENTHAL described a case of extensive purpura of the skin associated with haematuria, which was shown to be due to a tuberculous kidney; the existence, therefore, of generalised purpura with haematuria should not lead to the conclusion that the latter symptom is necessarily due to purpura.

2. Anaphylactoid purpura

This condition is regarded as being the result of a sensitivity reaction occurring in the endothelium of blood vessels, and is sometimes associated with bacterial infections. Microscopic red blood corpuscles in the urine or gross haematuria may be present; the condition is commonly associated with nephritis, which may account for the occasional haematuria, and also with joint manifestations (GAIRDNER). MARTIN described a case in which a patient, whose presenting symptom was haematuria, developed the full Schonlein-Henoch syndrome eight months later and who died with widespread purpuric lesions of the kidneys, ureters and bladder.

II. Primary purpura of the urinary tract

1. Renal purpura

Haematuria from one or both kidneys may sometimes be unexplained, however carefully the case is investigated. In some such cases of unilateral bleeding in which the usual investigations have been negative, the kidney has been removed for persistent haemorrhage and a tiny lesion, such as an angioma of a renal calyx or papilla, or focal nephritis or papillitis, has been found when the kidney has been carefully sectioned and examined, and which would explain the haematuria. In a few similar cases of persistent renal bleeding, no such lesion has been found, but purpuric spots or ecchymoses have been observed in the mucous membrane of the renal passages; similar purpuric spots may have been seen in the vesical mucosa during cystoscopy. Such cases provide the evidence for the existence of a true primary renal purpura, since they may be unaccompanied either by purpuric manifestations elsewhere in the body, or by the usual haematological findings which are present in cases of symptomatic purpura affecting the urinary tract. Before making a diagnosis of true purpura it is always necessary to exclude those occasional cases of acute pyelo-cystitis, renal calculus or tuberculosis in which haemorrhage has been a prominent and even a misleading symptom. Possibly some of the cases which are diagnosed as essential haematuria, are cases of renal purpura.

MARTIN suggests that such cases of primary renal purpura may be the result of a hypersensitivity reaction to certain antigens in the capillary endothelium, (RICH). MARTIN suggests that the commonest antigen is likely to be a bacterial one; he quotes such a case which resulted from a bacillus coli infection, the patient also having recurring attacks of cystitis; the infection in the urinary tract was the only focus of infection which was found in the body and he believed that the renal purpura resulted from a local toxic reaction or from bacterial hypersensitivity; the changes of chronic pyelonephritis were found histologically in the kidney after its removal. Whether such cases should be called primary renal purpura or whether they are clinical variants of chronic pyelonephritis, however, is uncertain. KIDD similarly noted the association of renal purpura with a streptococcal renal infection. RHODES reported haematuria, as well as cutaneous petechiae at the site of injection, following the prophylactic injection of tetanus antitoxin; the symptoms, which subsided and did not recur, may well have been allergic in origin: no pathological evidence was reported in this case. Haematuria following the ingestion of certain articles of food may be allergic in origin and may be manifested by renal purpura (MILLER and UHLE, LAZARUS), but little is accurately known about such conditions, autopsy evidence usually being lacking.

2. Purpura of the bladder

The diagnosis of purpura of the bladder can only be made after a patient, who is known to have had haematuria, is cystoscoped soon after the bleeding has occurred. Tiny, circular petechial haemorrhages or larger submucous extravasations of blood would suggest the diagnosis. Haemorrhage from one ureter associated with these findings may lead (if the pyelograms are normal) to the suspicion of a similar lesion in the pelvis or calyces of the kidney.

The cystoscopic findings must be differentiated from haemorrhages resulting from distension of the bladder or from slight trauma following the introduction of the cystoscope. The condition must also be distinguished from haemorrhagic cystitis. WARD reported a case of purpura of the kidney and the bladder, in a girl of seventeen, who presented with haematuria; cystoscopy revealed patchy congestion of the bladder and bleeding from the right ureter, for which nephrectomy was later carried out; the renal pelvis showed purpuric patches but no other lesion. According to KIDD, vesical purpura gives rise to pain.

True vesical purpura, though useful as a descriptive diagnosis, would appear, except in the thrombocytopenic and the anaphylactoid group, and possibly when associated with true renal purpura, to lack as yet a firm pathological basis upon which the diagnosis can be made.

F. Foreign body in the bladder

Foreign bodies in the bladder usually assume clinical importance soon after they reach the bladder, since they may give rise to symptoms of dysuria, frequency, haematuria and sometimes urinary retention. A few foreign bodies give rise to minimal or no symptoms and are therefore allowed to remain in the bladder, but in such cases vesical calculi, resulting from encrustation of the foreign body, form later and give rise to symptoms; many vesical calculi, especially in women, are, in fact, formed around foreign bodies previously introduced. In CRENSHAW'S series, of 606 patients (577 male and 29 female) with vesical stone, in 21 males and in 8 females the stone had formed around a foreign body (silk or other sutures, rubber tissue, the head of a catheter or bougie, hairpin, needle, knife blade, or a fragment of bone).

I. Classification

Foreign bodies in the bladder fall into the following groups, classified by the mode of their introduction:

1. Along the urethra:
   a) Self-introduced
   b) As a result of urological procedures.
2. By way of penetrating wounds of the bladder.
3. Following open surgical operations on the bladder.
4. Migratory foreign bodies.
5. By way of the intestine.
6. From pelvic dermoid cysts, tumours or teratoma.

1. Along the urethra

a) Self-introduced

Foreign bodies may be self-introduced into the urethra in either sex, either for the purpose of masturbation, or as a practical joke, or in children out of curiosity or perversity. Foreign bodies introduced in that way may not pass, in the male, beyond the membranous urethra, or alternatively the proximal end may protrude into the bladder. Some foreign bodies appear to cause few symptoms immediately after their introduction, the patient appearing for treatment years later. In PHILIP'S case there were no symptoms of note until six years after the introduction of a four-inch needle, when the patient developed frequency and later urinary retention and urethral bleeding; radiography showed that two-thirds of the needle had broken away and had become surrounded by calcareous debris, while the distal third had penetrated the bladder wall and had stuck into the back of the pubic bone. Most cases, however, are reported soon after the foreign body has been introduced. If they have been in the bladder for any length of time, most wooden or metallic foreign bodies become encrusted with phosphates. A great variety of foreign bodies self-introduced into the bladder have been reported: Lead and slate pencils, knitting needles, nails, hairpins, button-hooks, feathers, paint brushes, kidney beans, watch chains, leather shoe strings, straw, chewing gum, chalk, slippery elm, paraffin, tallow, bougies and catheters. Specific references include the following: Pencils (COOK, BINTCLIFFE); needles (PHILLIP); clinical thermometer (FENGER et al.); perfume bottle (DUNCAN); condom (forming nucleus of a vesical calculus) (BADENOCH and CAMPBELL); toothbrush handle (HIME); decapitated snake (GEYERMAN); squirrel's tail (GILL); thermometer (ASPINALL); snail (HARRIS);

sponge (MAUTERER); a wasp (LINS). Blades of grass, ears of corn and a caudal vertebrae of a squirrel have been removed from the male bladder (BADENOCH).

The most unusual foreign body of all (not self-introduced) is the Candiru fish which invades the bladder of native Indians of the Amazon basin (Vandellia cirrhosa). This fish, which is dangerous to humans, is a little less than three inches in length and 3—4 mm. in calibre, with cat-like teeth and a broad, rounded head, and sometimes enters the urethra of bathers in the Amazon,

propels its body into the bladder and attaches itself to the wall of the bladder by some spines on its neck. Thus it acts as a foreign body and gives rise to cystitis. The natives wear a protection for the genital organs when bathing. In order to try to deal with this, the natives drink a hot brew of the buitach apple (Genipa americana L.) which they believe has the property of dissolving the skeleton of the fish.

Pieces of chewing-gum have been used for the purpose of masturbation by stimulation of the deep urethra. The method, according to MOORE, is to roll the gum into a pencil-like structure which is introduced by the patient for the purpose of producing an orgasm by stimulation of the deep urethra. During the act, the part of the gum nearest the bladder breaks off and is sucked into the bladder

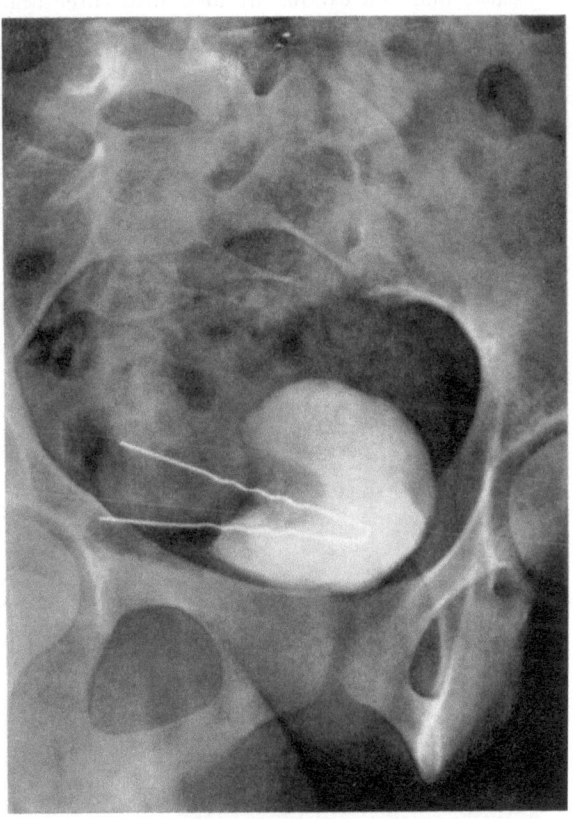

Fig. 12. Hairpin in bladder (self-introduced). Opaque medium has been introduced into the bladder. The ends of the hairpin have passed through the bladder wall

by muscular contraction. A first, the gum causes little trouble, but if infection or incrustation with calcium phosphate follows, clinical symptoms appear. The stones are quite firm, the outer coat consisting of a mixture of phosphate and oxalate. Similar cases have been reported by EWELL, GARSH-WILER and BALCH, GILL, GOODMAN, MACKENZIE.

Soft, pliable materials composed mainly of wax are commonly used; candle wax (TURNER); paraffin candle (KATZEN); bougies made of beeswax and wax tapers (GOLDSTEIN). The wax foreign bodies are sucked into the bladder and at first they float; only when they are coated by a deposit of salts sufficient to cause them to sink in the urine, do they come to the floor of the bladder and gradually cause symptoms. According to GOLDSTEIN the period between the introduction of the wax and the onset of symptoms is in direct relation to the specific gravity of the material; thus it is approximately 2 years in the case of paraffin wax, 1 year for beeswax, 7 months for chewing gum.

KATZEN's case, a man aged 21, illustrates the clinical picture and treatment in this group. A paraffin candle two inches in length and a quarter of an inch in thickness had been introduced along the urethra as a practical joke. The patient felt the candle slip into the bladder and he shortly developed frequency, dysuria, straining and terminal haematuria. Cystoscopy showed an intense cystitis, and the candle, divided into three segments attached to one another by the wick, was found floating near the vault of the bladder. The surgeon was unable to grasp the segment by cystoscopic forceps, and the continued use of this broke off many bits of the candle. In order to effect the removal of the wax, 100 cc. of warm, sterile liquid petrolatum instilled into the bladder through a rubber catheter was used. The suprapubic region was massaged and the patient instructed to refrain from micturition as long as possible. The next day some more of the liquid was instilled and also on a third day; this was followed by much reduction of the urinary distress, the wax evidently being dissolved; the wick of the candle was later passed and on cystoscopy five weeks later, a normal bladder was seen.

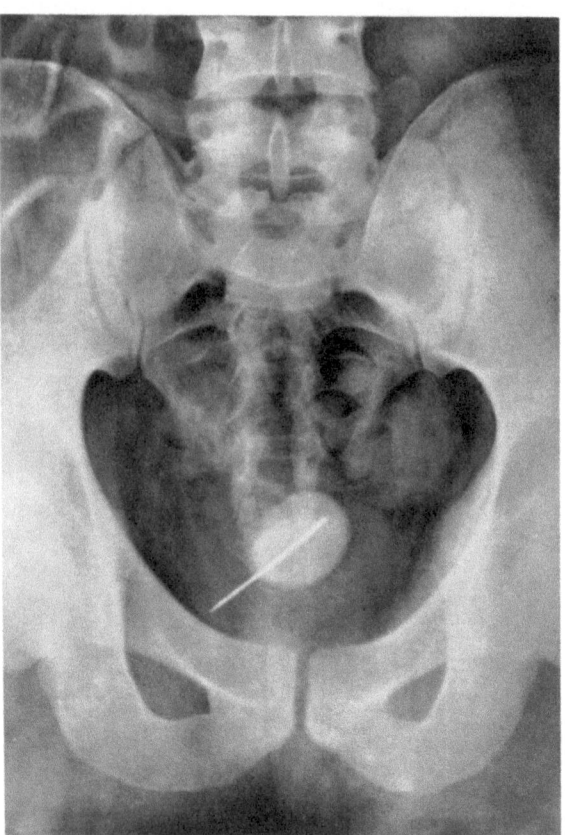

Fig. 13. Stone in bladder which has formed around a needle, which was self-introduced. The needle was partly outside the bladder

Paraffin acts as an irritant to the bladder causing a haemorrhagic cystitis within twenty-four hours. The cystoscopic lithotrite or Young's rongeur is sometimes successful in removing wax, but quite often the wax breaks up into fragments and the procedure fails. Other solvents have been tried. LOHNSTEIN was the first to describe the use of chemical solvents and he used an instillation of 15 ml. of benzene, retained in the bladder for 45 minutes; this was repeated later. The procedure resulted, however, in benzene poisoning as well as much vesical irritation. CAPLES dissolved a paraffin wax candle by instilling into the bladder liquid petrolatum at 110° F containing 33$\frac{1}{3}$ per cent gasoline, using seven injections over a period of three days. HOTTINGER, in order to dissolve beeswax in the bladder of a boy, removed part of it through an operating cystoscope and dissolved the remainder with xylene, first injecting water into the bladder and following with 12 ml. of xylene which was retained for four hours.

Xylene was also used by Geyer who used 50 ml. which was repeated five days later. Turner first introduced sterile water followed by 12 ml. of xylol for thirty minutes, the patient being in the recumbent position. The use of liquid petrolatum was first described by Harris who injected 100 ml. of warm liquid petrolatum into a boy's bladder to try to remove a paraffin candle. He gave 4 instillations at twenty-four hour intervals; most of the wax was dissolved after two injections and when cystoscoped several days later there was no evidence of foreign body.

Joelson recommended the removal of wax foreign bodies through the McCarthy resectoscope after distending the bladder with air instead of water; the foreign body then came to rest on the floor of the bladder and could be reached by the instrument and removed piecemeal by means of the Lowsley cystoscopic rongeur. Schulte devised a novel method of securing the removal of a piece of paraffin from the bladder of a male, aged 21, which had been introduced six months previously, the patient refusing further operative treatment after the removal of a part of the paraffin rod by a cystoscopic rongeur. The patient was told that if he could urinate whilst standing on his head the paraffin might pass naturally. He returned two weeks later having taken fluids in con-

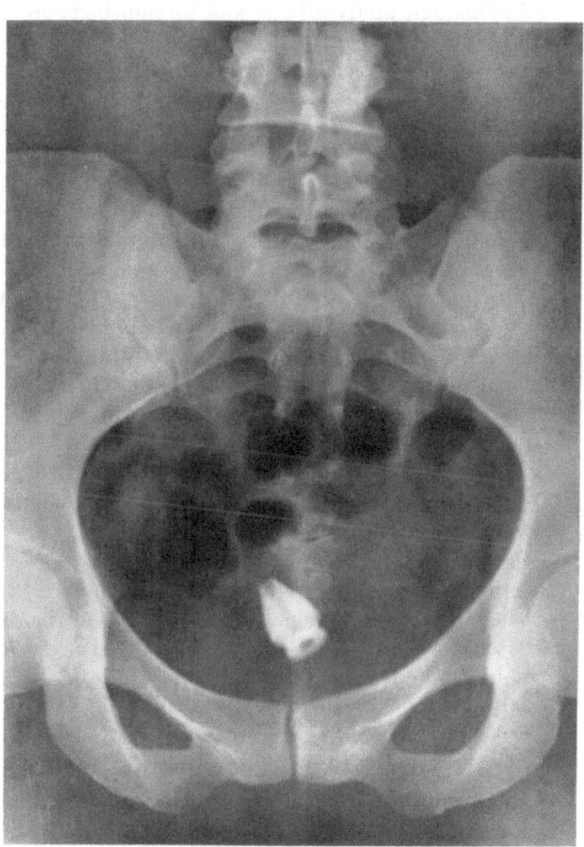

Fig. 14. Metal adaptor for a syringe in bladder

siderable quantity; when his bladder was distended he had suspended himself by his knees over a trapeze with his head down and he then urinated; the paraffin was passed spontaneously and measured $22 \times 9 \times 6$ mm; on later cystoscopy the bladder was normal.

In women, foreign bodies are not uncommonly passed along the urethra in mistake for the cervix uteri, in an attempt to procure abortion. Pieces of wood (commonly slippery elm), needles, wire or other pointed or blunt pieces of metal are frequently used. Often the foreign body lodges in the urethra whence it may be passed spontaneously or be easily removed. Such foreign bodies may remain partly in the urethra and partly in the bladder when a mushroom-shaped phosphatic stone ultimately forms, the stalk remaining in the urethra and the expanded, flattened, upper portion resting in the most dependent part of the bladder. The foreign body may penetrate the pelvic cellular tissues.

13*

The patient gives a history of dysuria, frequency or incontinence and a blood-stained discharge, or actual haematuria. In cases in which the foreign body still rests in the urethra, there may be a heap of granulation tissue around the urethral orifice. The foreign body in such cases can be palpated through the anterior vaginal wall. Cystoscopy often shows an intense cystitis. In some cases removal by the operating cystoscope is possible though the risk of injury to the urethra is not inconsiderable; more often suprapubic removal is necessary.

Tcherlock analysed cases of foreign body in the bladder in women between 1910 and 1929, 45 in all; most had been introduced in an attempt to procure abortion, while curiosity and masturbation accounted for most of the remainder. Charles and also Farncombe reported foreign bodies composed of slippery elm; McClinton recorded the case of a piece of wood in the bladder which had been introduced two years before. Brewer and Marcus reported the case of a woman who passed a fountain pen into the bladder, which gradually became surrounded by a calculus.

Rarely foreign bodies are self-introduced into the bladder by children. Thus Waller and Adney recorded the case of a female child, aged 5, with pain on micturition, who was found to have a stone in the bladder surrounding a bobby-pin; on cystoscopy the two ends of the pin could be seen sticking out of the stone and pressing against the bladder wall. Similarly in a boy of fifteen with a urinary tract infection, who admitted having introduced two long pieces of plastic tubing, the foreign bodies were removed suprapubically (Butters).

b) As a result of urological procedures

A part of a urological instrument may be broken off during the course of urethral instrumentation and may remain in the bladder; less commonly, the whole of such an instrument may be accidentally so retained. The end of a self-retaining catheter, in-dwelling in the urethra for a long time, may, if the rubber perished, become detached, and remain in the bladder when the rest of the catheter is withdrawn. Parts of a damaged ureteric catheter, bougie or lithotrite may be broken during instrumentation and be left in the bladder. A glass catheter may break during use and portions of it may be retained in the bladder.

2. By way of penetrating wounds of the bladder

Rifle bullets and shell fragments as a result of war injuries may lodge in the bladder after penetrating any part of the parietes of the abdomen, thighs or the perineum; portions of clothing or bone may enter the bladder at the same time as a metallic foreign body. As a result of a complicated fracture of the bony pelvis, especially the pubic rami or the symphysis pubis, sequestra may pass into the bladder cavity at the time of the original injury; such cases will be diagnosed either at the time of the wound toilet carried out soon after the injury, or following radiography and cystoscopy carried out later to elucidate the cause of dysuria, frequency and haematuria.

3. Following open surgical operations on the bladder

Sutures of silkworm gut, linen thread or other non-absorbable material used during an operation on the bladder, may eventually separate from the bladder wall, project into or actually become lodged in the bladder cavity and give rise to symptoms, or form the nucleus for a stone. Rubber tubes used for the purpose of open drainage of the bladder, or the end of a De Pezzer catheter, or needles used during an operation, may be left in the bladder. Radon seeds, the commonest foreign body to be deliberately left in the bladder wall, appear to remain there quite harmlessly and do not become encrusted with phosphate.

4. Migratory foreign bodies

A foreign body such as a bullet or shell fragment may come to rest in the pelvis near to the bladder and give rise at first to no urinary symptoms. Gradually such an object may move nearer to, and finally ulcerate into the bladder. Haematuria may result from the congestion of the vesical mucosa before ulceration has occurred, or it may occur when the foreign body finally penetrates the mucosa.

FULLERTON reported a case in which a fragment of shell eroded into the bladder twelve days after the injury. WALKER and KAUFMAN reported the penetration of a bullet into the bladder five years after the wound had been inflicted; when removed the bullet was not encrusted by phosphates. In a case recorded by BORS and BOWIE a shell fragment appeared in the bladder five months after it had entered the body. In LATTIMER's case the patient sustained an injury by a shell fragment which came to lie between the upper part of the rectum and the bladder, where it was located radiographically. Eight weeks later, urgency, tenesmus and urinary infection suggested that the piece of metal had penetrated the bladder, and, in fact, its presence was confirmed cystoscopically: it was removed through an operating cystoscope.

THOMSON-WALKER (reported by BORS and BOWIE) reported a case of Spencer Wells in which an artery forceps, which had been left behind during a pelvic operation, ultimately passed into the bladder.

Sequestra from the pelvic bones or even the femur, often resulting from a complicated injury, which have been deprived of their blood supply, may migrate through the pelvic cellular tissues and the bladder wall and finally reach the cavity of the bladder. In SADEK's case, a boy of 14, the separated upper femoral epiphysis ultimately passed into the bladder, causing haematuria and pyuria, and was removed suprapubically two and a half years after the original fracture.

Steel pins driven through the great trochanter into the femoral neck to repair a fracture, may gradually move inwards through the acetabulum into the pelvis, and months later may pierce the bladder. They then begin to cause symptoms of pain, dysuria, frequency and haematuria, and one end of the pin may be seen on cystoscopy to be penetrating the bladder; such migration may be unaccompanied by leakage of urine into the pelvis, though the patient may have an intermittent temperature. Such cases have been reported by RAVENSWAY and by BRANHAM and RICHEY.

Strips of fascia lata introduced into the pelvis as a sling around the female urethra near the bladder neck for the treatment of urinary incontinence, have been known to cut through the bladder wall insidiously, and ultimately be recognised cystoscopically in the bladder.

Gauze swabs or packs, left behind at operation in the pelvis, may ulcerate into the bladder cavity after having given rise to no symptoms for many months.

5. By way of the intestine

Foreign bodies, faeces or faecoliths from the intestine, may reach the bladder along a fistulous track resulting from a vesico-colic fistula complicating diverticulitis or malignant growth of the colon, vesico-intestinal fistula due to tuberculosis or Crohn's disease of the smal intestine, or a vesico-appendical fistula. A fistulous connection between the intestine and the bladder may develop as a result of the penetration of a sharp foreign body which is lodged in the intestine through the walls of both these viscera. Vesico-intestinal fistulae are not common and usually the debris which is discharged into the bladder, being small in amount, is voided in the urine. Rarely, solid matter entering the bladder may be recognisable (for example, fruit seeds, NITSCHKE), or may simply act as the nidus for a vesical calculus. An extreme example of foreign bodies swallowed by a patient, reaching the bladder through an ileo-vesical fistula is BOND's case, a man of 58, from whose bladder there were removed suprapubically, 90 nails (weighing 350 gm.), 20 gm. of cobbler's nails, some carpet tacks, one roofing nail, several pieces of glass, a piece of enamel from a tooth, a small animal bone and 2 three-inch screws.

Occasionally, a pointed foreign body which has been swallowed and which has passed down the gut, later penetrates the wall of the intestine at a certain point; if such penetration takes place into the peritoneal cavity, general peritonitis or a localised abscess follows; alternatively the penetration may take place into the bladder, the dome of which has become adherent to the loop of gut containing the foreign body. YOUNG reported the extraction of a needle from the bladder through an operating cystoscope 9 years after it had been swallowed. BARON and LIPSHUTZ reported the case of a man, aged 33, who, in order to arrest haemorrhage from a tooth socket, used the pressure from a pledget of cotton wool mounted on a long bristle, both of which were accidentally swallowed during sleep. Ten days later he had haematuria, and cystoscopy demonstrated the end of the bristle protruding into the bladder; the bristle was extracted by a Young's cystoscopic flexible biopsy forceps.

6. From pelvic dermoids, tumours, or teratomata

Teeth, pieces of bone or hair may reach the bladder from ovarian dermoid cysts or teratomata, which first become attached to, and later ulcerate into the bladder, discharging some of their contents. Foetal bones from an old unrecognised abdominal ectopic pregnancy may similarly be discharged into the bladder. The passage of such bodies per urethram, or pain, frequency and haematuria preceding their passage, may be the first symptom of the primary disease. Treatment will require the removal of the foreign bodies from the bladder, the resection of the primary lesion and repair of the bladder wall.

A case of a tooth in the bladder from a ruptured ovarian dermoid was recorded by GROSS and of hair from a similar source by KEYES.

FORSHAW reported the case of a woman, aged 38, who following short periods of dysuria, had passed small bones in the urine during a period of three years. She gave a history of an apparent pregnancy nine years before, and at term she had been seized with severe abdominal pain and uterine bleeding for one day only, and in the next three months the abdominal swelling gradually subsided. When she finally attended hospital, she was found to have a hard swelling deep to the anterior vaginal wall which was palpable bimanually, and when a metal sound was passed into the bladder a hard mass was detected. Following a diagnosis of an old ruptured ectopic pregnancy which had ruptured into the bladder, numerous small bones including portions of a mandible and of a skull and some small ribs, were removed from the bladder by the suprapubic route, the patient making a good recovery.

II. Symptoms

The foreign body, unless quite smooth such as a rifle bullet, usually becomes encrusted by several layers of calcium phosphate or of ammonium magnesium phosphate. Sometimes a foreign body remains in the bladder for prolonged periods giving rise to no symptoms. A foreign body may give rise to the clinical symptoms of vesical stone, namely pain and frequency of micturition and there may be intermittent haematuria; if it engages in the urethral opening there may be strangury or even retention or incontinence. It may be possible to palpate the foreign body per vaginam, per rectum or bimanually. Radiography may reveal at a glance the nature of the foreign body and show the extent of its incrustation; wax and liquid paraffin are non-opaque. A stone in the bladder of a woman, which is uncommon apart from those which are formed round foreign bodies, often gives a shadow with a central lacuna which represents the foreign body, which is commonly a piece of wood. Cystoscopy will confirm its existence and its nature.

III. Treatment

When possible, foreign bodies should be removed per urethram. This may be possible in women, especially if the foreign body has not completely passed from the urethra into the bladder. The surgeon may be able to seize the foreign body with a pair of forceps: portions of a catheter may be seized and extracted with a lithotrite. The loop of a hairpin may be engaged by a wire carrying a terminal hook passed down the operating channel of a cystoscope. A cysto-scopic rongeur forceps may be used to extract some foreign bodies; using this instrument one end of some elongated objects in the bladder can be grasped and removed as the rongeur is withdrawn.

The various methods which have been used to deal with foreign bodies made of paraffin wax, have already been referred to.

If the foreign body is pointed or encrusted with calculous material, or if a part of it has penetrated the bladder wall, it should be removed through a suprapubic cystotomy opening and the bladder closed by primary suture if there is no associated infection, a catheter being placed in the urethra. If there is a severe cystitis the bladder should be drained suprapubically for a few days, the appropriate antibiotic or sulphonamide being administered.

G. Priapism

Priapism is a condition of prolonged, persistent and usually painful erection of the penis, unaccompanied by sexual desire. The condition is uncommon. Extensive reviews of the subject can be found in the papers by HINMAN, and ABESHOUSE and TANKIN.

During erection of the penis there is an increased in-flow of blood and a partial occlusion of the venous outflow. The veins of the corpora cavernosa pass through the crus and the ischio-cavernosus muscle. The veins of the corpus spongiosum leave by way of the bulb through the bulbo-cavernosus muscle. The efferent venous outflow from the glans penis is by way of the dorsal vein of the penis. The muscles causing erection are supplied by the pudendal nerves and when they contract, the act of erection is completed and there is compression of the efferent veins which they surround. The venous pressure is greatly increased in the erectile tissues. In priapism, the obstructed blood becomes thick and grumous (pseudo-thrombosis) or there may be localised thrombosis in addition.

I. Pathology

While many causes, which are enumerated below, are occasionally responsible for priapism, the exact local pathology in the penis is not always easy to determine. An abnormal erection is the first step, but this must be accompanied by a change in the viscosity of the blood which leads to actual thrombosis in the sinuses of the erectile tissue of the corpus cavernosum penis. The principal findings at operation following incision of the penis are quite often only thick, grumous blood with no clots, though the presence of clots in the emergent veins of the penis could not be excluded in such cases; obvious clots escape from the incision in some cases. The mechanism of the production of the viscous grumous blood is still a matter for speculation and the various views are summarised in the paper by ABESHOUSE and TANKIN.

II. Classification

A classification of the causes of priapism, based on earlier papers by SCHEUER, HINMAN and others, is given by ABESHOUSE and TANKIN of which the following table is a slight modification:

1. *Neurogenic origin.*

   a) Ascending peripheral stimuli from diseases of the lower genito-urinary tract.
   b) Direct stimuli from the centre of the spinal cord or stimulation of the nervi erigentes or pudic nerve: spinal cord tumour, tabes, spina bifida, meningocele.
   c) Descending stimuli from the brain or spinal cord:
      α) Traumatic: fracture of the skull or vertebrae.
      β) Neoplastic: tumour of the brain or spinal cord.
      γ) Inflammatory: tabes dorsalis; disseminated sclerosis; meningitis.
      δ) Vascular: cerebral or subarachnoid haemorrhage.
      ε) Functional: psychoneurosis; sexual neurasthenia.

2. *Local mechanical origin.*

   a) Vascular:
      α) Phlebothrombosis, pseudo-thrombosis, or thrombophlebitis of the sinuses in the cavernous tissue or of the deep pelvic veins.
      β) Gangrene of the corpora cavernosa or of the penis.
   b) Traumatic:
      α) Penetrating wounds of the urethra or corpora; crush injury from falling astride; injury during coitus; transurethral prostatic resections.
      β) Fracture, dislocation, strangulation or torsion of penis.
      γ) Foreign body in the urethra.
   c) Neoplastic:
      α) Primary malignant tumour of the urethra, penis or corpora.
      β) Secondary malignant tumour of the corpora.
   d) Inflammatory:
      α) Primary infections or abscess in or about the penis, urethra or corpora.
      β) Secondary or metastatic infections from distant focus.

3. *Systemic origin*

   a) General diseases: Rheumatic fever, malaria, pneumonia, typhoid.
   b) Vascular: Leukaemia, sickle-cell anaemia, haemophilia
   c) Metabolic: Gout, diabetes.

4. *Idiopathic*

1. Priapism due to neurogenic causes

This variety is less common than the mechanical or combined variety. Possibly some cases are not recognised as having a nervous origin.

Some inflammatory conditions of the lower genito-urinary tract, such as urethritis, prostatitis, phimosis with venereal warts, and urethral polypi, are complicated by erections of short duration which are probably of reflex origin and which are associated with pain but not with sexual desire. Cases of priapism associated with urethritis have been reported by KIMBROUGH et al., and by NOE. Vesical calculi in children may be complicated by priapism (SCHEUER). Such erections may be transitory and of short duration but they have a tendency to recur, and in those respects they differ from true priapism in which the erection is persistent and prolonged. These symptoms, however, precede the onset of true priapism and may therefore predispose to that condition. More troublesome nocturnal erections associated with pain, but still of a temporary character, may complicate tabes and sexual neurasthenia. Cases of gonorrhoeal or non-specific urethritis may be complicated by prolonged erections which usually subside following the administration of sedatives and antibiotics, and which should not be confused with true priapism (ABESHOUSE and TANKIN).

Thirty-two of HINMAN's 170 cases of true priapism were attributed to descending nervous impulses from lesions of the brain and spinal cord, among which were a cerebellar tumour, a gunshot injury of the brain, birth injury resulting from forceps delivery, epilepsy, extra-dural haemorrhage, cerebro-spinal syphilis and nasal polypi. Fifteen of HINMAN's cases were associated with injuries, gunshot wounds or diseases of the spinal cord, such as myelitis or sarcoma; thirteen resulted from a fracture of the spine, 11 being in the cervical region, 1 at the level of the third lumbar vertebra and 1 at the level of the second dorsal vertebra. In cases resulting from fracture, the priapism came on at the time of the injury and usually subsided after a few days; in many cases the erections were not complete. OLLINGER reported a case of priapism associated with spinal tuberculosis complicated by paraplegia.

Direct stimulation of the erectile centre in the spinal cord may give rise to priapism, and tabes dorsalis is the most common cause of this type; cases have been recorded by FOURNIER, BOUCHER and D'ABUNDO. Localised lesions of the cord or of the vertebrae which occasionally cause priapism include spina bifida and meningocele, tumours of the cord or of the vertebrae. A lesion of the conus medullaris causing priapism was recorded by WILSON and MAUS. Priapism may be an early lesion in disseminated sclerosis (WILGUS and FELL, BERKEY and WHITE).

Death following hanging by the neck, or strangulation, results in priapism, the mechanism of which is not clear but it probably arises from the direct stimulation of the cervical cord. CALLOMON believed that the priapism in such cases results from the accumulation of carbon dioxide, resulting in irritation of the medullary vaso-dilator centre. The penis becomes suddenly and forcibly erect at the moment of strangulation and remains in that state for an hour or longer.

In some idiopathic cases of priapism, psychic factors are believed to play a part and close questioning of the patient may reveal such a factor which otherwise does not come to light (ABESHOUSE and TANKIN). SCHEUER recorded 9 cases of priapism in which a history of hysteria or sexual neurasthenia was obtained. These obscure causes should be borne in mind in otherwise unexplained cases of priapism.

2. Priapism due to local venous thrombosis

Thrombosis of the veins draining the corpora cavernosa may well be the most common cause of priapism and may be the initiating cause when that condition complicates general infective, toxic, traumatic and even nervous causes. Conversely, thrombosis of some of the deep pelvic veins is probably a common complication of diseases of the prostate, urethra and rectum, yet priapism only rarely occurs in association with these conditions.

The mechanism of the priapism is a thrombosis of the veins of the penis which interferes with venous return of blood. BAILEY states that the clotting probably begins behind the erector muscles in the veins which connect the venous plexus of the phallus with the pelvic venous plexus, in which are included the prostatic, vesical and haemorrhoidal plexuses. The corpora cavernosa are usually involved alone in the thrombosis. If the thrombosis is more widespread, the corpus spongiosum and the glans penis share in the erection; the thrombosis then involves the venae profundae which drain the bulb of the corpus spongiosum and open into that part of the pelvic venous plexus which communicates with the middle haemorrhoidal vein.

3. Priapism due to general disease and intoxications

Though priapism occasionally complicates certain systemic diseases, the precise cause is often difficult to determine and in many it is probably a local venous thrombosis, and is therefore mechanical in origin. In this group priapism has been recorded as a complication of tuberculosis, syphilis, typhoid (SCHEUER). malaria, pneumonia and rheumatic fever (IMBERT, PATEL). In syphilis, however, there may be a local cause in the form of a primary intra-urethral chancre (LOEHE, MASLOW).

Priapism has complicated gout, HINMAN reporting 4 cases, and also diabetes, as in the two cases reported by KLEHMET. Here again vascular causes were more than likely the immediate cause. Alcoholic intoxication has been responsible for priapism in several recorded cases, and here nervous as well as toxic factors have probably been responsible.

Four cases of HINMAN'S series resulted from intoxication, 3 following the self-administration of cantharides and 1 complicating diabetes. Other cases have been reported following poisoning by turpentine, muscarin and carbon monoxide (ABESHOUSE and TANKIN). FINKLER reported the development of priapism in a eunuchoid man following the administration of 100 mg. of testosterone propionate in four equally divided doses over a period of seven days; recovery followed cessation of the treatment and the condition did not recur after a second course of the treatment. A case of priapism complicating lead poisoning was described by EMODI.

4. Priapism due to primary or secondary malignant disease

Malignant priapism is caused by infiltration of the shaft or the base of the penis by a malignant tumour. Such tumours may be primary tumours of the penis, urethra, or corpora cavernosa, or metastatic from a malignant growth elsewhere in the body, usually in a neighouring viscus.

IKEDA et al collected from the literature 8 cases of primary malignant priapism. A further case was described by ABESHOUSE and TANKIN. The diagnosis between a primary and a secondary malignant tumour may be difficult even with repeated biopsis.

MAURER recorded the first case of primary malignant priapism caused by an angio-sarcoma or intra-vascular endothelioma of the penis associated with enlarged glands in both groins, which became ulcerated and gangrenous. COLMER reported a case of diffuse infiltration of the corpora cavernosa and the corpus spongiosum from malignant endothelioma; there were generalised metastases in the pelvis. CREITE reported a case of carcinoma of the penis of an atypical character in a child of two for which amputation of the penis was performed. ALLENBACH'S case was a cylindrical-celled carcinoma of the posterior urethra just below the bladder neck, a stony-hard tumour being palpable on rectal examination; there were metastases in the viscera. GOBBIE reported a case of perithelioma involving the corpora cavernosa in a boy of ten months. Other cases were recorded by FRONTZ and ALYEA (spindle-celled sarcoma), YAMAMOTO (endothelioma) and GAYET (epidermoid carcinoma). The ages of these patients varied from ten months to 58 years. The priapism was usually painful. Haematuria was observed in one case. Painful micturition and also urinary retention were occasional symptoms. Usually there was a complete erection though sometimes the penis was in a state of semi-erection.

Persistent erection of the clitoris in the female resulting from an infiltrating carcinoma of the cervix uteri was described by GUIBAL and PAVI.

Of 15 cases of secondary malignant priapism collected by IKEDA et al., in 6 the primary tumour was in the prostate (carcinoma 5, myxosarcoma 1), in 4 in the urinary bladder (carcinoma), in 2 in the testis (teratoma), in 2 in the rectum (carcinoma), and in one each in the liver (hepatoma) and in the kidney (hypernephroma). The ages of the patients ranged from 9 to 78 years, 1 being in the first decade, 3 in the fourth, 2 in the sixth, 6 in the seventh, 1 in the eighth, while in 2 the age was not stated. Sometimes there was complete replacement of the corpora cavernosa by growth, in others only partial replacement; in some cases there was involvement of both the corpora cavernosa and the corpus spongiosum. The initial symptoms were difficulty in micturition; in 3 cases there was haematuria and in 3 urinary retention; some cases had frequency of micturition. The onset of priapism usually occurred when the local disease was in an advanced state. ABESHOUSE and TANKIN found a further three cases of secondary malignant priapism in the literature.

5. Priapism due to diseases of the blood

a) Leukaemia

Leukaemia is one of the commonest causes of priapism, but priapism is an uncommon symptom of leukaemia. HINMAN found that leukaemia was the principal cause of priapism in 45 of 125 cases and considered that mechanical causes alone or combined with nervous causes were responsible; he believed that the abnormality of the blood may have initiated a reflex erection which was perpetuated by a subsequent thrombosis; in 70 per cent of these cases the priapism lasted for 20 to 60 days and in many the presence of thrombosis in the penis was confirmed at operation: in others, however, the attacks were of short duration and were probably not associated with thrombosis.

Priapism may be a manifestation in the acute and the chronic forms of myeloid and lymphatic leukaemia, especially the former. ACHARD recorded from the literature 50 cases in which priapism was the first sign of myelogenous leukaemia.

In cases of priapism associated with leukaemia, CRAVER states that leukaemic thrombi have been found in the corpora cavernosa, while in other cases the blood is merely thickened and there is no true clot. RUH and also WARTHIN also showed by autopsy studies that there was a thrombosis of the corpora cavernosa and in the dorsal veins of the penis in leukaemic cases with priapism. It has been suggested that the enlargement of the spleen may produce pressure on the abdominal veins or on the nervi erigentes, but yet in lymphoid leukaemia in which priapism also occurs occasionally, the spleen is seldom large enough to produce such an obstruction. Spontaneous remission of the priapism may occur.

b) Sickle-cell anaemia

Priapism may complicate sickle-cell anaemia, the first case having been reported by Diggs and Ching in 1934. Further cases have been recorded by Getzoff, Dawson, Rosokoff and Brodie, and Levant and Stept. According to Campbell and Cummins, the association with sickle-cell anaemia had been recorded in 16 cases to 1951; of 181 instances of sickle-cell anaemia admitted to their hospital during a ten-year period, priapism was the chief complaint in 5 cases. The age of the patients varied between 5 and 33, patients with this disease rarely living beyond the age of 40. Priapism may be a very early symptom but most patients have anaemia, fever, jaundice, vague pains in the abdomen and in the joints, and general weakness. The diagnosis of sickle-cell anaemia depends upon the discovery of sickled red blood cells and an associated anaemia. The mechanism of the thrombosis or pseudo-thrombosis of the penis in this condition is not, however, accurately known. Capillary engorgement of the tissues from the resistance encountered in the capillaries from the spiked and elongated red blood cells, was suggested as a cause of the priapism by Diggs and Ching. Getzoff suggests that the relative stasis of the blood is followed by a reduced oxygen tension and a temporary reduction of the pH of the blood, which results in a further release of those factors which influence sickling. Consequent upon these changes the red blood cells exhibit a greater tendency to rouleaux formation, which is followed by agglutination and thrombosis in the venous channels. Murphy and Shapiro found that the number of sickle cells in the circulating blood preceding the crisis was progressively raised.

c) Haemophilia

Guttman and Vorster (quoted by Scheuer) reported priapism in a case of haemophilia.

6. Priapism due to trauma

In Hinman's series, 7 cases of priapism were due to trauma. Many varied injuries, which include falling astride an object (Callomon, Defesche), blows on the perineum, gunshot injuries to the penis and ruptured urethra, may cause priapism, which is preceded by extravasation of blood into the erectile tissues, which is followed by clotting.

Foreign bodies within the urethra causing mucosal damage (Gutierriez) and strangulation of the penis (Dakin) may cause priapism.

Hughes quoted a traumatic case in which thrombophlebitis of the deep pelvic veins had resulted from a war injury in which a piece of cloth had been carried into the depth of the wound. Aaron and Robbins recorded a case of priapism in a man of 67 following perurethral resection of the prostate; the condition gradually subsided.

7. Idiopathic cases

A few cases have been recorded in which none of the above-mentioned causes for the priapism was found (Freedman; Rolnick, Cottrell and Lloyd; Lewis and Schwarez). The most likely cause for such cases was probably a limited thrombosis of the efferent or pelvic veins.

III. Age incidence

HINMAN, in a collected series of 170 cases of priapism, found 97 between the twentieth and the fiftieth years, 18 cases under the age of 30, and 14 cases over the age of 50. The youngest case, which occurred in a congenital syphilitic, presented soon after birth, while the oldest was in a man of 75.

In 34 cases collected by BAILEY, the condition was also found to occur at any age from the very young to the very old. Young patients usually suffered from sickle-cell anaemia or from leukaemia, whereas neoplasm was the most important cause in older patients. The decades from 20 to 40 showed the highest incidence though many cases were found in the age-group 40 to 70.

IV. Clinical picture

The cause of the priapism brings about very little variation in the clinical picture. In some cases there is a history of previous attacks of priapism of a few hours duration which have been followed by complete resolution. The onset of the main attack is usually sudden and often occurs during the night or after coitus.

The penis is rigid, tense and enlarged to three or four times its normal size. Because of venous congestion it is bluish in colour. It remains erect at an acute angle with the pubis. On examination the corpora cavernosa are the seat of a solid induration whereas the glans and the corpus spongiosum are not usually turgid. These differences can be determined by careful palpation with the finger and thumb of the component parts of the penis and of the urethra in the perineum. Sometimes one corpus cavernosum is more markedly involved than the other, resulting in deviation of the penis towards that side. In cases of priapism resulting from a fracture of the cervical spine HINMAN states that a complete erection follows manual stimulation of the glans, but that this gradually becomes modified, leaving the corpora alone erect; in other nervous cases the entire penis is involved.

In some cases pain, so severe as to require the repeated use of sedatives, is present, and it is increased by palpation of the penis or by pressure of the clothing. In priapism of nervous origin, the pain may be slight. No relief of pain can be obtained from coitus, even though ejaculation occurs.

When the corpus spongiosum is not involved (in 90 per cent of cases, according to DEFESCHE), dysuria or urinary retention are infrequent. In nearly half the reported cases, however, there is urinary frequency or difficulty and sometimes there is urinary retention, needing catheterisation.

The condition usually persists until surgical relief is given, though some cases subside spontaneously in a few days. According to HINMAN, in 85 per cent of cases of priapism due to nervous injury or disease, the duration of the condition was less than ten days. In serious cases of spinal injury the patient may die quickly, the local condition showing no improvement, while in other cases the condition has subsided gradually. In over 65 per cent of cases of priapism attributed to mechanical causes, the priapism lasted from 20 to 60 days, and in others it persisted for longer periods. In 80 per cent of the leukaemic cases the priapism lasted for long periods. Although priapism is not itself a fatal condition, it may occur during the late stages of leukaemia, or of injuries to the brain and spinal cord. Gangrene of the penis may occur if the priapism cannot be relieved by simple surgical measures. Urinary extravasation has been recorded, with death from septicaemia (BAILEY).

Impotence following the relief of the condition, has been reported by many observers. On the other hand, HINMAN states that inability to have erections following the subsidence of the priapism was noted in only 17 cases of 170 in his series, and since many of the cases had only been followed up for a few months, he thought that eventual recovery of function in a very high proportion seemed likely. In some cases recovery of sexual potency was partial, and in others which had been associated with thrombosis, the penis was distorted during erection. Recovery of normal erection is usual when the priapism is the result of peripheral reflex stimulation from lesions in the urinary tract, but unusual when it complicates a neoplastic or degenerative lesion of the brain or spinal cord. Impotence also commonly complicates priapism resulting from leukaemia.

V. Treatment

It is sometimes not possible to decide whether priapism is due to a lesion of the central nervous system or to a local condition, and in such cases, if a spinal anaesthetic is given, the priapism is relieved when the cauda equina is paralysed; neurogenic priapism almost never requires the use of any of the operative procedures discussed below.

1. Operative procedures

Division of the ischiocavernosus muscles is said by BAILEY never to have cured the condition. Similarly division of the pudendal nerves is referred to in the literature, but no reports of satisfactory results from such treatment have been found. Ligature of the dorsal arteries of the penis near the pubic symphysis at the base of the penis, has been suggested but it would appear to favour gangrene of the penis and is therefore unsound; there is no evidence of its having given relief.

a) Aspiration of the corpora cavernosa

MACKAY and COLSTON reported good results in 3 cases by aspirating stagnant, thick, grumous blood with a wide bore needle, from the shaft of the penis. The procedure was done under a local anaesthetic, each corpus cavernosum being aspirated in turn; following aspiration, saline solution was injected in an attempt to break up and remove blood clot. The second corpus cavernosum may be aspirated through the same needle puncture, by thrusting the needle through the septum. It is possible by this method to produce complete deflation, following which a bandage is applied. If there is recurrence of the priapism the aspiration must be repeated two or even three times. The procedure is most likely to be successful if it is used soon after the condition has developed, before extensive clots have formed. There have been some successes using this method but many failures. It is the method of treatment which should be tried first.

b) Incision

If repeated aspiration fails to give relief in two or three days, a short incision should be made using a fine-pointed knife, into both corpora cavernosa, in the centre of the shaft of the penis, the incision opening the fibrous sheath of the corpus cavernosum widely, and extending well into the spongy tissue, penetrating to the centre of the corpus and of the crus; the thick, grumous blood is expressed, and every effort made to extract clots, either by the use of forceps or by external pressure; irrigation with saline through the incision may help. The wound is left open to promote free drainage, or a small rubber drain is inserted; suturing of the wound is often complicated by haematoma formation. Infection is a possible complication but should be anticipated by the administration of antibiotics.

Lowsley and Gonzalez described a case of persistent priapism in which the corpora cavernosa were first exposed by dissection through a perineal incision, and were then incised, a tube being stitched into each corpus. There was rapid relief of the priapism but the condition recurred after some hours. The wounds were re-opened and the drainage tubes were re-inserted along with a catheter through which a solution of saline and heparin was irrigated. The condition gradually subsided.

c) Malignant priapism

Radical amputation of the penis is required for primary malignant priapism and a temporary suprapubic cystotomy is often desirable. For secondary malignant priapism a palliative amputation of the penis may be needed, since radiotherapy cannot be expected to produce a rapid improvement.

2. Other measures

Radiotherapy to the spleen in leukaemic cases, though desirable on its own account, cannot be expected to bring about rapid improvement in the priapism, for which the local measures described above should be used. Local radiotherapy to the penis has been recommended (Kaplan, Barney).

Riches described a case of a man of 49 in whom priapism which had lasted for 34 days, subsided gradually after rectal diathermy; the electrode was left in the rectum at first for ten minutes, the time being gradually increased to twenty minutes, and the rectal temperature being raised to 104⁰ F. The penis returned to normal after ten treatments given thrice-weekly; it was thought the condition was one of pseudo-thrombosis and that the diathermy reduced the viscosity of the blood thereby promoting a normal venous return.

Hinman records a case of persistent priapism in a man of 45 who had a positive Wassermann reaction, which responded to treatment by Salvarsan; the penis became flaccid but during the subsequent period of observation there was no return of normal erections. It was believed that the condition had resulted from a luetic lesion of the nervous system.

Treatment by dicoumarin or heparinisation may occasionally be effective in vascular cases. Smith reported a case of priapism successfully treated by dicoumarol, and a further case was recorded by Fraser. Brody, Lahr and Carroll used heparinisation for four days, followed by forcible massage of the penis under anaesthesia; there was rapid subsidence of the priapism and relief of pain.

H. Peyronie's disease

Peyronie's disease of the penis is a fibrous hyperplasia of the sheaths of the corpora cavernosa penis and was first described by François de la Peyronie, a French surgeon, in 1743. The condition is referred to in the literature under a variety of names: plastic induration of the penis, fibrous cavernositis, fibrous sclerosis, and fibrous plaque of the penis.

I. Aetiology

Lowsley states that two-thirds of the reported cases have occurred in men between the ages of 40 and 60 years; of 50 personal cases reported by him, the ages ranged from 27 to 69, the average age being 50 years; 72 per cent were in the fourth and fifth decades. Burford, Glenn and Burford, in a series of 71 cases, recorded 72 per cent in the age-group 50 to 70 and 92 per cent between 40 and 70. Some cases of plastic induration of the corpora cavernosa penis in young individuals are recorded in the literature, between the ages of 19 and 34 (Durant, Callomon, Rejka, Martenstein, Lowsley, Fricke and Varney).

The condition is probably much more common than the reported series would suggest. However, Polkey in 1929 collected 549 cases in the literature, to which Burford in 1940 collected 40 further cases and described 40 cases of his own.

II. Pathology

The cause of the disease is not known. The condition has been ascribed to syphilis, injury, gout, arterio-sclerosis and chronic inflammation of unknown aetiology. It has been attributed to slight trauma resulting from coitus, such as the rupture of small blood vessels in the corpora cavernosa, with the subsequent replacement of the haematoma by fibrous tissue. Later, deposition of calcium salts in the resulting plaque can occur. Fricke and Varney noted 2 cases in which the fibrous plaque had developed following radiotherapy for cancer of the rectum.

Some observers have noted a connection with Dupuytren's contracture; in Fogh-Andersen's 8 cases treated by operation, 1 patient had also been operated on for Dupuytren's contracture. The occasional association of Peyronie's disease with Dupuytren's contracture was also noted by Heite and Siebrecht who referred to 10 such cases. Waller and Dreese reported the association of the two conditions in 10 further patients between the ages of 48 and 67. No specific reason for the association is known. The condition has also been ascribed to lymphogranulomatosis inguinalis (May).

The condition affects the fibrous sheath of the corpora cavernosa either on their superficial aspects or in the septum between them. The corpus spongiosum is not affected. It most commonly begins at one side of the shaft of the penis forming a localised, well-defined, firm plaque which involves the fibrous sheath of the corpora, and the deep fascia (Buck's) of the penis. The plaque may extend for a considerable length along the shaft of the penis and in some cases seems to encircle it; sometimes more than one plaque is present. In the cases described by Lowsley and Boyce the plaques were of various sizes, being commonly 3 cm. in length by 1.25 cm. wide, while their thickness varied from 3 to 7 mm. In more than half the cases one or more plaques extended from the region of the coronal sulcus to the point of separation of the corpora cavernosa beneath the arch of the pubis. When multiple plaques were found, they were usually seen to be connected by some thickening of the intervening fascia. The condition is probably self-limiting, suggesting that there is some basic abnormality in the fibrous tissue in which it arises.

Calcific deposits may occur in the plaque in old or advanced cases; bone formation demonstrable radiographically has been reported (Fogh-Andersen). A deposit of calcium was present in 20 per cent of the cases of Lowsley and Boyce, bone formation in 2 cases and cartilage in one case. Malignancy and ulceration never complicate the condition.

Microscopically the plaque consists of dense fibrous tissue; there are bundles of collagen fibres, with fibroblasts between the bundles; blood vessels and elastic fibres are few. There is a complete absence of inflammatory cellular infiltration (KATZ-GALATZI). LOWSLEY and BOYCE state that around the small blood vessels are often found groups of spindle-like cells first described by MARCHAND which are believed to be undifferentiated mesenchymal cells still persisting in the adult. Hyaline degeneration is present in a majority of the cases. The microscopic appearance is similar to that of keloid or of Dupuytren's contracture (CALLOMON).

III. Clinical picture and diagnosis

By the time the patient has discovered the lump on the surface of the penis, he has usually found that erections are accompanied by discomfort or pain, and that the penis gradually becomes curved towards the side of the plaque, especially when erect, so that intercourse is rendered difficult or impossible. Many patients develop impotence. Often the patient becomes worried about his sexual difficulties and nervous or even mental symptoms arise. Urinary difficulty is exceptional. The initial lesion often develops rapidly for several months and then more slowly, or it may remain stationary. Most patients attend for treatment within a few months of the onset of symptoms. In LOWSLEY's series the average duration of symptoms was 23 months.

Physical examination reveals one or more flat, circumscribed, hard, fibrous plaques on the corpora cavernosa penis, most commonly on the dorsum, the skin overlying them being smooth and movable. In some cases the plaque extends from the coronal sulcus to the point at which the corpora cavernosa separate. Rarely the plaque is almost concealed by the pubic arch. Occasionally the shape of the penis is distorted. Sometimes the plaque lies on the ventral aspect of the penis between the septum and the corpus spongiosum urethrae. The proportion in which the various symptoms and signs are present can be seen by reference to the papers by SCARDINO and SCOTT, and LOWSLEY; not all the cases manifest all the symptoms.

Firm or hard periurethral indurations in the shaft of the penis occasionally complicate gonorrhoeal urethritis or old urethral strictures; since they are secondary to urethral inflammation they are closely related to the urethra, and hence are mainly in the corpus spongiosum, but the swelling may appear to extend into the corpus cavernosum. Instrumentation usually reveals some narrowing of the urethra which is never found in Peyronie's disease. Such indurations may slowly disappear spontaneously, or as a response to treatment with sulphonamides or antibiotics.

Carcinoma of the urethra begins in the lumen of the urethra and gives rise to haemorrhage and ultimately urinary obstruction. Gradually the mass extends into the corpus spongiosum and later to the corpora cavernosa.

IV. Treatment

In a few cases the fibrous plaques have disappeared spontaneously. Active treatment has given uncertain results and this fact accounts for the many and varied methods which have been tried. Local massage, diathermy, electrolysis and ultra-violet radiation, the local application of counter-irritant drugs such as iodine, mercury, camphor and iodoform, and injections of fibrolysin and of hyaluronidase have all been tried and are probably without effect.

The methods of treatment reported to have given a proportion of good results are radiotherapy, the administration of vitamin E, local injections with cortisone and surgical excision of the plaques with replacement by a free fat graft. As long ago as 1914 DREYER used, with success, heavily screened radium as a local treatment. The difficulties of treatment were seen in POLKEY's series of 419 cases; he found that only 76 patients were cured and 39 benefited by the various forms of treatment described.

1. Radiotherapy

FRICKE and VARNEY reported the results of radiotherapy in 141 patients treated at the Mayo Clinic from 1938 to 1943 inclusive. The average age of the patients was 53.2 years, the two youngest being 27 years old and the two oldest, 70. One treatment only was given to 86 patients, whereas 6 had four treatments each, the average number of treatments per patient being 1.5. Heavily filtered radium at a distance from the skin was applied over the entire shaft of the penis, which was secured in an extended position by a loop of adhesive tape applied to the sides of the penis for the period of the treatment. The scrotum was protected by a lead plate. The radium was left in position for twelve hours. Of the 112 cases treated, there was a poor result in 50 cases and a fair result in 20; the result was good in 28 cases and excellent in 14. The age of the patient and the duration of symptoms did not appear to influence the end-result.

BURFORD, GLENN and BURFORD treated 29 patients by a radium plaque, 13 receiving two or more treatments. The plaques, held in place by adhesive tape, were applied directly over the fibrous lesion, the testicles being protected by a lead plate. They used this method as a primary treatment and also in recurrent cases. 72.4 per cent of the cases were either cured or improved. FRICKE and OLDS considered that failure of some cases to respond to radium treatment has often been due to inadequate dosage. Some good results following radium treatment were also reported by SOILAND.

DAHL described the different radiotherapeutic techniques in use at Radium-hemmet, Stockholm. Some of the earlier methods of treatment resulted in extensive telangiectasis, atrophy of the skin with a tendency to ulceration, or slight symptoms of urethritis. The technique used for the treatment of more recent cases consisted of the surface application of radium tubes in glass capsules, the dosage being adapted to the tolerance of the tissues. Tubes containing 8 mg. of radium, placed in glass capsules, are applied, parallel to each other in one or two rows, to the surface of the penis. If the fibrous induration involves the deeper layers of the sheath of the corpus cavernosum, or if it is fairly extensive or is situated in the perineal part of the corpora cavernosa, deep X-ray treatment was used instead of radium; the details of dosage are given in DAHL's paper. In some cases in which the first method of treatment was unsatisfactory, supplementary treatment by radium or deep X-rays was given six months after the initial treatment. Lead protection must always be arranged for the testicles during treatment and the penis is fixed against the skin of the lower abdominal wall. Of 96 cases (79 of which were treated by radium, with deep X-rays as supplementary treatment, and 17 of which were treated by deep radiotherapy), 19 showed pronounced improvement, 34 showed moderate improvement and in 43 the results were poor.

CINIEWICZ reported the results in 64 patients treated by irradiation of the penis using X-rays in all but 3 cases. Of cases followed up there was no improvement in 15, slight improvement in 6 and in 21 there was marked or considerable improvement, including shrinkage of the plaque, disappearance of pain, and diminution of the curvature of the penis during erection.

PETERSEN reported the disappearance of symptoms in 9 of 27 cases treated by deep radiotherapy or radium, and improvement in a further 37 per cent. FUREY found radiotherapy useful in only 2 of 13 cases.

2. Vitamin E

Vitamin E has been used with varying succes. It is available either as a natural preparation (wheat-germ oil) containing α-, β-, γ-, and sometimes δ-tocopherol, while other preparations contain synthetic α-tocopherol.

Following reports by STEINBERG of some encouraging results in the treatment of Dupuytren's contracture by vitamin E, SCOTT and SCARDINO used this substance for the treatment of some cases of Peyronie's disease and reported some good results. These workers reported the treatment of 33 patients in two series of 23 and 10 respectively, treated with 300 mg. of mixed tocopherols or 200 mg. of synthetic α-tocopherols; no toxic manifestations were encountered. Their papers classify the various symptoms which responded to continued treatment in some of the cases. In an overall evaluation of the treatment of 27 cases which were followed up the results were good in 13, fair in 9, and no response in 5. In the cases with good and fair results, there was a disappearence or marked decrease in the penile plaques and of the curvature, as well as disappearance of pain on erection and a return to normal sexual intercourse. Other workers, KATZ-GALATZI, STEINBERG, and AURIG and SUSSE, have also recorded some successful results. NIKOLOWSKI reported improvement or complete healing in one-fourth of cases treated. Doses of 200 mg. daily for several weeks, followed by 50 mg. daily for some months, have been given (KATZ-GALATZI).

DAHL reported detailed results in 18 cases. Mixed tocopherols were used alone or almost exclusively in 9 cases; in the other 9 cases synthetic α-tocopherol was administered at first followed later by mixed tocopherols. A dose of 300 mg. daily by mouth was used. No adverse side-effects were noted though NIKO-LOWSKI considers that spermatogenesis may be impaired by vitamin E and advises repeated examinations of the seminal fluid. Treatment was continued for at least six months or for as long as improvement could be observed, which in one case was as long as seventeen months. In 4 cases there was pronounced improvement, in 6 moderate improvement and in 8 the result was poor. In 17 cases, penile curvature, which was present from the onset, disappeared under treatment in 2 cases and diminished in 4. Pain on erection which was present in 13 cases, disappeared in 4 and decreased in 3 cases. Pain was thus the symptom which responded to treatment most frequently. Penile plaques disappeared in 3 cases and improved in 7 cases. He found some evidence that vitamin E may give further improvement in cases which had been treated by irradiation. AURIG and SUSSE believe that treatment by radiotherapy is more effective following, rather than preceding, treatment by vitamin E; they suggest that the vitamin E increases the vascularity of the fibrous tissue and that this leads to a more favourable response to irradiation. The results may not be apparent until some months after the treatment has been concluded.

3. Cortisone

Cortisone given by local injection into the plaques was first used by TEASLEY who reported good results in a few cases.

BODNER, HOWARD and KAPLAN treated 17 patients, between the ages of 35 and 46 years, with cortisone injected locally. The patients were in three groups, one of which was treated with cortisone alone, another with cortisone and hyaluronidase and the third group with hydrocortisone. In one-third of the cases (6 of 17) the results were good, while in 9 of 17 the condition was improved. The proportion of cases improved in the three groups was approximately the same. The dose was 25 mg. (given with novocaine) of the selected drug once weekly for ten weeks. Those patients who received hyaluronidase received a dose of 150 turbidity reducing units. In the early injections it was difficult to inject the hard plaque but after two or three treatments, less pressure was required; it was found to be best to begin the injections at the lateral aspect of the plaque. Hyaluronidase, when given alone, was unsuccessful. Hydrocortisone was probably slightly more effective than cortisone. The symptom of pain was relieved before there was any visible improvement in the curvature of the penis.

FUREY treated 13 cases by injection of cortisone once or twice weekly into the plaques, under local anaesthesia of the whole penis. The treatment helped the curvature in 30 per cent and relieved the pain in 61 per cent of the cases. An average of 6 to 10 injections of 25 mg. each of either cortisone or hydrocortisone was necessary before there was much significant change, the best effects being noted after 15 to 20 injections. Extensive lesions and those in which there was calcification responded worst while those with a single non-calcified nodule responded best. He regards the method as safe and easy. A metal syringe specially adapted for the injection of fibrous plaques is recommended by RIBA.

4. Surgical treatment

LOWSLEY reported some good results following excision of the plaque and replacement by a fat graft.

The steps of LOWSLEY's operation are as follows: A rubber tourniquet is passed round the base of the penis. A longitudinal incision is made over the plaque, which is then dissected away. Sometimes the induration extends into the corpus cavernosum and care must then be taken to inflict the least possible injury. When the plaque extends to the ventral surface of the corpus cavernosum it may be necessary to free the corpus spongiosum urethrae in order to remove the fibrous tissue completely. A block of fat removed from the subcutaneous tissue of the lower abdominal wall is then sutured by fine catgut stitches to that part of the corpus cavernosum from which the plaques have been removed. There should be no approximation of the cut edges of the sheath of the corpora, indeed the fat must be sutured so as to keep them apart; if this is not done, a dense fibrous scar may almost reproduce the original condition. The penile fascia is carefully sutured over the graft. The tourniquet is released and any bleeding vessels are sealed. The skin incision is then closed by sutures. A Foley catheter is placed in the urethra and a compression bandage is applied to the penis. In order to prevent painful erections, stilboestrol in doses of 12 mg. daily is administered for some days before and some days after operation.

If the induration involves the ventral portion of the intercavernous septum, the initial incision may be made on the ventral surface of the penis in the middle line. The dissection is carried around the urethra to expose the fibrous plaque which is then dissected away. A fat graft is introduced and the urethra sutured over it. Some cases may require a dorsal as well as a ventral incision. Special precautions are needed to guard against infection from accidental displacement of the dressings. Antibiotics should be administered.

Of 17 patients in LOWSLEY'S series in which the plaque was excised without the use of a fat graft, 9 were cured and 4 were markedly improved. Of 33 patients in whom excision was combined with a fat graft, 20 were cured and 6 were markedly improved; 7 patients required a further operation because the formation of scar tissue had partially reproduced the deformity.

In all the 8 cases operated on by FOGH-ANDERSEN, the graft took without complications and in the majority of patients the penis returned almost to normal after some months. Follow-up of these patients to three and a half years after operation revealed a satisfactory result in the majority of cases, the best effects being observed in younger patients.

J. Hydrocele

The term hydrocele refers to a collection of serous or other fluid in the tunica vaginalis or in a part of the processus vaginalis which is still patent. The term hydrocele of the epididymis is used by some writers to describe cysts of the epididymis or spermatocele, which are here described in a separate article.

I. Classification

1. By anatomy

a) Primary hydrocele of the tunica vaginalis

α) Idiopathic hydrocele of the tunica vaginalis

The tunica vaginalis is distended with fluid. Rarely, part of the tunica vaginalis is obliterated by adhesions between the visceral and the parietal parts of the sac resulting in a localised or encysted hydrocele of the tunica vaginalis.

β) Congenital hydrocele

The tunica vaginalis is distended with fluid, and in addition the processus vaginalis is patent, communicating with the peritoneal cavity by a small opening which is large enough to allow fluid to pass between the peritoneal cavity and the tunica, but not large enough to allow omentum and intestine to herniate downwards.

γ) Infantile hydrocele

The tunica vaginalis, which is distended with fluid, has an upward prolongation into the spermatic cord, which is part of the funicular process; there is no communication with the peritoneal cavity.

δ) Hydrocele associated with incomplete descent of the testis

This is similar to primary hydrocele except that it occurs in association with mal-descent of the testis, and it calls for no special description.

ε) Bilocular hydrocele

The distended tunica vaginalis is connected by an isthmus in the upper part of the scrotum or in the inguinal canal with a second abnormal loculus.

b) Hydrocele of the testis

This term is applied to a collection of fluid which may form between layers of the tunica albuginea testis; this is excessively rare and calls for no special description.

c) Hydrocele of the spermatic cord
α) Encysted hydrocele of the cord

This is a cystic distension of an unobliterated part of the processus vaginalis, the resulting swelling usually occupying the central portion of the spermatic cord at some distance above the testis.

β) Diffuse hydrocele of the cord

This condition, the name of which is misleading, is really a lymphangiectasis or oedema of the spermatic cord.

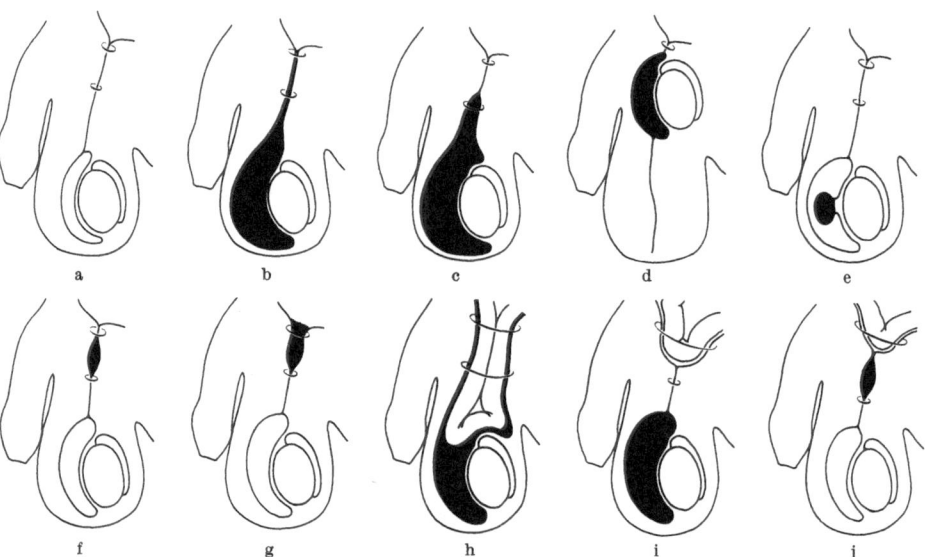

Fig. 15a—j. Drawings to represent the various types of hydrocele [after M. F. Campbell, Surg. Gynec. Obstet. **45**, 193 (1927), by courtesy of the Editors]. a Normal arrangements of the tunica vaginalis. b Congenital hydrocele with patent processus vaginalis. c Infantile hydrocele. d Hydrocele in association with incomplete descent of testis. e Hydrocele of the testis. f Encysted hydrocele of the cord. g Hydrocele of a hernial sac. h Congenital hydrocele with hernia. i Hydrocele of the tunica vaginalis with an associated hernia. j Encysted hydrocele of the cord with an associated hernia

d) Hydrocele of a hernial sac

Here there is a clear effusion into an inguinal hernial sac, which no longer admits any of the abdominal contents, the neck of the sac having been obliterated.

e) Hydrocele of the canal of Nuck

This is a hydrocele forming in the female in an unobliterated part of the processus vaginalis or canal of Nuck.

2. By causation
a) Primary
α) Acute

Trauma to the tunica vaginalis occasionally gives rise to an acute primary hydrocele.

An acute primary pneumococcal vaginalitis with effusion has been referred to by de Quervain. In congenital hydroceles, acute infection and also tuberculosis may theoretically spread by direct extension to the tunica vaginalis from acute or tuberculous peritonitis respectively; such cases are excessively rare.

β) Chronic

This is the common idiopathic hydrocele of the tunica vaginalis.

b) Secondary
α) Acute

A secondary hydrocele may accompany acute infections of the testis and the epididymis, resulting from gonorrhoea, or from blood-borne infections such as typhoid, pneumonia, secondary syphilis or mumps; it may complicate torsion of the testis or its appendages.

β) Chronic

This may complicate the more slowly developing inflammatory or neoplastic diseases of the testis or epididymis. It is not uncommon in association with gummatous orchitis, which itself is now rare and it occasionally complicates tuberculous or non-specific epididymitis. A chronic hydrocele occurring weeks or months after operations for inguinal hernia or varicocele may also be called a secondary hydrocele.

3. By contents

The common idiopathic hydrocele usually contains clear fluid.

a) Chylous hydrocele

In this condition the tunica vaginalis is filled with milky or chylous fluid and is a complication of filariasis caused by the filaria sanguinis hominis (Bancrofti).

b) Meconium hydrocele

This occurs rarely in new-born infants.

c) Pyocele

The tunica vaginalis is distended with pus.

II. Age incidence

The following tables give the age incidence of hydrocele.

CAMPBELL Age of patients presenting with hydrocele 456 cases		OBNEY Primary hydrocele 791 cases	
Age	Number	Age	Number
Under 6	12	0— 9	23
6—14	2	10—19	35
15—19	35	20—29	51
20—29	126	30—39	107
30—39	74	40—49	134
40—49	89	50—59	184
50—59	79	60—69	171
60—69	32	70—79	80
70—79	6	80 and over	6
81	1		791
Not recorded	3		
	456		

CAMPBELL found that bilateral hydroceles were present in 4 per cent. of cases.

III. Pathology

1. Causation

No cause in known for the idiopathic chronic hydrocele and patients rarely volunteer any point in their past medical history which throws light upon its origin. Earlier writers sought to incriminate previous attacks of epididymo-orchitis resulting, for example, from gonorrhoea (CAMPBELL), though hydrocele still remains a common condition while gonorrhoea is much less common now then formerly.

Rarely there is a history of trauma though it is often vague and inconclusive and only applies to a small proportion of the cases. CAMPBELL, however, found a history of an actual blow to the testis in 34 of 502 cases.

Another theory is that a hydrocele is caused by a passive effusion into the tunica vaginalis from some unknown cause. The presence of an excess of fluid in the tunica vaginalis suggests that there may be some abnormality of the absorptive mechanism in the wall of the sac. The parietal part of the tunica vaginalis has a lymphatic plexus comparable to the subserous lymphatic plexuses of the diaphragm and of the intercostal pleura and which drains into the same regional lymph glands as the lymphatics of the body of the testis (ALLEN); possibly some derangement of this lymphatic field is responsible for the accumulation of fluid in excess of the small normal requirements. That the lining of the tunica does normally possess absorptive properties is shown by the experiments of CECA, who reported that iodide appeared in the urine 15 to 19 minutes after it had been injected into the tunica vaginalis. TORRACA similarly reported that phenolsulphon-phthalein appeared in the urine of the dog in 14 to 23 minutes after injection, though in that animal there is a

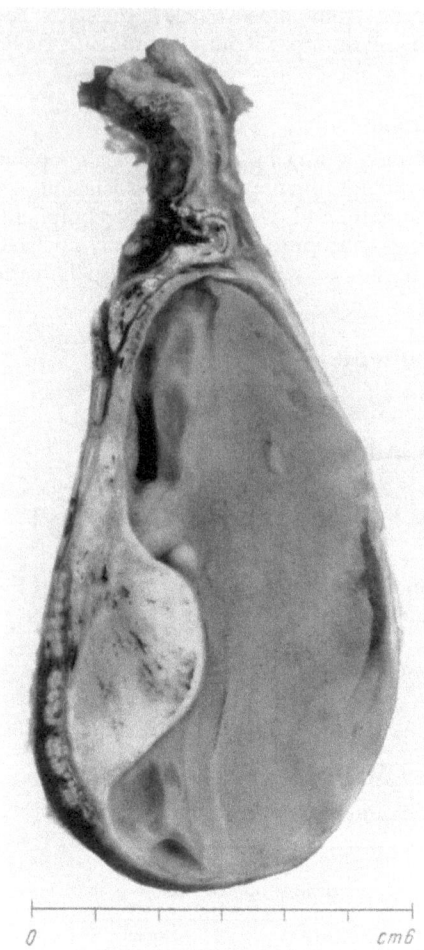

Fig. 16. Hydrocele of the tunica vaginalis. The wall of the tunica is slightly thickened, and its dilated cavity is filled with opalescent gelatinous material—the hydrocele fluid coagulated by the formalin fixative. The testis is slightly distorted by the pressure of the fluid and is distinctly atrophic. The big nodule projecting into the cavity above the testis is the upper end of the epididymis; the smaller nodule below it is a hydatid of Morgagni.
Incidental post-mortem finding in a man of 35

free communication between the tunica vaginalis and the peritoneum. HUGGINS and ENTZ found that phenolsulphonphthalein injected into the tunica vaginalis of patients was excreted in the urine in 47 minutes; similarly MONARCI injected 10 per cent potassium iodide into 4 patients with idiopathic hydrocele and found that some of the iodide appeared in the urine five to ten days later.

2. Pathological anatomy

In the common chronic idiopathic hydrocele the walls of the hydrocele sac, which are merely the distended walls of the tunica vaginalis, consist of fibrous tissue lined by a layer of endothelium. In many recent hydroceles and even in some of long standing, the coverings are thin from pressure of the fluid. If the sac is injected with paraffin, long finger-like processes may sometimes be seen extending into the connective tissue of the scrotum (THOMSON-WALKER); sometimes the cavity is partly subdivided by septa into compartments. In old-standing hydroceles, especially in those which have been repeatedly tapped, the coverings are thick and leathery and the inner surface of the sac may be covered with fibrinous plaques which occasionally undergo calcification; sometimes the walls may be as much as half-an-inch in thickness; these thickened sacs are often as hard and as dense as cartilage. Occasionally villous or fibrous processes are found projecting from the walls; small fibrous or cartilaginous bodies may lie free within the cavity.

The contained fluid is clear and pale yellow and contains 4 to 6 per cent of albumin, and a trace of glucose (GREENE). In old-standing cases the fluid contains a shimmering cloud or even a heavy deposit of cholesterin crystals, and microscopically endothelial cells and a few leucocytes are found. The fluid contains fibrinogen and therefore coagulates on the addition of blood which contains fibrin ferment. Several ounces of fluid are commonly present, but giant hydroceles occasionally occur. THOMSON-WALKER reported one case of a hydrocele of thirty-two years' standing from which 4 quarts of fluid were withdrawn, a fortnight later 3 quarts and six weeks later 6 quarts; MAISONNEUVE recorded a case of a hydrocele containing 21 litres of fluid.

In old hydroceles pressure atrophy of the testis may occur, the tunica albuginea being considerably thickened. In some cases the epididymis is lifted away from the testis because of the pressure of fluid in the digital fossa. The vasa efferentia are thinned and possibly there may be some interference with the normal discharge of spermatozoa. The spermatic cord may be thickened from hypertrophy of the cremaster.

IV. Primary hydrocele of the tunica vaginalis

1. Clinical picture

a) Symptoms and signs

The condition usually arises as a slowly increasing painless distension of the tunica vaginalis though a few have an acute onset following trauma. The patient may experience a dragging sensation as the swelling increases in size. In large hydroceles, the skin covering the penis may be dragged forwards so that the penis is lost in the sac, causing some inconvenience in micturition.

The swelling, which is confined to the scrotum, is ovoid, round or pear-shaped, not tender, smooth and its upper limit can easily be determined. Fluctuation can be elicited except in the very tensest swellings. There is no impulse on coughing. The swelling is translucent to light except in cases with very thick or calcified walls or in those which have been the seat of haemorrhage from previous tappings; translucency may, however, be present in the hernia of a child if it contains a coil of distended intestine. The position of the body of the testis can usually be determined by the difference in its consistency, and by the eliciting of testicular tenderness, in the lower posterior part of the sac; transillumination may assist in locating the position of the testis. Occasional cases of inversion of the testis must be noted. Occasionally a hernial protrusion in the wall of the sac exists in large hydroceles; there is then a round opening which will admit a finger, bounded by a hard ring of fibrous tissue, through which the serous lining of the tunica prolapses, resulting in the production of a secondary swelling beneath the skin of the scrotum, which is reducible on pressure; WALKER records a case in which the testicle was able to prolapse through such an opening, though it could easily be replaced into the main sac, and pushed out again by manipulation.

It is the common experience that inguinal hernia and hydrocele are frequently found together. OBNEY refers to this association and found that in 14,442 cases of inguinal herniae, there were 730 or 5.05 per cent, which had an associated hydrocele, an association which was especially noticeable in the older age groups. A large hydrocele may extend into the upper part of the scrotum as far as the external ring, making difficult the diagnosis of a small associated hernia.

b) Complications

Complications are not commonly found. Local trauma, as in attempts at tapping, may result in haemorrhage into the distended tunica vaginalis, converting the hydrocele into a haematocele; it is then no longer possible to transilluminate the swelling. Haemorrhage occasionally occurs spontaneously.

Traumatic rupture of a hydrocele of the tunica vaginalis occurs occasionally. The patient experiences a sensation of something giving way and later there is the appearance of oedema of the scrotum and penis; later the perineum, the opposite side of the scrotum and the lower abdominal wall are involved in the extravasation. There is considerable scrotal tenderness and ecchymosis and a considerable haematoma may accumulate. In the absence of treatment, the extravasated fluid will be gradually absorbed after three or four weeks and the hydrocele will refill; during this time there may be slight temperature. Occasionally gangrene of the skin is an additional complication (HINMAN). Rarely, spontaneous rupture of the sac may occur with the same sequence of symptoms. WOLF and GOMILA recorded such a case in a hypertensive male aged 62, in a long-standing hydrocele sac which had been tapped twelve months previously, the man presenting with an extensive scrotal haematoma and widespread ecchymosis in the penis, pre-pubic area and the upper parts of both thighs, which were shown at operation to be the result of a tear in the wall of the hydrocele sac.

Suppuration in a hydrocele sac rarely occurs unless infection is introduced from without, as from external injury from tapping, or from other causes.

2. Diagnosis

An idiopathic hydrocele must be differentiated from an inguinal hernia, haematocele, cysts of the epididymis and solid swellings of the body of the testis and the epididymis (malignant neoplasms, gummata and epididymitis).

An inguinal hernia descends from the inguinal canal into the scrotum, is associated with an impulse on coughing, is usually reducible and the upper pole of the hernial sac cannot be palpated since its walls are continuous with the peritoneum; in an irreducible hernia, the normal body of the testis can be distinguished by palpation from the hernial sac.

A haematocele is a firm, boggy or even hard swelling in the tunica vaginalis, cannot be transilluminated, and imparts a sensation of weight to the palpating hand. If there has been recent haemorrhage into the tunica the scrotal skin may be discoloured. If the swelling is explored by a needle, blood may be withdrawn. A long-standing chronic haematocele may be smooth or nodular, is heavy, does not transilluminate, and the skin may be attached to the underlying haematoma.

A cyst of the epididymis (or spermatocele) is a cystic swelling unilocular or multilocular, located in the upper pole of the epididymis and close to the body of the testis but quite separate from it. In the small and medium sized cysts of the epididymis, no difficulty should arise in diagnosis but in the larger ones, which may be the size of an orange, the cyst becomes very closely applied to the upper pole of the body of the testis and the tunica vaginalis, and may mimic a hydrocele of the tunica vaginalis; it is always possible, however, to recognise a groove between the body of the testis below and the cyst above. The fluid aspirated from a cyst of the epididymis may contain spermatozoa.

Solid swellings of the testis and the epididymis should be readily differentiated from hydrocele by the absence of fluctuation and transillumination and by palpation of the epididymis and the body of the testis separately.

3. Treatment

A hydrocele may be treated by simple tapping, by aspiration of fluid contents followed by the injection of sclerosing fluids, or by radical operation. Rarely, patients who refuse treatment of any kind use a scrotal support to counter the dragging discomfort.

a) Tapping

The tapping of a hydrocele can only be regarded as a palliative procedure except perhaps occasionally in infants, when a single aspiration of the sac may not be followed by recurrence. Acupuncture of the sac at several places in such patients to allow the fluid to drain into the scrotal cellular tissues, was reported by WELLS.

Tapping is used for the treatment of patients for whom an operation is inconvenient, in very old men, in whom it may have to be repeated at intervals, and for patients who refuse radical surgical treatment; it is also used for diagnostic purposes in secondary hydroceles to enable the body of the testis and the epididymis to be accurately palpated.

Tapping may be carried out with a fine trocar and cannula, or the fluid may be aspirated with a syringe mounting a thickish needle. Before tapping or aspiration, the surgeon must satisfy himself either by palpation or transillumination, of the position of the body of the testis and epididymis, in order to avoid injury to those structures; the rare inversion of the testis, when that organ lies in relation to the anterior instead of the posterior wall of the tunica vaginalis, must be borne in mind. In order to tap the hydrocele with a trocar and cannula, the sac is rendered tense by pressure of the surgeon's left hand, an avascular portion of the skin of the scrotum is selected on the anterior surface some distance from the body of the testis and a small amount of local anaesthetic is infiltrated into the skin at that point. With the tunica still rendered tense, the trocar and cannula are driven obliquely through the skin and coverings into the hydrocele sac and the fluid is allowed to escape. The puncture is sealed with a tiny dressing of cotton wool and collodion. In aspirating a hydrocele the point through which the aspiration is to be done is selected as described above, the needle being introduced obliquely and in an upward direction so as not to injure the tunica albuginea and the fluid is then withdrawn.

b) Injection of sclerosing fluids

This is advocated as a wholly satisfactory method of treatment by some surgeons who claim a high percentage of good results; the method may be used for patients not desirous of having operative treatment.

Historically, it is of interest that Celsus in Roman times injected saltpetre into hydrocele sacs (FOOTE), and in the eighteenth century, spirits of wine was similarly used by SHARP, and port wine and decoction of red rose leaves by EARLE (WILSON). The practice of using port wine for injection persisted up to the time of SYME of Edinburgh (PATERSON). The later use of iodine and of carbolic was often followed by a severe inflammatory reaction, and even sloughing of the scrotal tissues and of the testicle; loculation of the hydrocele sac was a common end-result (ROSS). WHITBY reported successful treatment by injection using 3 minims of iodised phenol (4 parts of iodine and 2 parts of phenol) for every ounce of fluid withdrawn from the sac; the scrotum was massaged for five minutes after injection but the patient usually experienced a burning sensation for a few minutes.

α) The solution

The ideal sclerosing solution, which should be bactericidal, must be painless after injection, must not cause general toxic reactions nor give rise to haemorrhage nor to too fierce a local reaction, and it must modify or possibly destroy the endothelial lining of the tunica vaginalis if recurrence is to be avoided.

The sclerosing solutions at present in use are quinine urethane and sodium morrhuate. A solution of quinine urethane (quinine hydrochloride 4 gms., urethane 2 gms., water to 100 cc.) is usually injected in amounts of 10 cc. at a time, though in large chronic hydroceles up to 15 cc. may be used without any ill-effects (ROSS). DIAMOND used a solution of 2 per cent quinine hydrochloride and urethane and 30 per cent lithium salicylate containing 1 per cent tutocaine. Alternative fluids are 5 mls. of a 5 per cent solution of sodium morrhuate, or 5 mls. of ethamoline.

β) Contra-indications

Treatment by injection may be used for the common idiopathic hydrocele but it should not be used for the congenital hydrocele in children, encysted hydrocele of the cord, large bilocular hydroceles, hydrocele associated with hernia, nor secondary hydroceles in which treatment should be directed to the primary cause. Even very large sacs may respond, though two or three injections may be needed as in one case reported by JAMES in which the sac measured $10^3/_4$ inches in length and contained 64 ounces of fluid.

γ) Technique

The fluid in the sac is first removed as completely as possible in order to avoid dilution of the sclerosing agent, using either a fine trocar and cannula or a syringe and needle. The sclerosing fluid is then injected into the tunica vaginalis, gentle massage of the scrotum for a few minutes ensuring its diffusion over the whole of the surface of the cavity. The patient should rest for some hours following the injection and should wear a scrotal support for two or three weeks. There is commonly some outpouring of reactionary fluid during the next two or three weeks, which may be absorbed spontaneously, or may need a further single aspiration. RHIND found that in many cases the reactionary fluid which accumulated following aspiration and injection, reached a maximum after two weeks and was then gradually absorbed during two months. One cause of failure following injection appears to be the reaccumulation of fluid in the upper part of the sac, the lower part being thickened and empty; this may be an effect of gravity or failure to massage the sclerosing fluid after injection. In some cases the treatment has to be repeated on more than one occasion at intervals until obliteration of the sac has occurred. The fluid at a second aspiration is usually turbid. A second injection should never be given if there are signs of inflammation in the scrotal contents.

Following injection the tunica vaginalis contains a small amount of dark amber fluid and some long loose strands of organised fibrin attached to the walls of the sac; the endothelium of the tunica may be still intact and the blood vessels normal, but the subserous layer becomes thickened and infiltrated with organising fibrous tissue (EWELL, SARGENT and MARQUARDT).

δ) Results of injection

MILBERT reported good results up to fourteen months after injection in 19 cases of hydrocele and 7 of spermatocele; multiple injections were given in some cases. EWELL et al. reported 165 cases of hydrocele treated by injection with quinine urethane, with only 7 failures or recurrences (4.3 per cent). SMITH (quoted by EWELL et al.) had a recurrence rate of 6 per cent. ROBERTSON treated with good results 7 patients with large hydroceles using sodium morrhuate. DIAMOND reported the successful treatment in all of 76 patients with hydrocele, by one twin-injection of lithium salicylate and quinine hydrochloride and urethane. FOOTE reported excellent results in 95 per cent of cases using quinine urethane alone or with lithocaine, and NICHOLSON reported success in all of 40 cases which were followed up after treatment with quinine urethane; in 6 cases two, and in 4 cases three injections were needed. WILSON reported 18 cases (12 hydroceles and 6 spermatoceles) treated by injection; in 16, one injection sufficed, in 1 case two, and in another three injections were needed; in 15 patients followed up between 8 and 20 months after injection, the results were excellent while in one case there was a re-accumulation of fluid in spite of three injections. Ross reported the results of 40 cases followed up in a series of 45 treated by quinine hydrochloride; there were 2 recurrences, one being in an old man of 76 with huge bilateral hydroceles. RHIND reviewed 128 cases of hydrocele and spermatocele treated by injection; 82 were considered satisfactory in that the tunica vaginalis was entirely free from fluid or contained only a small quantity, while 20 cases were assessed as failures.

c) Operative treatment
α) Excision of the parietal layer of the tunica vaginalis

The best operation is excision of the parietal layer of the tunica vaginalis. The alternative operation of eversion of the hydrocele sac gives a greater percentage of recurrences and is therefore not recommended.

In order to excise the parietal layer of the sac, an incision is made starting at the external abdominal ring and carried downwards through the skin of the upper part of the scrotum. Some surgeons prefer to make the incision entirely over the hydrocele. The incision is deepened through the coverings of the tunica vaginalis until the blue wall of the sac is demonstrated. Pressure over the hydrocele sac by the hands of the assistant helps in the demonstration of the successive coverings of the sac. The sac itself is then freed from the overlying coverings over as large an area as possible, using blunt dissection until the visceral layer of the tunica covering the epididymis and the body of the testis is reached, when the parietal layer is cut away, the fluid being aspirated first or allowed to escape. There is usually some bleeding from medium-sized vessels, or some oozing from the cut edge of the tunica, all of which must be carefully controlled. Large vessels may be ligatured and oozing points in the edge sealed by diathermy. An alternative method is to insert a continuous catgut stitch through the cut edge of the tunica, which must be drawn taut to secure the complete haemostasis. The outer coverings of the hydrocele sac are drawn together by catgut stitches and the wound closed, leaving a drainage tube in the cavity.

Following operation there is usually a little oedema and local thickening which may persist for three or four weeks after the wound has healed. The patient should wear a suspensory bandage during early convalescence.

β) Inversion of the hydrocele sac

In the operation of simple eversion of the sac, a similar incision is used to that described above and the sac freed. The tunica vaginalis is opened and the sac is turned inside out, its cut edges being sutured behind the testicle and the lowest part of the cord.

γ) Complications of operation

Haematoma formation has already been referred to. If a large haematoma forms, it may be necessary to re-open the wound and evacuate the clot.

Infection may result from imperfect technique and if this occurs there may be some sloughing of the thin skin of the scrotum. Atrophy of the testicle has been reported but probably the blood supply of the testis has then been interfered with.

Recurrence of the hydrocele rarely follows a complete excision of the parietal layer of the sac, but if pockets are left unexcised, there may be recurrence from a new endothelialisation of the scrotal pocket. YOUNG (quoting SANFORD) gives a recurrence rate of 5 per cent. There is a significant recurrence rate following operation for those hydroceles which are associated with an inguinal hernia. Thus OBNEY found 44 recurrences (6.0 per cent) of 730 hydrocele operations done at the time of a hernia operation, but no recurrence for operation for uncomplicated hydrocele; the possible causes of such recurrences are the stripping of the spermatic cord during the repair of the hernia, thus interfering with the lymphatics and their absorptive mechanism, and too tight suturing of the external abdominal ring, thus interfering with the venous return.

V. Congenital hydrocele

Congenital hydroceles of infants, which are not necessarily present at birth, give rise to a cystic, pyriform, translucent swelling, which can be reduced by pressure with the fingers, or will slowly disappear spontaneously when the child is recumbent and slowly reappear when the child resumes the erect position. An impulse on coughing may be detected. Since the processus vaginalis is patent, a hernia may later develop.

Spontaneous cure may take place in the neonatal period, or soon after but if this does not occur the sac should be excised and the processus vaginalis closed, when the child is a little older.

VI. Infantile hydrocele

This condition may be bilateral and may be seen in neonatal children or in infants. There is a cystic scrotal swelling which is translucent to light; the extension along the cord to the peritoneum may be palpable. There is no impulse on coughing.

If the swelling does not disappear spontaneously in infancy, it should be treated surgically, excising the parietal part of the tunica vaginalis and the upward extension of the processus vaginalis. Aspiration of the fluid is usually followed by recurrence.

VII. Bilocular hydrocele

This condition is also alternatively called properitoneal or interstitial hydrocele. The hydrocele sac consists of two chambers, one of which lies in the scrotum around the testis, as in idiopathic hydrocele, and the other is directed upwards towards the abdomen; there is a neck or isthmus which separates the two loculi. It is believed that the abdominal loculus arises from distension from an unobliterated portion of the processus vaginalis, which increases in size after the initial distension of the tunica vaginalis. At operation a short fibrous cord may be found, extending from a point on the wall of the abdominal portion, to the peritoneum. Such hydroceles are usually associated with incomplete descent of the testicle and are usually on the right side. The distended upper loculus may pass either in front of or behind the peritoneum, the latter being the

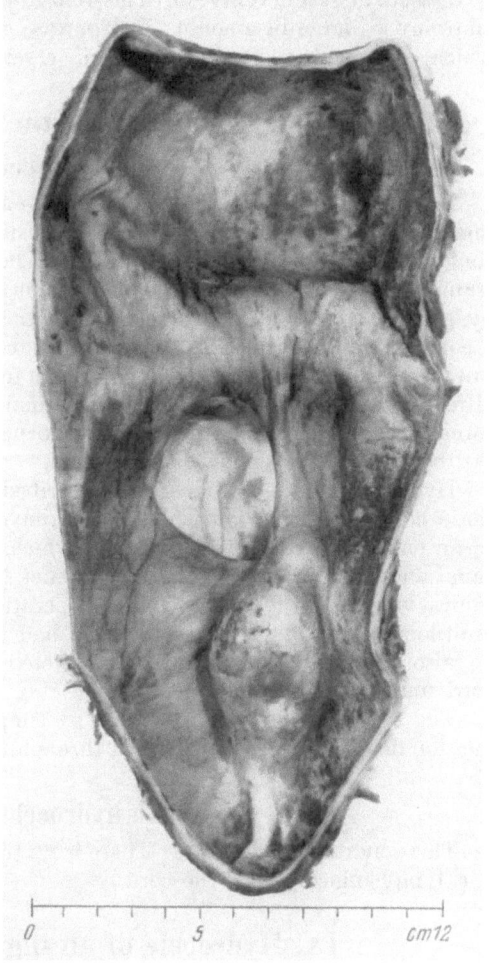

Fig. 17. Multilocular hydrocele. Patient admitted to hospital with swelling of left side of scrotum of eighteen months duration. Six months after scrotal swelling first noticed, an abdominal swelling appeared in the suprapubic region. Both the scrotal and abdominal swellings enlarged painlessly until they reached the dimensions shown in the photograph

more common; alternatively the abdominal portion of the sac may lie in the interstitial position between the muscle layers of the abdominal wall or, very rarely, superficial to the muscles, and to such cases the term interstitial hydrocele is sometimes applied. The full extent of the sac can be determined radiologically after injecting a solution of sodium iodide. The abdominal part of the swelling, which the patient may discover for himself, is often more prominent than the scrotal portion. Cases have been reported by HÜBOTTER, BICKLE, LASBREY, and HOLMES.

For these rare hydroceles the incision must be made over the inguinal canal and extended downwards to the scrotum. The inguinal canal must be opened as in the operative approach for the radical cure of inguinal hernia. Preliminary aspiration of some of the fluid may be desirable. The scrotal part of the sac should be defined first, then the neck, and the extraperitoneal sac should finally be dissected out and removed. The fluid in the sac is usually pale strawcoloured, and may be large in amount; in HOLMES' case there were seven pints of fluid, which contained masses of cholesterin crystals.

VIII. Hydrocele of the spermatic cord
1. Encysted hydrocele of the cord

This is a rounded, cystic swelling in the spermatic cord between the external inguinal ring and the testis, occurring usually in children (where it is often congenital), but occasionally in adults (when there is sometimes a history of trauma). The swelling is irreducible and not tender and can be drawn downwards by gentle traction upon the testicle, to the upper pole of which it maintains a constant relation; if it is accessible, it is translucent to light, though this test cannot be applied to swellings lying in the inguinal canal. The swelling must be differentiated from an irreducible inguinal hernia, but the examining hand cannot get above the latter, though the upper pole of an encysted hydrocele can usually be determined on palpation.

Hydrocele of the cord as just described constitutes the great majority of these cases but some unusual variants have been described. CAMPBELL refers to an encysted hydrocele of the cord which occupied the whole of the inguinal canal and was associated with and distinct from a rather large hydrocele of the tunica vaginalis. He also found at operation in an eleven-month old boy a multilocular encysted hydrocele which had four separate cystic compartments; he also recorded a pedunculated encysted intra-abdominal hydrocele of the cord found near the internal inguinal ring in a boy of two and a half years.

The swelling should be removed by simple dissection of the cyst, using an incision directly over it and carried through the coverings down to the cyst wall.

2. Diffuse hydrocele of the cord

This condition is very rare. There is no true hydrocele sac since the condition is a lymphangiectasis of the cord.

IX. Hydrocele of an inguinal hernial sac

The patient may give a history of having had a hernia for some years. Later the neck of the sac becomes obliterated at its upper end; the sac then becomes distended with fluid and is translucent to light. The diagnosis from an irreducible epiplocele may be impossible except at operation, which is usually necessary.

X. Hydrocele of the canal of Nuck

The processus vaginalis has become obliterated in its upper portion but a cystic swelling develops in the lower part; the condition is comparable with encysted hydrocele of the cord in the male. The patient, who may be a child or a young adult, notices a painless cystic swelling in the inguinal region. The diagnosis from an irreducible inguinal hernia may be difficult and possibly only determined at operation. Operative removal of the cyst is usually desirable.

XI. Secondary hydrocele

Acute secondary hydroceles may give rise to only a small collection of fluid in the sac and may pass unnoticed by comparison with the symptoms and signs of the underlying primary condition which is usually an epididymo-orchitis. The effusion is usually straw-coloured and not abundant, while in the severer grades of acute epididymo-orchitis, infected or purulent fluid may be present. The fluid in the tunica is usually absorbed when the primary condition resolves, though sometimes it may need to be aspirated, or even drained by an incision if suppuration occurs.

Chronic secondary hydroceles are rarely large, but since they may sometimes mask the diagnosis of the initiating disease, the fluid may need to be aspirated. The treatment is that of the primary disease. The hydrocele which may follow operation for hernia or varicocele requires the same treatment as the primary idiopathic hydrocele.

XII. Chylous hydrocele

This condition is due to obstruction of some lymphatics draining the scrotum, by the filaria sanguinis hominis (Bancrofti) and the rupture of other lymphatic vessels coursing round the wall of the tunica vaginalis. The fluid in the hydrocele sac is milky, and on standing a layer of fat forms on the surface. The appearance of the hydrocele may be preceded by attacks of fever and pain. Other evidence of filariasis, such as elephantiasis of the legs and of the scrotum may be present. The cyst is not translucent to light and when it is tapped the sac rapidly fills up again. Excision of the sac, often with part of the greatly enlarged scrotum, is the only treatment.

XIII. Meconium hydrocele

If in a newborn infant with the rare meconium peritonitis resulting from a perforation of the intestine in foetal life, there is a patent processus vaginalis, meconium may fill the tunica vaginalis producing a considerable scrotal swelling. The meconium may undergo calcification (FRIES and TALBOT, WILLIAMS).

XIV. Pyocele

This condition may follow an abscess in the body of the testis or in the epididymis and occasionally complicates pyaemia. In the congenital hydrocele of infants, an upward extension of the purulent exudate may lead to peritonitis, and conversely, a pyocele may then be formed by direct extension from suppurative peritonitis; CAMPBELL refers to 4 such cases in childred aged 2 and 4 months, in one of which the condition was bilateral. Treatment of a pyocele is incision and drainage of the tunica vaginalis and the administration of suitable antibiotics.

K. Haematocele

A haematocele is a collection of recent or altered blood in the tunica vaginalis or in the spermatic cord. The anatomical arrangements of the tunica vaginalis and of the cord have been referred to in the article on hydrocele.

I. Haematocele of the tunica vaginalis

1. Aetiology

The aetiology and varieties of haematocele are:

a) *Haematocele resulting from gross trauma:*

α) Crush injuries to the scrotum.
β) Penetrating external wounds involving the tunica vaginalis.
γ) Following the tapping of a hydrocele.
δ) Following surgical operations on the testis or the epididymis.

b) *Symptomatic haematocele:*

α) Torsion of the testis, or of the appendages of the testis or the epididymis.
β) Neoplasms of the testis.
γ) Orchitis or epididymo-orchitis.
δ) Gumma of the testis.
ε) Spontaneous haemorrhage into a hydrocele.

c) *Chronic haematocele:*

α) Spontaneous and idiopathic.
β) Symptomatic, complicating certain blood dyscrasias such as purpura, scurvy, the anaemias.

2. Pathology

Following the accumulation of blood in the tunica vaginalis, for example, after injury, clotting occurs and serum separates out, leaving shaggy masses of altered blood which are attached to the parietal part of the tunica and to the body of the testis and the epididymis. If no operation is done, the clot is gradually organised, and becomes converted into fibrous tissue; the tunica vaginalis will eventually become thickened.

The term haematocele almost always refers to haemorrhage into the tunica vaginalis of the testis or into an idiopathic hydrocele of that cavity. The large bilocular hydroceles may, however, also be converted into haematoceles by the slow accumulation of blood in them, as in the case reported by PYRAH.

When haemorrhage occurs into the sac of a hydrocele, there already exists a cavity of moderate or large size and bleeding may occur to fill or even distend the sac. In most symptomatic haematoceles, the bleeding may not be severe, or there may be merely blood-stained fluid.

In chronic haematocele, layer upon layer of old altered blood-clot, which may have been changed almost beyond recognition, are found filling and distending the tunica; evidence of more recent haemorrhage may be found among the outer layers. The testis may have undergone partial or almost total atrophy. Calcification, occasionally found in long-standing cases (KICKHAM, WEYRAUCH), may occur in plaques in the layers of old blood-clot.

3. Clinical picture

a) Haematocele resulting from gross trauma

A relatively simple crush injury to the scrotum, a kick, a blow, for example from a cricket ball, or a penetrating wound of the tunica vaginalis with a knife or shell fragment, may result in a haematocele. In such cases the injury may be followed by slow or rapid distension of the tunica vaginalis with blood, the increasing pressure of which may ultimately arrest the bleeding; bleeding may occur through the external wound. The haematocele may suppurate, and this may be followed by sinus formation.

The patient usually experiences pain, the severity of which depends upon the rate of the bleeding. The physical signs are the presence of a tender, boggy or firm swelling in the tunica vaginalis, not translucent to light; there may be subcutaneous ecchymosis. The position of the body of the testis becomes obscure, but its position may be demonstrated by eliciting testicular sensation below and behind the swelling. If a trocar is inserted into the swelling, blood may escape.

In traumatic cases of moderate severity, a scrotal haematoma and haematocele may co-exist, and since there are abundalnt smal vessels in the subcutaneous tissue of the scrotum, bleeding into the various layers readily occurs; moreover, the laxity of the tissues encourages extensive bleeding. An associated haematoma in the scrotum may collect beneath the dartos tunic, rarely in the septum of the scrotum (in which case diagnosis is difficult), or immediately outside the tunica vaginalis between the tunica and its fibrous coverings (LOWSLEY and KIRWIN).

In severe crush injuries, a haematocele may be but one part of a more extensive iujury; thus there may be extensive ecchymosis of the perineum, scrotum and penis, the blood extending upwards along the spermatic cord to the lower part of the abdominal wall and the upper two inches of the anterior part of the thighs. The testis itself may be damaged or even avulsed from the cord.

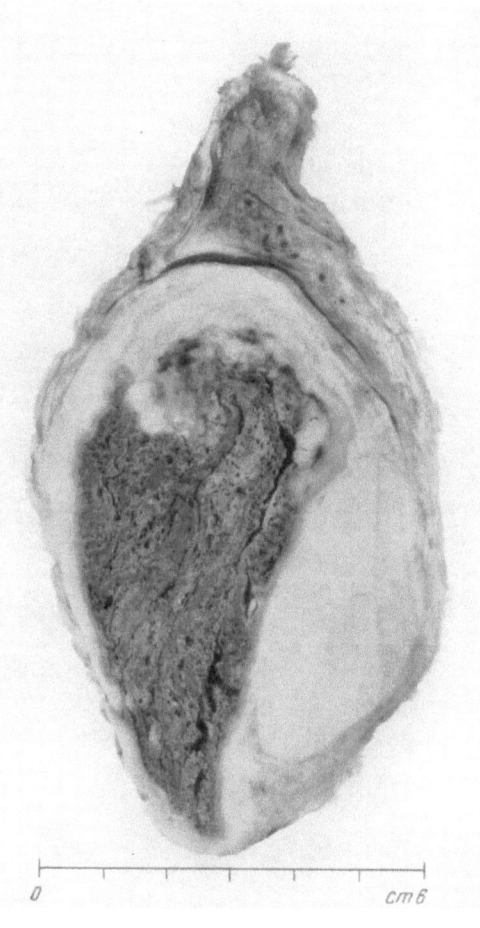

Fig. 18. Post-inflammatory haematocele of the tunica vaginalis. The testis is compressed and distorted by a large collection of blood-clot in the cavity of the tunica vaginalis, and the wall of the tunica shows gross fibrous thickening, especially at its upper end, where the boundary between the wall and the clot is not very clearly defined. The wall shows two distinct zones of tissue, the outermost white and opaque, the innermost grey and translucent, and in places there is a thin, opaque, cream-coloured zone between these two. Microscopically the outer layer is composed of dense fibrous collagen, containing a fair number of fibroblasts and occasional inflammatory foci; in the inner layer the fibroblasts are nearly all necrotic and there are numerous empty lacunae in the hyaline collagen; the intermediate zone shows similar a structure, but the lacunae are filled with degenerate polymorphs. There are also big groups of polymorphs amongst the clot in the sac, which suggests that the lesion has been caused by a chronic bacterial infection

There may be an associated injury to the pelvic bones around the symphysis, and even to the pelvic viscera. In such complicated injuries the tunica vaginalis may be torn or punctured and contain only a little blood; or alternatively a gross haematocele, which may be bilateral, may appear as a major physical sign.

A haematocele resulting from the tapping of a hydrocele should be prevented by avoiding puncture of prominent scrotal veins, by visualising them before carrying out the puncture. A warning that a haematocele may possibly develop may be seen from the escape of blood-stained instead of completely clear fluid when the hydrocele is tapped. If a haematocele does follow such a puncture, it usually gradually forms in the succeeding days, when the scrotum slowly distends, with some accompanying discomfort or pain, and the production of a swelling of firm or hard consistency no longer translucent to light.

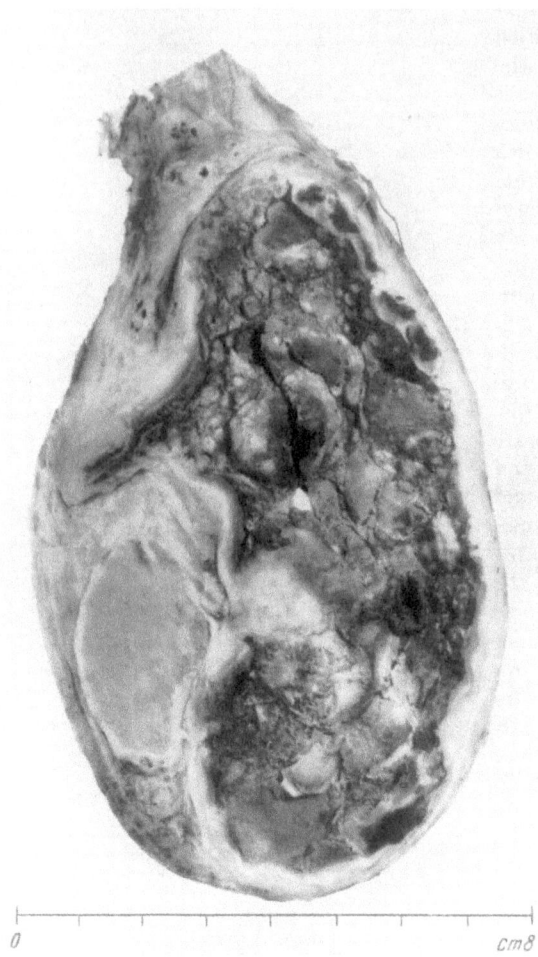

Similarly a haematocele or perhaps more strictly a haematoma, may form in the scrotum following an operation upon the epididymis or testis, or following an operation for hydrocele or spermatocele, when small bleeding points have not been ligatured. Such a swelling has to be differentiated from the oedema of the scrotum which is a common immediate sequela of such operations. Small collections of blood may resolve spontaneously and become organised into fibrous tissue.

Fig. 19. Organising haematocele of the tunica vaginalis. The specimen is a haematocele of the tunica vaginalis, which was removed at operation from a man of 58, five weeks after he had injured his scrotum in a fall. He first noticed some pain and swelling in the scrotum about a week after the injury, and three weeks later the swelling had increased to the size of a grape-fruit. A large amount of disintegrating yellowish clot escaped from the grossly dilated cavity of the tunica when the specimen was incised, and the portion preserved shows only a shaggy nodular layer of fibrin, heavily pigmented with haemosiderin, in the lining of the sac; external to this fibrinous layer there is an irregular zone of translucent grey granulation tissue, which has been laid down recently on the inner surface of the original membrane, causing massive thickening of the tunica. The testis is also surrounded by a bulky mass of fibrous tissue, mostly of recent formation

b) Symptomatic haematocele

Haematocele always complicates torsion of the testicle and very often torsion of the small, pedunculated cysts or appendages of the testis or of the epididymis. A blood-stained effusion or an actual haematocele not uncommonly complicates neoplasms and gummata of the testis, and rarely orchitis or epididymo-orchitis from any cause. The symptoms and signs are those of the underlying disease, together with those of the distended tunica vaginalis which is not translucent to light.

Spontaneous acute haematocele may rarely be seen in old or debilitated subjects as a complication of arteriosclerosis, scurvy, diabetes or syphilis (WHIT-NEY). HOWARD described spontaneous haematocele among African natives. A case of spontaneous rupture of a haematocele was recorded by BARRINGTON.

c) Chronic haematocele

Slow, even insidious bleeding may occur into the tunica vaginalis from vessels of the tunica itself, in the absence of injury. The patient notices a unilateral swelling in the scrotum, which undergoes slow and progressive enlargement over weeks, months or even years. The increase in size is not usually accompanied by pain, though gradually its very size may cause local discomfort and mechanical inconvenience. Because it is relatively symptomless, many patients do not report such a swelling for months or years after its onset, especially if the condition is idiopathic and not associated with a blood disorder which would give rise to other symptoms.

On physical examination, a chronic haematocele may vary in size from an ovoid swelling a few centimetres or more in diameter, to a large swelling as big as a grape-fruit or larger. At first it is rounded, irregular or ovoid and not tender on palpation. Later, probably because parts of the tunica vaginalis are at first stretched and then penetrated by the increasing mass, the swelling becomes lobulated or nodular: the demarcation between the testis and the epididymis disappears and the swelling may extend upwards to the lower part of the spermatic cord. The swelling, which is not usually tender, imparts a sensation of weight to the examining hand, and in consistency it varies from very firm to stony hard. In the larger swellings, the skin of the scrotum may be stretched and thinned over lobules of the mass; rarely there may be ecchymosis of the overlying skin and the skin may give way locally from necrosis, leaving a raw granulating surface covering the underlying blood clot; or the haematocele may actually fungate through the skin. The mass, therefore, comes to mimic very closely a malignant growth arising in the body of the testis.

Rarely, a long-standing chronic haematocele may undergo a sudden painful increase in size, causing the patient to seek advice. Such a case was reported by COWDERY et al. WEYRAUCH also recorded the case of a man aged 78 who had had a scrotal mass for twenty-five years with no history of injury; examination revealed a stony-hard mass which was adherent to the skin in some places and parts of the mass crepitated on palpation from the grating against each other of calcified masses, which were also demonstrable on radiography. Similar crepitation in longstanding haematoceles, a very rare sign, was also recorded by WATSON and CUNNINGHAM.

4. Diagnosis

In acute haematocele, the diagnosis usually presents no difficulty since a swelling resulting from an injury is present, which distends the tunica vaginalis and which is not translucent to light. Sometimes patients with acute epididymitis report an antecedent injury, real or imaginary, but there should be no difficulty in diagnosis. If the haematocele is a swelling which the surgeon expects to resolve on conservative treatment, and which fails to do so in two or three weeks, the possibility of haemorrhage into a soft malignant growth of the testis must be considered, as in NELIGAN'S case; in such cases an exploratory operation is called for.

A haematoma of the scrotum not involving the tunica vaginalis, and possibly difficult to differentiate from a haematocele, is found occasionally, as in VERNON's case in which a painful swelling appeared in the scrotum following a muscular strain but unassociated with a direct blow to the scrotum: in such cases it is usually possible to define the body of the testis separate from the haematoma.

Chronic haematocele may present difficulty in diagnosis. The absence of the usual landmarks which differentiate the body of the testis and the epididymis from the swelling may cause it to be mistaken for epididymitis. Hydrocele of the tunica vaginalis is usually softer in consistency than haematocele and may fluctuate on palpation, and it is translucent, though this sign may not be present in thick-walled, long-standing hydroceles. A chronic spontaneous haematocele is very often mistaken for a malignant growth of the testis and indeed there may be no method of distinguishing the two until after orchidectomy. Fungation of the scrotal mass, which may suggest malignant disease, may also be associated with haematocele; both conditions present with a mass which may be believed to arise in the body of the testis, which slowly increases in size and which presents a stony-hard or fibrous swelling which imparts a sensation of weight to the examining hand; calcification may be present with both conditions. Operative exploration, and often orchidectomy, are needed to establish the diagnosis, and, in fact, this is the appropriate treatment for both conditions. Gummata, which are rare now-a-days, arise in the body of the testis and may be as large as a duck's egg. The swelling may be nodular, and is not usually tender. The discovery of other signs of tertiary syphilis in different parts of the body and a positive Wassermann reaction should establish the diagnosis.

Rarely a haematocele of the tunica vaginalis may be mistaken for a haematocele of the epididymis. In such cases there is a history of injury, followed by the appearance of a firm or cystic swelling in the epididymis; such cysts are usually thick-walled and when examined months or years later may contain the remains of old blood-clot. In TAYLOR's case of a boy of fourteen who sustained an injury of one testicle, which was followed by a swelling which diminished in size but did not disappear, operation fourteen years later following a further increase in size, revealed a haematocele of the epididymis. Similar cases were recorded by VALENCE, MATIGNON and VENNET, and O'CONOR.

5. Treatment

a) Small recent haematoceles will require bed rest and sedatives, support to the scrotum, and toilet of the penetrating wound if one be present; they will usually gradually resolve, with organisation of the clot. The treatment of the larger haematoceles consists in opening the tunica vaginalis, at the same time excising any penetrating wound which is present, evacuating the clot, ligating bleeding points and closing the wound; a temporary drainage tube is desirable unless the wound is quite dry.

The treatment of haematoceles associated with more severe injury will include a general toilet of the injured area, with arrest of bleeding, the removal of the blood within the tunica vaginalis or a testis which is deemed to be too badly damaged for recovery, and also the appropriate treatment of fractured bones and damaged viscera.

b) A haematocele complicating the simple tapping of a hydrocele usually constitutes an indication for operative treatment, the tunica vaginalis being widely opened, the clot evacuated and the parietal layers of the tunica being excised as in the ordinary operation for hydrocele. When a haematocele or haematoma follows operation for hydrocele, temporary treatment by a scrotal support and cooling lotions to relieve discomfort are usually all that is needed. The treatment for larger haematomata consists in opening part of the wound, evacuating the clot, ligating bleeding points if such can be found, and closing the wound.

c) The treatment of a chronic haematocele is usually orchidectomy with removal of the haematocele. It is usually impracticable to conserve the body of the testis, which is buried in layers of clot and which may even have undergone a considerable degree of atrophy; an attempt to preserve the testis may result in considerable local bleeding. Since many chronic haematoceles occur in elderly patients there is usually no objection to orchidectomy.

II. Haematocele of the spermatic cord

Haematocele of the spermatic cord is very rare. There are two varieties.

1. Encysted haematocele of the spermatic cord

Encysted haematocele of the spermatic cord may result from a haemorrhage into an encysted hydrocele of the cord. It can only be diagnosed by an exploratory operation. The treatment consists in the removal of the haematocele by dissection.

2. Diffuse haematocele of the spermatic cord

Diffuse haematocele of the spermatic cord results from rupture of one of the veins of the pampiniform plexus and may follow a blow or a violent muscular exertion. A case of spontaneous haematoma of the spermatic cord was reported by BLISS and by FLORIO. The patient experiences a severe pain in the inguinal region and shortly afterwards an elongated swelling appears in the cord extending from the inguinal canal to the testis; bruising of the scrotum may be present. A small haemorrhage of this kind may absorb spontaneously and treatment is therefore usually conservative; a larger one requires an incision with removal of the clot and surgical treatment of the varicocele.

L. Varicocele

The term varicocele denotes a varicose condition of the veins of the pampiniform plexus of the spermatic cord.

IVANISSEVICH demonstrated in 1918 that the venous drainage of the testis and of the epididymis was effected through a deep and a superficial venous system. The deep veins (or the primary venous drainage) consist of the internal spermatic vein draining the pampiniform plexus, the vein of the vas deferens and the external spermatic vein, the last two anastomosing with the internal spermatic vein near the external inguinal ring. The superficial veins (or secondary venous drainage) comprise the superficial and deep inferior epigastric veins, the superficial internal circumflex vein and the scrotal tributaries of the superficial and deep external and internal pudendal veins. There is a free communication by collateral anastomosing branches, between veins of each group, and also between the primary and secondary venous systems, by way of the cremasteric branches of the external spermatic vein, at the level of the external inguinal ring (Fig. 20). ROBB has confirmed by dissection, and also by radiography after the injection of the pampiniform plexus with diodone, that secondary venous channels pass in close relation to the vas deferens, to the internal iliac veins, and that communications also exist between the pampiniform plexus and the pudendal and circumflex systems of veins.

The veins of the pampiniform plexus itself form an anastomosing group in front of the vas deferens within the spermatic cord, and pass upwards through the external abdominal ring into the inguinal canal, where they form two veins lying in front of the vas and one posterior to it. At the level of the internal abdominal ring or sometimes higher, these veins join to form one venous trunk,

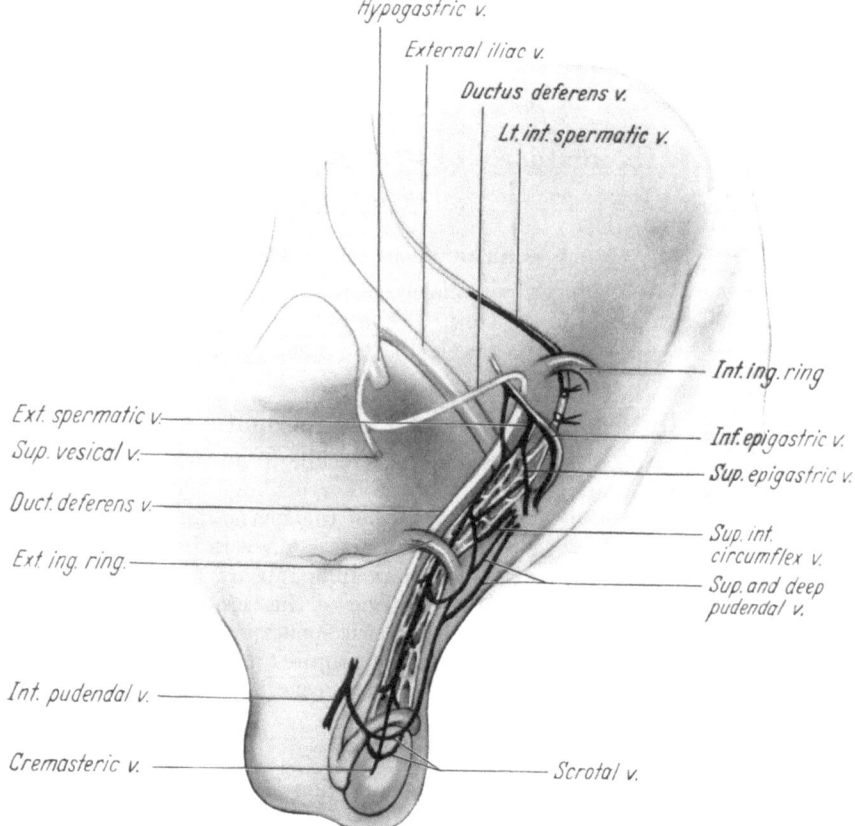

Fig. 20. Diagram of the venous drainage of the scrotal contents. [Adapted from JAVERT and CLARK, Surg. Gynec. Obstet. **79**, 644 (1944), and also from OLSON and STONE, New Engl. J. Med. **240**, 877 (1949).] (By courtesy of the Editors.) *Primary system*—internal spermatic, ductus deferens and external spermatic veins (in white). *Secondary system*—superficial epigastric, superficial internal circumflex and cremasteric veins, and through scrotal tributaries the superficial and deep pudendal and the internal pudendal veins (in black)

the internal spermatic vein. On the left side, the vein lies on the posterior abdominal wall behind the sigmoid colon and in front of the psoas muscle, and joins the left renal vein at a right angle. On the right side the vein passes upwards in relation to the inner border of the caecum and along the surface of the psoas muscle before joining the inferior vena cava at an acute angle.

I. Primary, spontaneous or idiopathic varicocele
1. Pathology

Given the complicated anatomical arrangements of the testicular veins, it is still not clearly understood why a varicocele should develop. Possibly there is some congenital abnormality of the walls of the veins. The internal spermatic veins have no valves except near their points of union with the inferior vena cava. Since both veins are long, any incompetence of their valves will predispose to the formation of a varicocele by allowing a retrograde flow of blood into the pampiniform plexus with resulting over-distension of the veins, especially when the man is in the erect position. The right-angled mode of union between the left spermatic vein and the left renal vein is similarly disadvantageous and will predispose still further to varicocele. Pressure of a chronically loaded pelvic colon may be a contributory cause.

When fully formed, the veins constituting the varicocele are greatly increased in length and in their calibre and become very tortuous (Fig. 21). Local venous dilatations are found and phleboliths may be found in the walls of the veins at those points. The connective tissue between the distended veins undergoes hyperplasia. The testis in relation to a varicocele is frequently smaller and softer than its fellow on the normal side and it may be that the venous congestion resulting from the varicocele affects adversely the nutrition of the testicular tubules, and also by interfering with the normal temperature-regulating mechanisms of the organ damages their delicate epithelial cells. The clinical importance of this possibility is referred to later.

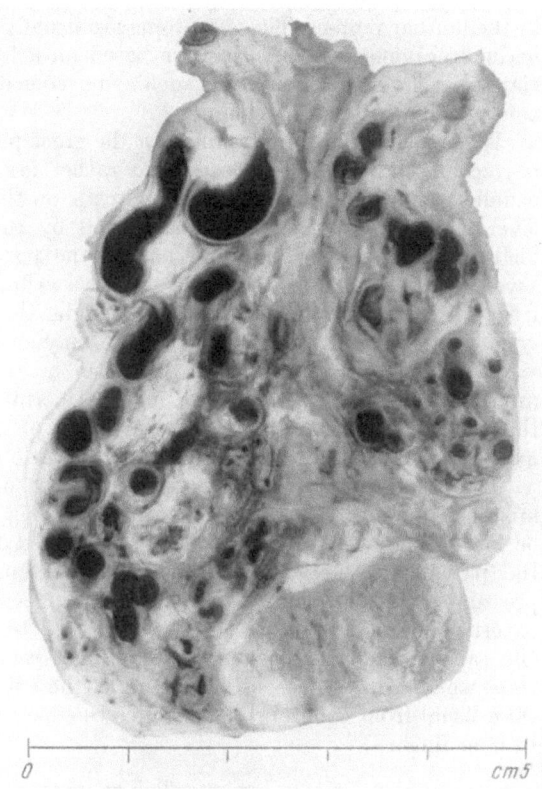

Fig. 21. Autopsy specimen of a thrombosed varicocele. The greatly distended veins of the pampiniform plexus show the usual appearance of varicose veins, they are thick-walled, dilated and tortuous; most of them are filled with laminated, ante-mortem thrombus, which maintains the channels in the over-distended condition that was habitual during life, whereas the vessels without thrombi have collapsed. Incidental post-mortem finding in a man of 51

2. Clinical features and diagnosis

This variety of varicocele, the commonest, is found in adolescents and young adults and is of insidious onset. There is a great preponderance of cases on the left side. In a survey of 1,500 young men of military age Lewis found 16.5 per cent with varicoceles; of these, he classified 15 per cent as being severe, and having symptoms; 55 per cent as small and symptomatic; 30 per cent as of moderate size with some, but not severe, symptoms. Campbell (1928) reviewing 500 cases of idiopathic varicocele, found 3 cases in 12-year old boys, and 1 in a 59-year old man; 90 per cent of the cases were between the ages of 15 and 35, the period of greatest sexual potentiality; nearly all were unmarried, and he suggests that possibly inadequate sexual relief perpetuated a constant congestion of the genitalia. The incidence, however, in the male population as a whole is very small.

Very commonly the patient is unaware of any abnormality until he is medically examined before entering the Armed Forces, since he has experienced no symptoms. Other patients complain of a feeling of weight in the affected testicle, or of a persistent ache which is worse towards the end of the day, and sometimes of neuralgic pain. In some cases the discomfort radiates to the groin and even to the lumbar region. The symptoms are usually relieved by rest. Some patients become nervous and introspective based on a fear of loss of virilism, and associated psychological symptoms such as nervous depression, headaches and sexual neurasthenia, are not uncommon.

When the patient is examined in the erect position the condition can usually be recognised by the elongated and rather lax scrotum, by the fact that the usually somewhat smaller and softer testis on the affected side hangs at a lower level than that on the normal side, and by the enlarged, tortuous and easily visible coil of veins above and behind the testicle which renders the scrotum asymmetrical. On palpation the venous swelling feels like a "bag of worms"; it is not tender. When the patient coughs there is a thrill-like impulse. The venous swelling is much reduced and may even disappear when the patient is recumbent and reappears slowly when he resumes the erect position. In a number of cases there is a small inguinal hernia associated with the varicocele; thus BENEVENTI found such a hernia in 8 of 28 cases which he had operated upon, and JAVERT and CLARK found a sac in one-third of 32 cases.

The diagnosis of varicocele need never be in doubt and it should never be mistaken for any other of the scrotal swellings such as epididymitis, hydrocele or hernia. A varicocele may be distinguished from an omentocele by placing the patient in the horizontal position and emptying the swelling by gentle pressure on the scrotum; the examining finger is then firmly pressed upon the external inguinal ring and the patient asked to stand. In the case of a hernia, the sac does not fill so long as the finger remains pressed upon the external ring, whereas in the case of a varicocele, the veins readily fill to a greater extent than usual from collateral veins, since the spermatic veins are partly obstructed by the finger.

3. Treatment

A symptomless varicocele, strictly speaking, usually requires no treatment. Patients who become aware of their varicocele and who complain of aching or discomfort in the testicle should first wear a closely applied suspensory bandage which may give relief. The relation of varicocele to subfertility, however, has provided a new reason for operative treatment in subfertile men. The regulations of the Armed Forces and of some industrial concerns providing services in tropical countries, have rendered essential the operative treatment of varicocele in such patients, even when no symptoms are complained of. Operation is indicated for persistent neuralgia or pain in the affected testicle. Treatment by injection as an alternative to operation has been recommended.

CAMPBELL (1928) described three clinical groups of idiopathic varicocele in which the indications for operation vary. The first type, or the asymptomatic varicocele, which may be large or small, is best left untreated since operation may actually induce symptoms and may lead to a neurosis. Most patients in this group are not aware of the existence of a varicocele until it has been shown to them by the examining surgeon. Some of these patients, unfortunately, have to submit to operation in order to conform to Service regulations. The second type, in which the varicocele is of medium or large size, complain of a sensation of weight and local discomfort in the scrotum; operation is usually successful for patients in this group. Many cases fall into a third group in which the varicocele is small, yet the patient is greatly concerned about it, and complains of local discomfort, and some are the subjects of a sex anxiety neurosis. Operation is not uniformly successful in this group. Most surgeons would agree with these views.

a) Treatment by injection

Treatment by injection has been described by HANSCHELL, by PORRITT and by McEVEDY. The solutions used were quinine urethane (1 cc. of a solution of quinine hydrochloride 4 gm., urethane 2 gm., distilled water 30 cc.), or a 5 per cent solution of sodium morrhuate. The injection is given with the patient standing. The vein selected for injection should be steadied by the surgeon's left hand and after the injection it should be compressed for a short time to avoid the formation of a haematoma or the escape of the injected fluid into the tissues. After a successful injection, the veins become hard and tender and may so remain for several days. Usually one, but occasionally two injections are required. The results of treatment by injection are uncertain and occasional haematoma formation and local discomfort are disadvantages.

b) Operative treatment

The operative treatment of varicocele is intended to interrupt most of the venous channels in the spermatic cord, leaving, however, a few of the more normal veins or some of the collateral veins to carry back the venous blood, thereby relieving back-pressure on the testis and eliminating the varicocele. The older technique has been to make a mass division between ligatures of most of the veins of the pampiniform plexus, hoping that just the right amount of venous tissue could thus be eliminated to dispose of the varicocele. Since this method was seen to be followed by a number of failures to cure the varicocele, other surgeons advocated ligature of the main internal spermatic vein itself, either in the upper part of the inguinal canal (where the main vein may still not be formed and may still be only represented by two biggish tributaries) or above the inguinal canal. Following current interest in the operation for varicocele for the subfertile male, even this operation has been shown to be by no means always successful because the presence of large varicose veins which leave the spermatic cord low down to enter the pelvis behind the symphysis pubis, may perpetuate the varicocele. HANLEY therefore advocates the selective division of veins in the inguinal canal itself, such division to include the ligature of many of the large collateral veins referred to, as well as most of the veins of the pampiniform plexus. PALOMO recommented ligature at a high level of the internal spermatic vein and the testicular artery.

α) The low operation

The older operation (the so-called low operation) consists in exposing the spermatic cord by an incision in the upper part of the scrotum or over the external abdominal ring, extended into the upper part of the scrotum. The spermatic cord is freed and its coverings incised to expose the pampiniform plexus. The large anterior group of veins lying in front of the vas is isolated and divided between ligatures, excising one inch of the plexus. The ligatures securing the divided ends are tied together so as to shorten the cord and elevate the testicle.

Care is taken to preserve the spermatic artery, the vessels immediately surrounding the vas, and also the smaller posterior group of veins. The wound is then closed.

This operation very often relieves the aching and discomfort though some patients with neuralgia of the testis remain unrelieved. The operation itself is not, however, without complications; haematoma formation, painful thrombosis, testicular atrophy, hydrocele, and recurrent varicocele have all been described. DOUGLAS in a follow-up of 106 cases operated on for varicocele, found that 48 per cent were normal several months later, 39 per cent had hydrocele and 4 cases had atrophy of the testis. Hydrocele following the operation was found in 30 per cent of cases by BLOODGOOD.

The interruption of the venous return, resulting from the low operation, which may interfere with or disrupt in a casual or disorderly fashion the venous anastomoses between the primary or deep, and the secondary or superficial groups of veins, which are found at or near the level of the external inguinal ring, is probably unsound and may lead to bad venous drainage of the scrotal contents and a recurrence of the varicocele (OLSON and STONE). Moreover, since there is some slight risk during the low operation for varicocele of producing atrophy of the testis by damaging not only the main testicular artery but also the anastomosing vasal and cremasteric vessels (which HARRISON has shown by arteriography to exist), it is safer to ligate the internal spermatic vein or veins at a point where the testicular artery can be avoided with certainty, by careful dissection.

β) The high operation

An alternative procedure, therefore, is to divide the internal spermatic vein between ligatures, either at the upper end of the inguinal canal or after it has emerged from the canal into the retroperitoneal space. Retroperitoneal ligature of the internal spermatic vein for the treatment of varicocele was first recommended by IVANISSEVICH as far back as 1918, but the method did not secure general adoption until much later.

A short incision is made above and parallel to the inguinal ligament, the mid-point of the incision being over the internal inguinal ring. The aponeurosis of the external oblique is divided down to the external ring and upwards to expose the internal inguinal ring; the muscular fibres of the cremaster muscle at the proximal end of the canal and if necessary those of the internal oblique and transversalis are divided to expose the extraperitoneal fat; the fat and the peritoneum are displaced slightly upwards to expose the testicular vessels as they lie on the posterior abdominal wall at the outer side of the external iliac artery. The internal spermatic vein is dissected out, separated from the testicular artery, and divided between ligatures; sometimes the internal spermatic vein is formed within the upper end of the inguinal canal by the union of two large tributaries or alternatively these two veins may not unite until some distance above the canal, when they should both be ligated and divided. The distal ligated end of the divided spermatic vein may be drawn upwards and secured to the internal oblique muscle before suturing the cremaster, thus assisting in the suspension of the testis. There should be no danger, with careful dissection, of inflicting damage on the testicular artery. If a hernial sac is present, it should then be excised. The incised muscular fibres of the cremaster are then sutured. Following operation the patient should wear a suspensory bandage for a week. Early ambulation is advantageous.

JAVERT and CLARK reported cure of the varicocele and relief of symptoms in 100 per cent of 22 cases operated on by this method. RIBA reported cure of the varicocele in all, and relief of symptoms in 87 per cent of 23 cases. OLSON and STONE reported 20 good results and 1 poor result following operation on 21 cases of varicocele presenting with typical symptoms; of 4 cases who presented with atypical symptoms, which included aching pain or sharp stabbing pain, or localised scrotal pain with tender testes, only 1 case had a good result following operation. PRICE reported cure of the varicocele in 9 of 10 cases operated on, and complete relief of symptoms in 8 and partial relief in 1. No complications were reported from the operation. LEWIS reported satisfactory results in 42 cases operated on.

PALOMO advocated ligature of both the spermatic artery and of the internal spermatic vein for varicocele and claimed good results in all of 38 cases. The operation is done through an incision at the level of the internal inguinal ring. The arterial supply of the testis is chiefly through the spermatic artery. The smaller deferential artery, a branch of the inferior vesical artery, lying in intimate contact with the vas deferens, and a third artery, the cremasteric, a branch of the deep epigastric running in the external sheath of the cord to the lower pole of the testis and the epididymis, also assist in the arterial supply. There is an anastomosis between these three vessels and consequently ligature of the spermatic artery need not necessarily result in atrophy of the testis (PICQUE and WORMS). There would, however, appear to be at least some risk of atrophy of the testis, in spite of the anatomical arrangements described above.

In spite of many favourable reports of the high operation, some unsatisfactory results have been found by some surgeons, but it is especially the current interest in the relation of varicocele to male subfertility that has led to modification. BERNARDI had, however, at an earlier date, reported excellent results following an operation which involved not only ligating the internal spermatic vein, but also excising all the veins within the inguinal canal. The views of HANLEY are recorded in the paragraph on varicocele and subfertility.

II. Symptomatic or secondary varicocele

A varicocele, especially on the left side but sometimes on the right, occasionally occurs as a complication of renal neoplasms, hydronephrosis and pyonephrosis, either from the pressure of the resulting tumour upon the internal spermatic vein, or, in the case of renal neoplasms, from partial blockage of the vein by growth which has extended into it from a mass of growth inside the lumen of the renal vein. Pressure from a mass of malignant glands lying along the course of the vein may also cause a varicocele. A varicocele occasionally complicates pelvi-rectal cancer (SKINNER) and retroperitoneal sarcoma or teratoma (CAMPBELL 1928). Rarely, a varicocele may be the presenting symptom or sign of such conditions, hence the term "symptomatic" varicocele, which is sometimes applied to it. Symptomatic varicocele differs from idiopathic varicocele in its more rapid development, in its larger size and in not disappearing when the patient is recumbent. A case was described by PEPPER in which the sudden appearance of a varicocele preceded the clinical discovery of a renal sarcoma by six months.

Analysing the cases of symptomatic varicocele occurring along with tumours, malformations and inflammations of the kidney at the Massachusetts General Hospital, BATE found 14 cases of symptomatic varicocele in a total of 122 cases in which the clinical diagnosis had been malignant growth of the kidney. In 12, the varicocele occurred on the left side and in 2 it was bilateral. In one of these bilateral cases the growth extended into the right renal vein across the inferior vena cava.

A varicocele in infants and in older males calls for a full urological examination, including an intravenous pyelogram (Fig. 22) to determine if a lesion of the upper urinary tract is compressing the spermatic vein. CAMPBELL (1944) refers to a case of a thirteen-month old boy with a large, left-sided varicocele which disappeared following nephrectomy for a Wilm's tumour. In another case of a thirteen-year old boy with a varicocele, investigation revealed a symptomless left-sided hydronephrosis, and a diagnosis of a possible vascular obstruction

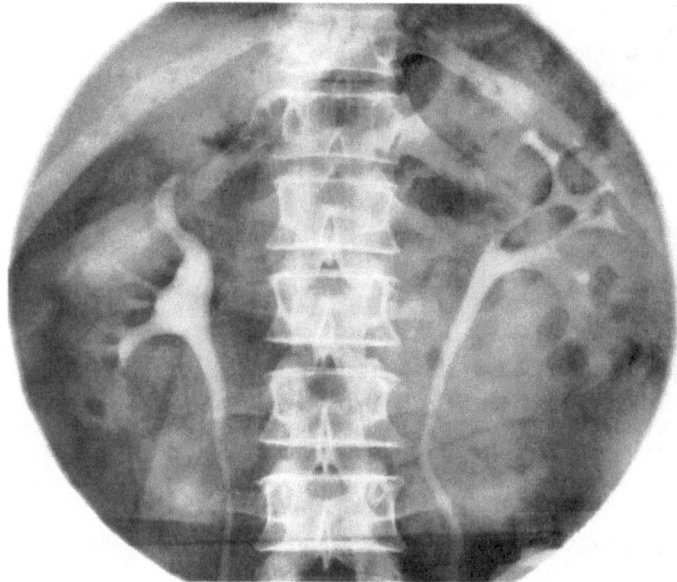

Fig. 22. Retrograde pyelogram of a carcinoma of the left kidney. The patient had a varicocele which disappeared following nephrectomy. The varicocele did not produce any symptoms of its own

of the upper ureter was made; at operation, an aberrant artery from the lower pole of the kidney to the aorta was found, which compressed not only the pelvoureteric junction but also the upper part of the left spermatic vein thereby causing a varicocele. Three months after operation the varicocele had disappeared. WHITE recorded the disappearance of a left-sided varicocele of six weeks' duration, immediately following the removal of a pyonephrotic kidney.

The management, therefore, of symptomatic varicocele calls for the recognition, diagnosis and treatment of the cause. No direct surgical attack upon the varicocele is called for.

III. Varicocele and subfertility

The presence of a varicocele, because of the resulting venous congestion, affects the temperature level of the testes and consequently the activity of the spermatic epithelium, and therefore fertility. The temperature level of the testes in man, as well as in many mammals, is maintained at a lower level then that of the body as a whole; the difference in man averages 2.2° C. (BADENOCH). The normal temperature-regulating mechanism of the testes, situated as they are in the scrotum, includes the contraction and relaxation of the dartos muscle and of the cremaster and the muscles of the scrotum, which respond to elevation of the local temperature by relaxation, and to cooling by contraction (HARRISON). In experimental work on the ram PHILLIPS and McKENZIE showed that elevation of the temperature of the testes by about 2^{0} C. effected by insulating the scrotum, resulted in degenerative changes in testicular epithelium; after treatment in that way exceeding two weeks, the seminiferous tubules contained no spermatozoa and the sperm count of the seminal fluid showed a rapid descrease. GUNN et al. confirmed these findings. GLOVER showed that the elevation of testicular temperature in men by 5^{0} C. for only twenty-four hours resulted in the appearance of tailless spermatozoa 19 to 24 days after the experiment.

It has been noted that some men suffering from varicocele are subfertile and it has seemed possible that the abnormal vascular arrangements associated with a varicocele may result in an elevation of the temperature of the testis because of venous congestion and an interference with the normal drainage of venous blood: moreover, the tortuous collection of veins filled with blood in close proximity to the testes, will act as a radiator at normal body temperature. The scrotal support which is worn by many patients with varicocele would also tend to conserve heat and interfere with the normal temperature-regulating mechanism. Such persistent elevation of the scrotal temperature may interfere with normal spermatogenesis (DAVIDSON).

TULLOCH found in the routine examination of subfertile males, that the incidence of varicocele is high. Analyses of the seminal fluid of male patients with varicoceles showed variations, from moderate degrees of subfertility, to a marked depression of the quality of the seminal fluid and even azoospermia. RUSSELL reported similar findings and believed that a varicocele persisting beyond the age of 30 was particularly apt to be associated with semen of poor quality. Using the technique of testicular biopsy, TULLOCH showed that in a case of varicocele, the lumina of the seminiferous tubules were filled with much cellular debris and that spermatozoa were absent, but that within three months of operative treatment for the varicocele, spermatozoa were present in the seminal fluid, and pregnancy had been induced in nine months.

HANLEY has shown that there may be a difference of up to 3^{0} C. between the intra-scrotal temperature and the rectal temperature in the normal male, and that such difference may be reduced to as little as 0.2^{0} C in the presence of a large varicocele. He found that the removal of a moderate-sized varicocele usually reduced the intra-scrotal temperature by about 1.0^{0} C. thus producing a temperature differential between that of the rectum and of the scrotum of about 2.5^{0} C. If such a temperature change can be achieved following operation, he found very often some improvement in fertility. Conversely in some cases following operation there was no change in the temperature differential though there was a change in the fertility. HANLEY considers, therefore, that although the elevation of the intra-scrotal temperature is the best guide to an improved fertility, it may not be the only factor involved; for example, the operation may lead to improved testicular blood-flow with a reduction of ischaemia or anoxaemia, a factor about which little is as yet known, and this may determine improved fertility.

Although insufficient data is yet available regarding the numerical relationship between sterility and varicocele, the available evidence suggests that in cases of male infertility, a varicocele discovered on routine examination should be regarded as a possible cause, and should be treated surgically. Some of the results are referred to in a later paragraph.

DAVIDSON reported the results of operation in 12 men suffering from oligozoospermia who had unilateral varicoceles. The sperm counts ranged from $2^1/_2$ to 11 millions per ml. in 9 cases, and from 24 to 26 millions per ml. in 3 cases. Most of the patients showed other characteristics which he regarded as being associated with faulty thermo-regulation of the testis, namely an increased semen volume, impaired motility and morphology and an excess of desquamated precursors of spermatozoa. Many of the patients had had other forms of treatment in an attempt to relieve the oligozoospermia. After the varicocele had been operated on, the results showed a high proportion of increased sperm counts. Three of the twelve cases who had a moderate oligozoospermia reached normal figures after operation. Four cases of marked oligozoospermia produced counts within the normal limits of 40 millions per ml. or over, and four others showed an increase over 20 millions per ml. Only one case failed to improve and he was later found to have a residual varicocele. The increase in the sperm count was found to have occurred in three to four months after operation. Five men in the group induced a pregnancy in their wives within six months of the operation; all these men had failed to induce conception from two to six years preoperatively. He points out that the scrotal support should be abandoned after the operation in order to allow the heat-regulating mechanism to work naturally. He considers that a varicocele must always be looked for in cases of subfertility, when a semen test shows a relatively large volume and a sperm density of less than 20 millions per ml., and that, if found, it should receive appropriate treatment.

TULLOCH reported 30 cases of varicocele who underwent operation because of subfertility. In one-third of the cases there was a return to normal fertility and a successful pregnancy ensued in the wife. In a further third of the cases there was a considerable improvement in the sperm count after operation but pregnancy had not yet been induced. In a further third of the cases there was no improvement and sometimes even a deterioration of the sperm count. He suggested that in those cases the process of destruction of the germinal epithelium had become irreversible before operation.

YOUNG reported 12 cases whose varicoceles had been operated on for infertility. Of 10 cases which were traced, the sperm count had improved in 9 and remained unchanged in 1. Two wives out of 9 had become pregnant.

HANLEY has operated on over 300 subfertile men for conditions (the commonest being varicocele) which were thought to be affecting testicular temperature control. A few had hydrocele. Of the cases of varicocele, he found that not all of them had the characteristic "bag of worms" but that in many the spermatic cord contained larger numbers of small venous channels with much perivenous fat, in the pampiniform plexus rather than the large type of varicose veins. He calls this type of varicocele a "spongy cord". In this group the usual operation for varicocele has not always improved the temperature differential but it has led in some cases to a quite dramatic improvement in sperm morphology, possibly by improving the blood supply to the cord.

HANLEY reported the results of operation in 78 men. In 53 cases there was an improvement in the sperm count (often a reasonable one) and there were many pregnancies. In 8 cases the results were classed as doubtful; here the temperature differential and the sperm density figures before and after operation were not significantly changed, although the morphology and motility of the sperms was improved; 5 wives, however, conceived. Seventeen cases were classed as failures but this was possibly due to the fact that the operation did not in all cases reduce the thickness or the vascularity of the spermatic cord and did not produce any increased temperature differential; 3 men in this group were later transferred to the group of cases who improved, following a second operation.

M. Cysts of the epididymis

Cysts of the epididymis arise either in the epididymis itself or as a cystic dilatation of one of the embryonic vestigial remains which are attached or lie close to, the epididymis. Terms which have been used to describe this condition have included spermatocele, hydrocele of the epididymis, and spermatic or spermatogenic cysts, and their number has led to some confusion. It is probably best to use the term cyst of the epididymis to describe them all, bearing in mind, however, that they are not all derived from the same cause.

It is uncertain who first described this condition. CURLING (1843) referred to cysts containing clear fluid which were separate from the tunica vaginalis and which were situated in the head of the epididymis and he designated them "encysted hydroceles of the epididymis". LISTON and also LLOYD in 1843 reported cysts of the epididymis which contained spermatozoa. The subject has been reviewed by WHITNEY, CAMPBELL and by EDINGTON; and ABESHOUSE in 1937 gave an excellent account of the subject.

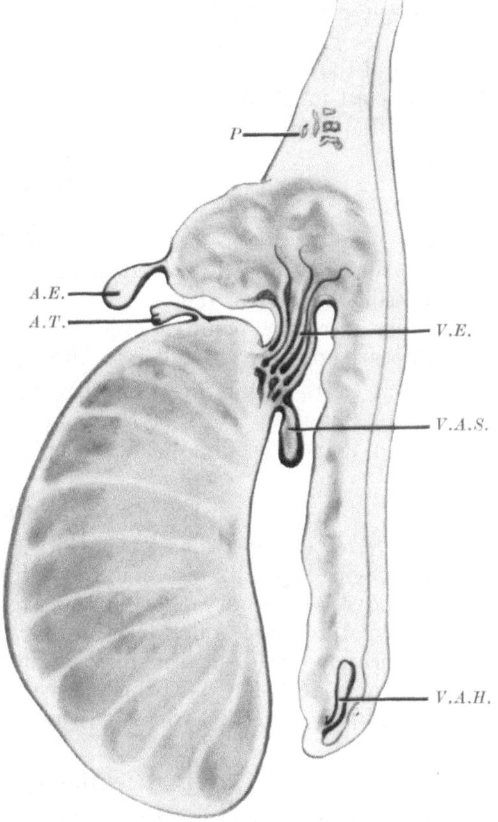

I. Surgical anatomy

The body of the testis is covered by a firm, fibrous capsule which sends prolongations into the substance of the testis dividing it into lobules, each of which contain seminiferous tubules which pass towards the apex of the testis where they unite to form the tubuli recti. The rete testis is formed by the tubuli recti and from these originate 12 to 15 vasa efferentia which pass through the tunica albuginea to form the coni vasculosi in the globus major of the epididymis; these ducts

Fig. 23. Schematic drawing of testis and epididymis showing the approximate position of the embryonic remains: *P* Paradidymis (organ of Giraldes); vestige of caudal part of mesonephros; *A.E.* Appendix epididymis (vesicular or stalked appendage); vestige of cranial end of mesonephros. *A.T.* Appendix testis (fimbriated or sessile appendage). *V.E.* Vasa efferentia. *VAS* Vas aberrans of rete testis. *V.A.H.* Vas abberrans of Haller

unite at the tip of the globus major to form the ductus epididymis or the main excretory duct of the epididymis.

The vestigial structures which are found in or near the epididymis and the testis include:

1. The appendix testis, or sessile or non-pedunculated hydatid of Morgagni. This is found at the upper pole of the testis or in the groove between the testis and the globus major of the epididymis, and it is derived from the upper end of the Mullerian duct.

2. The appendix epididymis or the pedunculated hydatid of Morgagni. This small structure is derived from the cephalic tubules of the epigentalis (Wolffian duct).

3. The organ of Giraldès, or paradidymis. This is a small body which is situated in the front of the vessels of the lower part of the spermatic cord at the level of the globus major and is composed of a number of blindly-ending groups of tubules. It does not communicate with the seminiferous tubules.

4. The vasa aberrantia. These are two groups of tubules derived from the Wolffian body. The first is the superior vas aberrans which arises from the rate testis and ends blindly in the globus major; when more than one such tubule is present, they are then known as Kobelt's tubes. The second is the inferior vas aberrans or vas aberrans of Haller; this is derived from the caudal end of the vas deferens itself or from the epididymis and it passes upwards along the cord for 2 to 5 cm. and ends blindly; it represents a persisting canal of the Wolffian body.

II. Pathology

1. Pathogenesis

When cysts are exposed at operation or specimens dissected, it is sometimes possible to arrive at a fairly certain conclusion as to their aetiology while in other cases the point of origin of the cysts cannot be determined. The cysts appear, however, to fall into three groups:

a) Cysts originating in embryonic remains.
b) Retention cysts of the epididymis.
c) Polycystic disease of the epididymis.

a) Cysts originating in embryonic remains

Cystic changes may occur in the sessile or non-pedunculated hydatid of Morgagni (appendix testis), in the pedunculated hydatid of Morgagni (or appendix epididymis), the organ of Giraldès (or paradidymis), and the vasa aberrantia. Spermatozoa are found in those cysts arising in structures which maintain a connection with the seminiferous system, namely the vas aberrans and the appendix testis. Cysts which have no such connection, namely the appendix epididymis, and the organ of Giraldès, do not contain spermatozoa.

b) Retention cysts of the epididymis

The majority of cases of cysts of the epididymis are believed to be retention cysts, resulting from obstruction by one or more of the spermatic tubules and such cysts may occur in any position in the seminal tract distal to the terminal portion of the vas deferens; they may result from stricture following old infection, from sclerosis arising from inflammation or even from injury, or possibly from faulty development of the vasa efferentia, which may fail to continue to form the coni vasculosi. These cysts occur most frequently as a result of the dilatation of the vasa efferentia and less commonly following dilatation of the coni vasculosi of the globus major (ABESHOUSE). They are usually single but may be multiple, they lie in the groove between the testis and the epididymis and they are usually small though they may be several centimetres in diameter.

HOCHENEGG has suggested an anatomical basis for the formation of cysts in the vasa efferentia on the basis of the anatomy of different parts of the spermatic tubules. The width of the seminiferous tubules of the testis varies between 0.1 and 0.2 mm.; the diameter desreases to 0.05 to 0.08 mm. in the rete testis; in the vasa efferentia it expands to 0.6 mm. and in the coni vasculosi it contracts to 0.2 mm. The vasa efferentia therefore represent a dilated part of the tube situated between two portions with constricted lumina; moreover the vasa efferentia are surrounded by loose connective tissue which does not prevent their dilatation, whereas the other parts of the spermatic tubule are encased in a firm capsule in either the testis of the epididymis, which would discourage cystic dilatation (ABESHOUSE).

c) Polycystic disease of the epididymis

This condition, to which the term cystic embryoma has also been applied, refers to a condition of multiple retention cysts, which are usually a millimetre or two in diameter and rarely exceed 0.5 cms., and which occupy a portion or a whole of the epididymis (CORLETTE). McCRAE states that every grade can be found from cases in which there is one cyst, to one in which the entire epididymis is involved. The cause of the condition is not known but it has been suggested that it is analogous to the cysts seen in polycystic disease of the kidney and may result from a non-union of the vasa efferentia derived from the mesonephros, with those tubules of the epididymis which are derived from the Wolffian duct. There is nothing to suggest that this condition is neoplastic in origin. In such cases the normal tissue of the epididymis is largely destroyed and replaced by cysts.

2. Causative factors

Since it is very rarely possible to obtain any history from a patient which may throw light upon the predisposing cause of the cyst, this is really a matter of conjecture, but various theories have been advanced.

In a few cases there has been a history of injury, such as a blow or a strain. Possibly an injury may cause some distraction between the more mobile testis and the more fixed epididymis, thus resulting in a tear of the vasa efferentia where they pierce the tunica albuginea: cysts may occur in consequence (HOCHENEGG). A history of injury was reported in 8 of 15 cases by KREBS, and by ABELL in 4 of 32 cases. Ligature of the vas deferens, commonly practised along with the operation of prostatectomy, does not lead to the development of cysts of the epididymis. In the great majority of cases there is no history of injury so that possibly a supposed history of injury may merely have drawn attention to a cyst which was already present.

CAVASSE and others considered that sexual abstinence or unsatisfied sexual desire were aetiological factors, arguing that such abstinence may result in dilatation of the spermatic tubules; in fact, any excess of spermatic fluid escapes as a seminal emission so that this theory has little to commend it.

It is inevitable that an occasional patient with a cyst of the epididymis gives a history of gonorrhoea, but the coincidence is rare and there is little to support the view that such an infection can lead to a cyst. Moreover, though a gonorrhoeal epididymitis used to be common, cysts of the epididymis are less common. The same arguments apply to the other forms of epididymitis, namely non-specific, tuberculous and pyogenic.

3. Pathological anatomy

The cysts are of two kinds, extra-vaginal and intra-vaginal, the former being by far the commoner.

Extra-vaginal cysts originate in the vasa efferentia and probably occasionally in the vasa aberrantia superioris. They lie outside the tunica vaginalis. They are attached to the rete testis in the groove between the testis and the epididymis and as they enlarge they may displace the epididymis backwards At first. there may be some projection into the vaginal cavity but as they grow upwards. they tend to enlarge into the tissues of the lowest part of the cord. A rarer primary type of extra-vaginal cyst is that which is derived from the paradidymis or organ of Giraldès and because of the anatomical position of that small organ. the cyst grows into the tissues of the spermatic cord without displacing the epididymis. Cysts of this group are usually unilocular though they may be multilocular if more than one cyst arises at the same time. Extra-vaginal cysts tend to be larger than the intra-vaginal cysts.

Intra-vaginal cysts arise most commonly in the vasa efferentia and rarely in the appendix testis or sessile hydatid of Morgagni. They present as small, spherical masses above the testicle and may be closely attached to the tunica albuginea testis or to the globus major of the epididymis. The cyst wall is usually very thin and it projects into the cavity of the tunica vaginalis.

Unilocular cysts are usually extra-vaginal but may be intravaginal. Some of the larger ones extend in an upward direction and reach a considerable size. Multilocular cysts are almost always extra-vaginal though there may be a communication between the loculi; they usually arise from the vasa efferentia. The cysts have thin walls and contain a clear fluid in which spermatozoa are usually found.

The cysts of polycystic disease are usually very thin-walled and contain a clear, colourless fluid. In the worst cases the cystic change extends throughout the epididymis and no normal epididymal tissue is recognisable. The testis usually appears to be of normal size and not atrophic.

The structure of the cyst wall consists of strands of fibrous connective tissue in which may be found a few elastic fibres. In the walls of the cysts which have their origin in the vasa efferentia and the duct of the epididymis, smooth muscle fibres may be found (ABESHOUSE, HERBUT). Long-standing cysts may be thick-walled but usually the walls are thin unless they have been subjected to repeated tappings when they are fibrosed and infiltrated with lymphocytes. The cyst wall may undergo hyaline degeneration; calcification has been reported occasionally (ENGLISH, ROCHE, MARSAN). Ossification was reported in one case by CAMPBELL. MATHÉ described an atypical cyst of the epididymis, the wall of which was lined with irregularly necrotic tissue which contained bone. The cyst is lined by a low columnar or cuboidal epithelium. In older cysts the epithelium may be of the flattened pavement type. Very rarely ciliated epithelium has been observed (ABESHOUSE).

Usually the body of the testis is normal but atrophic changes of part of the epididymis and of the testis have been recorded when the cyst is large and of long-standing. Hydrocele of the tunica vaginalis may co-exist with a cyst of the epididymis, and the common conditions of inguinal hernia and varicocele may also be found in association with it.

If there is a communication between the cyst and the seminiferous system, active spermatozoa will be found in the cyst and the fluid is opalescent; dead spermatozoa may be present in a cyst which formerly had such a communication. In small retention cysts and in the cysts of polycystic disease the fluid is clear and does not contain spermatozoa. The fluid is alkaline or neutral in reaction, has a specific gravity of 1000 to 1009, and contains 0.2 to 0.5 per cent albumin (ABESHOUSE). Lymphocytes, epithelial cells, fat globules, as well as spermatozoa, may be found if the fluid is centrifuged.

III. Aetiology

Though the series of reported cases are not numerous and though some writers have regarded the condition as uncommon, it is, in fact, a condition very commonly seen in the out-patient clinic in middle-aged and elderly men if the scrotum is regularly examined. The condition is often missed, either because the patient himself has not noticed any abnormality, or because, being symptomless, the patient has not reported it. ROLNICK found the incidence of cyst of the epididymis to be 1 per cent of 55,000 men examined during a period of four years. HOCHENEGG found 27 instances of the condition in 362 testicles of cadavers. Many short series, which have been reported as representing the incidence in a hospital over a period of years, are misleading as regards the incidence of the condition, since the patient commonly receives no surgical treatment.

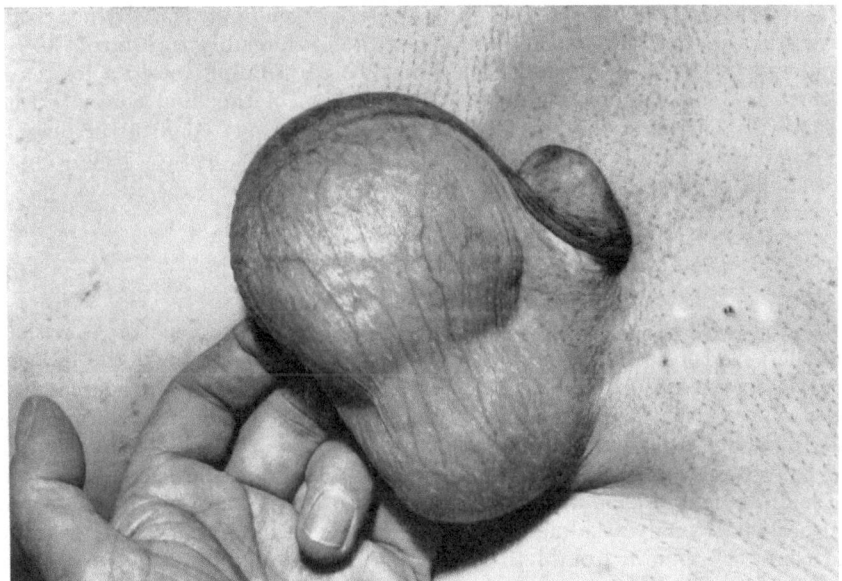

Fig. 24. Spermatocele in association with the right testis. The testis is at the lower pole of the swelling and groove separates it from the cyst

The condition is most frequently seen between the ages of 35 and 60 but the limits of 20 and 90 would probably include all cases. In WHITNEY's series of 90 cases, only 1 case was below the age of 20 and one above the age of 80, the average age being 49.8 years. In WAKELEY's series of 22 cases, there were 2 patients aged fourteen and eighteen respectively. The condition is usually unilateral, both sides being affected approximately equally. The condition is bilateral in between 10 and 18 per cent of cases (CAMPBELL, ABELL).

Cysts of the epididymis grow very slowly. and since they give no trouble and do not interfere with bodily function it is often not possible to say how long a cyst has been present when the patient reports for examination; many have been present for years before they are seen by the surgeon.

IV. Symptoms

The patient complains of a lump in the scrotum or one may be found by the surgeon on routine examination, the patient having experienced no symptoms. If the cyst is large, the patient may experience some dragging discomfort

and mechanical inconvenience from the clothing while a few patients complain
of pain which may be due to dragging on the testis and the spermatic cord. Pain
in the testicle following sexual excitment is said by CAVASSE to be pathognomonic
of cysts of the epididymis.

On examination, the body of the testis can be clearly differentiated below
and in front of the cyst, by its shape and by eliciting testicular tenderness. The
axis of the testis may be changed from the vertical to the transverse direction.
The cyst itself is ovoid or spherical, soft and fluctuant. In intravaginal cysts
a small, rounded mass can be felt at the upper pole of the testis to which it is
in fairly close relation. Large cysts may be readily separated from the testis
on palpation or they may be closely applied or almost moulded to the upper
pole of the testis; the testis can, however, always be differentiated from the
cyst by its difference in consistency, by the presence of testicular sensation and
by the discovery of a groove which is always present between the body of the
testis and the cyst. Cysts containing clear fluid are readily translucent to light,
while cysts which contain milky fluid may be transilluminant to a less extent.
If there is doubt about the diagnosis, aspiration of the fluid contents with a
needle will assist, and the presence of spermatozoa would confirm the diagnosis.

Haemorrhage into a cyst following injury, and infection (following, for
instance, tapping) are rare complications.

V. Differential diagnosis
1. Hydrocele

In hydrocele of the tunica vaginalis the swelling is ovoid or pear-shaped
and the testis projects into the sac. The position of the testis at the lower and
posterior part of the hydrocele sac may be recognised by the difference in the
consistency of that part of the wall of the sac even though the outline of the
body of the testis cannot be determined. In the case of cysts of the epididymis,
the body of the testis can be palpated separately from the cyst. A hydrocele
is usually more transilluminant to light than a cyst. The fluid from a vaginal
hydrocele is strongly alkaline in reaction, has a specific gravity of 1020 to 1026
and contains albumin from 4 to 7 per cent (ABESHOUSE) and does not contain
spermatozoa.

2. Encysted hydrocele of the cord

Encysted hydrocele of the cord is a rounded cystic swelling in the spermatic
cord, and lies some distance above the body of the testis and epididymis.

3. Haematocele

Usually there is a history of injury except in long-standing cases of chronic
haematocele. In a recent haematocele there is a boggy swelling of the tunica
vaginalis associated with bruising of the scrotum; aspiration will produce blood
or blood-stained fluid. A chronic haematocele is a firm or hard swelling of the
tunica vaginalis, quite different from a cyst of the epididymis.

4. Hernia

Hernia gives rise to a swelling in the scrotum which descends from the abdomen and its contents can be reduced into the peritoneal cavity. It is only
rarely that a cyst of the epididymis extends to the level of the external ring
and it never extends along the inguinal canal.

5. Varicocele

Varicocele should not be mistaken for a cyst of the epididymis; the dilated veins, which feel like a bag of worms and which reduce in size when the patient is recumbent, are quite different from the rounded, soft, persistent swelling of a cyst.

6. Nodules in the epididymis

Nodules in the epididymis resulting from fibrosis following past attacks of epididymitis, may be mistaken for small cysts; there is usually a history of an acute epididymitis in such cases.

7. Torsion of the testis, gummata and solid tumours of the testis

Torsion of the testis, gummata and solid tumours of the testis should not give rise to difficulty in diagnosis. Acute and chronic epididymitis give rise to characteristic swellings of a part or the whole of the epididymis which are quite different from the localised spherical swellings of a cyst.

VI. Treatment

Small cysts being harmless and symptomless, should receive no treatment. It should be explained to the patient that the condition is devoid of danger and that it will not be associated with any loss of function of the genital organs nor with any inflammatory nor malignant change. Larger cysts, if they cause dragging pain or discomfort, should be treated by operation. Some authorities advise surgical treatment because of the possibility of pressure atrophy of the testis.

1. Non-operative

Aspiration alone does not cure the cyst since it rapidly fills up again; moreover there is a slight danger of introducing infection.

Aspiration followed by injection with quinine urethane or sodium morrhuate has been performed, but the results of such treatment are often not permanent and do not appear to give the same percentage of successes as similar treatment for hydrocele of the tunica vaginalis (ROSS, MILBERT); the details of such treatment will be found in the section on "Hydrocele".

2. Operative

Surgical excision should be done for large cysts and for those which cause aching or pain. An incision is made through the scrotal skin overlying the cyst; an alternative is to make a low inguinal incision delivering the testis with the attached cyst into the wound, but this seems unnecessarily remote. The incision is carried through the coverings, to the wall of the cyst, which is then dissected out; it usually separates easily, and can often be removed intact. Occasionally a portion of the epididymis near to the cyst must be excised as well. The blood vessels entering the cyst wall from the epididymis, must be ligatured.

In polycystic disease, the part (or sometimes the whole) of the epididymis, affected by the cystic change, should be excised, but care should be taken to avoid injury to the blood supply of the testis.

Very rarely, orchidectomy may be indicated if the cyst is large and the testis is the seat of partial atrophy.

The results of operation are good and there should be no recurrence, though occasionally a new cyst develops. In the case of polycystic disease, recurrence may take place if the entire cystic zone is not excised.

N. Torsion of the testicle and of the appendages of the testis and the epididymis

I. Torsion of the testicle

By torsion of the testicle is meant the sudden twisting or rotation of the testicle either extravaginally or more commonly within the tunica vaginalis. The first case was probably described by DELASIARVE in 1840.

1. Aetiology

The condition is not uncommon. DONOVAN reported 163 cases. CAMPBELL stated that 500 cases of torsion had been recorded in the literature to 1951. Torsion of the testicle affects males of any age from the new born to old age, but it occurs most commonly in the younger age groups during adolescence and early adult life. ABESHOUSE reported the condition in a man aged sixty-eight. In O'CONOR's series the average age of 127 cases was fourteen years, and in ABESHOUSE's series the average age of 153 collected cases was 17.7 years. In 32 cases reported by SCUDDER, the average age was less then 24 in three-quarters of the cases. Of the 150 cases reported by WALLENSTEIN 90 were right-sided and 60 left-sided.

While somewhat more than half of the reported cases occur in patients having incompletely descended testes, yet because of the overall low incidence of that abnormality, such testes are clearly more prone to undergo torsion than is the normally descended organ. Incomplete descent was present in 75 of 127 cases of torsion collected by O'CONOR (three cases being examples of torsion of intra-abdominal testes), in 90 of 150 collected by WALLENSTEIN and 69 of 153 collected by ABESHOUSE. In OWEN and CAMPBELL's series one-half occurred in cases with incompletely descended testes.

Several cases of torsion of the spermatic cord in new-born infants have been reported in recent years. The first case was observed by TAYLOR who observed a case four hours after birth. Others have been recorded by RHYNE et al., GLASER and WALLIS, JAMES, BIORN and DAVIS. Thirteen cases have so far been reported. Probably the torsion has occurred in intra-uterine life. Birth trauma was believed to be the cause in CAMPBELL's (1951) case, though in that recorded by RHYNE et al. the child was delivered by Caesarean section. Many cases of torsion have been reported occurring in infants a few days or weeks after birth (LANGLEY).

Bilateral torsion is not uncommon, the second testicle undergoing torsion months or years later. Thus EWART and HOFFMAN found 26 cases of bilateral torsion in 489 cases from the literature; and MOULDER similarly found that 27 of 350 collected cases were bilateral.

2. Pathology

a) Pathological anatomy

Torsion of the testicle is of three kinds anatomically, namely torsion of the lower end of the spermatic cord outside the tunica vaginalis, the twist involving the entire testis and epididymis and the tunica vaginalis (extra-vaginal torsion); intra-vaginal torsion, the twist taking place at the lower end of the spermatic cord within the tunica vaginalis; and torsion through the mesorchium between testis and epididymis.

α) Extra-vaginal torsion

Extra-vaginal torsion implies the rotation of the testis, epididymis and the tunica vaginalis within their scrotal coverings, the rotation taking place at the spermatic cord outside the tunica vaginalis. More than a normal degree of mobility of these structures is needed if such torsion is to be possible. In infancy, as SIR ASTLEY COOPER was the first to point out, and CAMPBELL (1951) has confirmed, the attachment of the tunica vaginalis to the scrotal skin and the thin subcutaneous tissue is extremely loose, so that in fact the testis and epididymis along with the tunica vaginalis in the infant cadaver, can be almost lifted out with scarcely any dissection. Thus in the newly born and the infant, torsion of the spermatic cord can occur at a level higher than the tunica vaginalis, resulting in gangrene of the testis, the epididymis and the tunical coverings; in fact, extra-vaginal torsion is the usual form of torsion in the new born. It is not necessary for the gubernaculum to be ruptured in such cases though it may be. Extra-vaginal torsion must be excessively rare in patients other than infants. ALLEN and ANDREWS reported a true case of extra-vaginal torsion in a child of 10 months. YOUNG and DAVIS state that the condition can only occur other than in infants, when a severe external injury tears an undescended testis from its scrotal attachments. The testis, except in infants, does not possess a sufficient range of movement to allow rotation (except perhaps a partial rotation), to occur in any position other than intra-vaginally.

Fig. 25. Specimen showing intra-vaginal torsion of the testis. The testis and epididymis show a fairly uniform congestion resulting from occlusion of the spermatic veins. The twist in the cord occurs immediately above the epididymis

β) Intra-vaginal torsion

MUSCHAT has described in considerable detail the pathological anatomy of testicular torsion. In normal descent, the testis with the epididymis are drawn downwards behind the peritoneal space into the scrotum, helped by the traction of the gubernaculum, which keeps them there by the development of strong surrounding attachments. The testis becomes almost completely covered by peritoneum by invagination, while the epididymis is covered mainly on its lateral aspect, its posterior portion remaining outside the vaginal sac and adherent to the inner wall of the scrotum. Normally, when the tunica vaginalis is opened, it is found to be impossible to rotate the testis to any extent around its axis because of the strong attachment between the testis and the epididymis to the inner wall of the scrotum; rotation would only be possible if the testis were freed from the epididymis by dissection.

In order that the testis may rotate, it must be freely movable and free from any lateral attachments, and must be suspended by the spermatic cord as by a pedicle, within the cavity of the tunica vaginalis, circumstances which imply some disturbance in the normal descent of the organ. In cases of torsion the relations of the testis and epididymis to the tunica vaginalis do, in fact, differ from those of the normal. The testis and the epididymis begin to bulge into the tunica vaginalis until the latter completely surrounds the testis and epididymis, and when this occurs the lower portion of the spermatic cord is similarly invested with peritoneum above the testis, so that testis, epididymis and the distal part of the spermatic cord hang freely suspended by the cord, within the vaginal sac. There are then no lateral attachments between the scrotum, the testis and the epididymis. The reflection of the tunica vaginalis on to the testis in such cases of high investment of the cord, takes place only at the highest point of the vaginal sac.

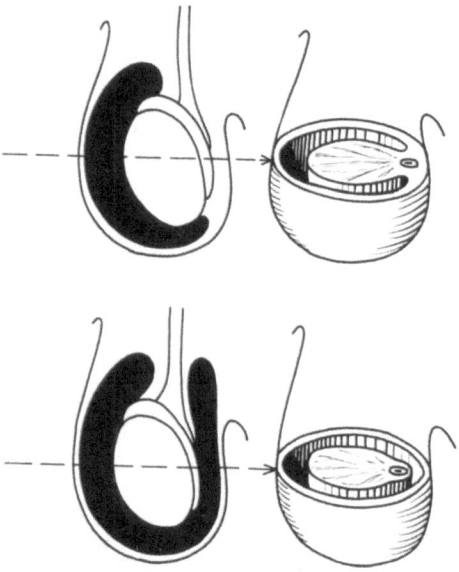

Fig. 26. Schematic drawing showing the relations of the tunica vaginalis to the testis and epididymis: 1. In normal cases the investment of the tunica is incomplete. 2. In cases with torsion, the testis and epididymis are very largely invested by the tunica vaginalis. [Adapted from WHEELER and CLARK, New Engl. J. Med. **247**, 974 (1952).] (By courtesy of the Editors)

MELTZER has pointed out that there are other abnormalities present in varying degree, in cases of torsion. These include a very roomy tunica vaginalis, the absence of the gubernaculum testis, and the posterior mesorchium, the absence of a scrotal ligament, and an abnormal attachment of the vessels and their coverings to the lower pole of the testis and to the globus minor of the epididymis, resulting in the formation of a narrow pedicle instead of a broad band called by HAMILTON, PEYTON and MANTELL the bell-clapper deformity or intra-tunicary pedicle; in addition there may be elongation of the globus minor, and loose connections between the testis and the epididymis. Normally, during their descent, the testis and epididymis rotate together on their axes through 180°, the lower pole of the testis becoming the permanent upper pole, and the upper pole, the permanent lower pole, the epididymis describing the same degree of rotation. Any interference with this physiological rotation may result in the relative positions of these organs becoming fixed in any stage of this rotation and probably more liable to undergo torsion.

MUSCHAT believes that the anatomy of the cremaster muscle in cases of torsion is the second aetiological factor. Although topographically the cremaster appears to be the same as in the normal individual, yet because of the high investment of the spermatic cord by the vaginal sac, its relative topography is changed. Instead of extending over the outer part of the vaginal sac as in the normal individual, in cases of high investment of the cord, the cremaster fibres are carried down into the inner part of the vaginal sac to the lowest point of the spermatic cord. A strong contraction of the muscle may result in rotation of a freely moveable testis and epididymis.

Some authors have stated that the left testis turns in a clock-wise direction, the right testis in a counter clock-wise direction (EWART and HOFFMAN, CAMPBELL); WHEELER and CLARK found that this was true in their series. The torsion may be from half a turn to four or five complete turns. The resulting twist usually causes infarction of the testis within a few hours of the onset if the testis is not untwisted by open operation; some cases of relapsing torsion, however, appear to resolve spontaneously.

γ) Torsion between the testis and the epididymis

The third variety of torsion, which is rare, occurs when the testis and epididymis are very loosely united to each other, giving an unusually long mesorchium, allowing rotation to take place at the junction of the testis and epididymis.

b) Pathological changes

At first the veins are completely occluded, giving rise to deep blue congestion and swelling of the twisted organ and later the arteries are also occluded. The tunica vaginalis contains blood or blood-stained fluid. If the torsion is not reduced, the testis and epididymis undergo massive necrosis, leading to complete atrophy and replacement by fibrous tissue. If a blood-borne infection follows, suppuration, abscess and sinus formation may occur.

In several of the cases reported in the new-born group, although the testicle is infarcted, satisfactory evidence of torsion was not found at operation, though in others the spermatic cord was found to be twisted extra-vaginally above the testicle, the tunica vaginalis being involved in the twist. It is possible that in this group there are two separate clinical entities, namely infarction of the testis without torsion (CAMPBELL, MacLEAN and RAVICH), and infarction resulting from torsion; some writers believe that the first group can be explained by assuming that torsion of the testis has taken place and caused gangrene of the testis, and that the organ has subsequently untwisted itself in some way. On the other hand, it would have been expected that the swelling of the testis which always accompanies the torsion would have led to the fixation of the mobile tunica vaginalis to the scrotum.

c) Rarer variants of torsion

A testis which is incompletely descended and which is also the seat of a malignant growth, may undergo torsion. In WALLENSTEIN'S 90 cases of torsion of incompletely descended testes, 4 contained malignant growths. In ABESHOUSE'S series, in which there were 84 cases of torsion of normally descended testes, 3 were found to contain malignant tumours.

Torsion of an undescended abdominal testis occurs occasionally, the first case having been reported by PEARLMAN. Several of the reported cases have been associated with a malignant testicular neoplasm. BELLER collected 10 cases of torsion of an intra-abdominal testis, some of which were the seat of malignant growth (GERSTER, STILES, LeCONTE, PEARLMAN, CHITTY). WHIT-TINGTON reported an unusual case of torsion of an abdominal testis which was associated with transposition of the abdominal viscera. CHARENDOFF et al. collected a further 7 cases from the literature and added one of his own, and of these the cases of KEY, VASTOLA, JOSSA, BURKE and BENNETT and SHAW were the seat of malignant tumours. In fact, 13 of the 20 recorded cases of torsion of an intra-abdominal testis were associated with malignancy, and one case was bilateral (KEY).

POPOV reported a case of torsion of the testis which presented with a swelling in the right groin in a patient of 50 who had previously been thought to be a female.

BOGGON and also WILSON and LITTLER reported torsion in a supernumerary testis in cases of polyorchidism. In the two cases of WILSON and LITTLER who were aged 35 and 44 respectively, the patients complained of a painful swelling in the groin which led to exploration, revealing the abnormal supernumerary testis. Each testis had a cord which joined up with the spermatic cord of the true testis of the same side but there was no vas deferens associated with it; both patients gave a history of previous transient attacks of pain, so that there had probably been earlier incidents of sub-acute torsion; the twisted accessory testis was removed in each case.

3. Clinical picture and diagnosis

The patient may be known to have an incompletely descended testis, or the affected testis may lie normally in the scrotum. Occasionally there is a history of previous short attacks of testicular pain, accompanied by local tenderness, lasting for half-an-hour or less and suggesting earlier attacks of torsion. He is suddenly seized with acute pain in the testicle, which sometimes radiates to the groin or even to the loin, and this is usually accompanied by nausea or vomiting and sometimes by lower abdominal pain. The affected testicle swells and becomes exquisitely tender and usually lies higher than normal in the scrotum; if the testicle is incompletely descended, the tender swelling usually lies in the inguinal canal or at the external inguinal ring, and the corresponding half of the scrotum is then empty. The skin overlying the swelling, whether in the scrotum or in the inguinal canal, after some hours becomes dusky and oedematous and later red. There is often some elevation of the temperature. After two or three days in the absence of surgical treatment, because of the onset of gangrene of the testis, the local pain and tenderness become less severe.

Torsion of the testicle must be differentiated from strangulated hernia, from acute epididymitis and from torsion of the appendages of the testis and of the epididymis, and if the testis is incompletely descended, from acute lymph-adenitis of the groin.

A strangulated hernia, which will give rise to spasmodic umbilical pain and vomiting, gives a swelling which descends from the abdomen and which can be differentiated on palpation from the testis in the scrotum; in torsion of the testis the upper border of the swelling can usually be readily defined even though there is some oedema. A strangulated hernia is not very often associated with an incompletely descended testis though it occurs occasionally, while torsion of the testis in such cases is probably more common; in this group it may not be possible to differentiate a strangulated epiplocele from torsion, without an exploratory operation, but a strangulated hernia containing intestine will give rise to symptoms of intestinal obstruction.

In acute epididymitis the normal body of the testis can be felt anterior to the swollen and tender epididymis, and normal testicular sensation can be elicited. There may be an associated prostatitis or urinary infection, which will give rise to frequency of micturition, dysuria and abnormalities in the urine. PREHN (1934) described a useful sign which helps in the differentiation; elevation of the swelling on a pillow placed between the patient's legs, gives some relief to the pain of epididymitis but not to that of torsion.

In torsion of the appendages of the testis or of the epididymis the pain is not as severe as that in cases of torsion of the testis; if seen in the early stage, there is a tender swelling in or near the globus major of the epididymis or at the upper pole of the testis, while if seen late there is a more diffuse, tender swelling apparently of the entire testis and epididymis and of the scrotal skin, and it may then only be possible to differentiate the condition from torsion of the testis by exploration.

In inguinal lymphadenitis, a primary focus of infection will be found in the skin of the foot, leg or anal region.

The chronic swellings of the scrotum, such as hydrocele, cyst of the epididymis and chronic epididymitis, should not be mistaken for torsion because of the absence of pain and tenderness.

In torsion of the testis in the new born, the scrotal swelling is discovered at birth and the infant does not seem to experience pain; the testicle is stony hard and irregular in outline and the overlying skin red and oedematous; evidently the torsion has occurred in utero and the gangrenous testis no longer gives rise to painful stimuli. DONOVAN similary reported the absence of pain in two infants with torsion aged respectively 8 and 10 months.

Torsion of an abdominal testis is usually associated with severe pain in the umbilical region or in the corresponding lower quadrant of the abdomen and with much local tenderness and muscle guarding; the temperature may be raised, thus mimicking acute appendicitis (WHITTINGTON). Moreover since the testis swells as a consequence of the torsion, a palpable lower abdominal swelling may be present, but if the corresponding side of the scrotum is seen to be empty, the correct diagnosis should be made. If a growth, such as a seminoma, complicates the twisted abdominal testis, a biggish palpable mass may be found, not necessarily on the same side as the half-empty scrotum, thus leading to difficulty in diagnosis (CHARENDOFF et al.).

A condition of partial or relapsing torsion may occur either in normal or incompletely descended testes. After a short period of pain and swelling, which may last a few minutes to an hour, the organ becomes spontaneously untwisted, the testicular circulation is restored to normal and the pain and local swelling disappear. Some patients have recurrent attacks at intervals of weeks or months before a final prolonged attack of torsion occurs. CAMPBELL records a case in which the attack of torsion appeared to have been stimulated by the administration of hormones which had been administered to induce descent of the organ. Possibly atrophy of the testicle may occur as a result of such repeated attacks though this entails serious interference with the circulation in at least one of the attacks. WALKER recorded a case of a man of 24 who had six attacks of torsion during three years, only one of which lasted more than twenty-four hours; the right testicle, however, was reduced to a small, round, hard body the size of a hazel-nut lying near the external abdominal ring.

4. Treatment

In any case of doubt, and in all cases in which the diagnosis of torsion has been made, operative exposure of the testicle is indicated. It is rare, and only in cases of very early diagnosis, that reduction of the torsion with restoration of the viability of the organ, is possible. According to DEMING and CLARKE if the circulation has been seriously interfered with for four hours, gangrene of the testis is inevitable; they report, however, preservation of the testis in 9 of 20 cases. If the testis should be viable after untwisting, it is then fixed to the inner wall of the tunica vaginalis by several sutures to prevent recurrent rotation. According to RIBA and SCHMIDLAPP, orchidectomy is necessary in at least 75 per cent of the cases. In practice, gangrene of the testis and of the epididymis is nearly always found at operation and orchidectomy is then the only possible treatment. In cases of torsion associated with an incompletely descended testis, any associated hernia should be dealt with at the same time. In intra-abdominal torsion, the abdomen must be opened usually through an incision above the inguinal ligament, the pedicle ligated and the twisted organ with any associated growth removed; sometimes an adherent loop of small intestine must be separated from the lump.

In none of the cases of torsion reported in the new born was the testis viable. Orchidectomy has been done in some cases as early as a few hours after birth, while in others operation has been deferred until several days later. GLASER and WALLIS described 2 cases in which they performed no operation, the infant not being distressed; in eight and twenty-four months later respectively the testes had undergone almost total atrophy; they suggest that orchidectomy in such cases is scarcely necessary.

The treatment of relpsing torsion is to operate, expose the testicle in the tunica vaginalis and deal with any associated hernia. The testicle and the epididymis are then secured by interrupted catgut sutures to the parietal part of the tunica vaginalis to prevent further rotation; the visceral part of the tunica covering the epididymis may be scarified so as to cause the latter to adhere over a broad area to the parietal part. Such treatment may be successful in preventing further attacks.

Since bilateral torsion occurs not uncommonly, and since it has been shown that the abnormal anatomical arrangements of the testis predisposing to torsion are usually present on both sides, there is good reason to advise orchidopexy on the sound side at a second operation, to prevent torsion.

II. Torsion of the appendages of the testis and the epididymis

1. Anatomy of the appendages

The appendix of the testis, alternatively called the sessile hydatid of Morgagni, or the sessile hydatid, is a small, rounded or ovoid, and usually pedunculated structure attached to the front of the upper pole of the testis or to the groove between the testis and the head of the epididymis. It is reddish in colour and soft in consistency. The pedicle may be short and stumpy or quite long, when it is perhaps more liable to undergo torsion (HOWSER and RIVER). Microscopically it consists of a stroma of delicate fibrous tissue with an abundant supply of capillaries and small arteries and veins. The surface of the swelling is covered with tall columnar epithelium with gland-like recesses. The stroma may contain tubules lined with cylindrical epithelium which is occasionally ciliated (KOELLIKER), and may show evidence of proliferation of fibroblasts and a few round cells, and also irregular spaces, the walls of which may contain deposits of calcium salts and the lumina may also show partially calcified necrotic tissue (RANDALL).

The appendix of the epididymis (alternatively called the pedunculated hydatid) is a small, pedunculated body attached to the front of the head of the epididymis; it is not so frequently present as the appendix of the testis.

The paradidymis or organ of Giraldès is a collection of tubules lying on the front of the lowest part of the spermatic cord above the head of the epididymis.

The vas aberrans is a small tubule or tubules which lie near the middle of the groove between the epididymis and the testis.

2. Torsion of the appendix testis

A case of either torsion of the appendix testis or of the appendix epididymis was first reported by OMBREDANNE in 1913. COLT reported the first successfully diagnosed case of torsion of the appendix testis in 1922, and this was followed by the cases of SHATTOCK and of WALTON, also in 1922. MOUCHET recognised the existence of torsion of the appendix of the testis as explaining some of the cases in infants and adolescents hitherto diagnosed as sub-acute epididymo-orchitis. In a series of several papers he reported 13 personal cases, which suggests that the condition is not excessively rare once its existence is known. He also described a mild type of the same lesion (les formes frustes de la torsion de l'hydatid sessile de Morgagni). Further cases were reported by RUTOLO, GARCIA and CUCULLU, LOPEZ, BROSTER and COYTE.

DIX, in a comprehensive review of the subject, collected 53 cases of torsion of the appendages from the literature including 2 of his own and 6 unpublished cases of MOUCHET; 48 of these cases were cases of torsion of the appendix testis (or hydatid of Morgagni), which is much the commonest in other series, and there were 5 others. Further reports were published by RANDALL, who found 17 additional cases in the literature between 1931 and 1937, which included one case of torsion of a cyst of the vas aberrans; by SCOTT, who collected 85 cases to 1940, and by VERMEULEN and HAGERTY, COPPRIDGE and ROBERTS, and HAMILTON et al. (who reported the simultaneous torsion of two appendices of the epididymis of the same testis). SEIDEL and YEAW in 1950 added 8 of their own cases (4 being cases of torsion of the appendix testis) which brought the total of published cases to that date to 107. AMBROSE and SKANDALAKIS reported a case of torsion of the appendix testis on one side which was followed six weeks later by a similar condition on the other; both sides were successfully operated on.

a) Aetiology

Any one of these structures may rotate on its pedicle and give rise to clinical symptoms, rotation of the appendix testis being not uncommon and rotations of the other appendages rare. The reason for such rotation is not known. Usually there is no definite history of injury; possibly some cases occur following a sudden movement since cases have been reported in boys after rising suddenly from bed or while playing a game. Whether the size of the hydatid influences torsion is not known but possibly the narrowness of the pedicle makes such an accident more likely to occur.

Torsion of the appendix testis occurs most commonly in boys up to or at the age of puberty. 27 of 42 cases in DIX's series occurred between the ages of 11 and 14, though 4 cases occurred under the age of 5. The two sides are almost equally affected.

Torsion of the appendix testis is occasionally associated with an appendix in an imperfectly descended testis lying either in the inguinal canal at the external inguinal ring, or in the groin (SHATTOCK, FOUCAULT, GARCIA and CUCULLU, MONCALVI).

b) Clinical picture and diagnosis

In some cases of torsion the patient gives a history of previous attacks of testicular pain which may have been less severe than that in the final attack (ROCHER and RIOUX, SHATTOCK, PAOLI). In SEIDEL and YEAW's cases, 2 of 7 gave a history of trauma. The patient experiences, sometimes after exercise, a sudden severe pain in the affected testicle, which may radiate to the inguinal region along the course of the spermatic cord; the pain, which is not of the same intensity as that associated with torsion of the testis, may become worse for the first day or two, when it usually abates with the onset of gangrene of the appendix, though in some cases it remains severe for many days up to the time of operation; in the "formes frustes" the pain may be very mild. Occasionally there is vomiting and there may be slight elevation of temperature.

On examination the findings vary with the inflammatory response in the tunica vaginalis and its coverings. A common finding is generalised redness and oedema of the affected side of the scrotum, associated with an apparently enlarged and tender testis and epididymis with a small secondary hydrocele; the entire testis and epididymis are then acutely tender on palpation and excessively so at the upper pole of the testis. In other cases, especially two or three days after the onset, and even when there is scrotal oedema, an acutely tender swelling can be distinguished at the upper pole of the testis and it may then be possible to palpate the body of the testis and the epididymis separately. In milder cases (as in those described by MOUCHET), there is no redness and oedema of the scrotum and no generalised swelling, but a small, tender mass can be palpated in relation to the upper pole of the testis.

In many recorded cases the correct diagnosis has been made following clinical examination, though many are wrongly diagnosed as acute or sub-acute epididymitis. Because of the severe redness and oedema of the scrotum, and the apparent considerable enlargement of the testis and epididymis, torsion of the testis has often been wrongly diagnosed; in such cases the true diagnosis may only be made at operation. In torsion of the testis, however, the pain and the constitutional signs are more severe than in torsion of the appendix testis.

c) Treatment

The testis is exposed through a scrotal or an inguinoscrotal incision. The tunica vaginalis, the coverings of which are oedematous, usually contains an appreciable quantity of clear or blood-stained fluid. The twisted appendix of the testis is seen as a rounded tumour, reddish, reddish-purple or black according to the degree of strangulation, in relation to the upper pole of the testis, its exact position partly depending on the length of the pedicle; the swelling is larger in size than an untwisted appendage. In some cases the pedicle does not at the time of operation show actual torsion of the pedicle and presumably in such cases it has untwisted itself. Cases have been reported in which the twisted hydatid became spontaneously untwisted when the tunica vaginalis was opened (MOUCHET, GARCIA and CUCULLU). When the twist is still present the number of turns has varied from one to four or five, and the torsion has been found to occur in either a clockwise or an anti-clockwise direction. The testis and epididymis may be normal or may show signs of some local inflammatory reaction. In SHATTOCK's case, there were two gangrenous testicular appendages.

The gangrenous swelling is excised and the base of the pedicle may be sutured with one or two firm catgut stitches, or left unsutured, bleeding points being coagulated by diathermy.

The operation is immediately successful in relieving pain and there should be no late complications. Probably in cases which are not operated on, the appendix undergoes total necrosis and ultimate fibrosis; the possibility, however, of a long period of pain before this occurs, as well as the acuteness of the immediate symptoms in most cases, makes early operation desirable in every case in which the condition is diagnosed.

3. Torsion of the other appendages

The clinical picture of torsion of the other appendages resembles very closely that of torsion of the appendix of the testis from which, indeed, it is clinically indistinguishable. Only a small number of these cases, however, have been recorded in the literature (CHATON, MICHEL, SOLIER & HUARD). A case of torsion of the appendix of the epididymis has been recorded by CHATON. Cases of torsion of small cysts of the vas aberrans have been recorded by RANDALL and PATCH, and two cases of torsion of the organ of Giraldès have been reported by SORREL.

References

A. Infarction of the kidney

ALDER-RACZ, A.: Ein Fall von Embolie der Nierenarterie. Z. Urol. **27**, 40 (1933). — ALLEN. E. V.: Symposium on treatment of hypertension. Proc. Mayo Clin. **27**, 495 (1952). — ASCHNER, P. W.: The clinical importance of aseptic infarction of the kidney. Amer. J. med. Sci. **164**, 386 (1922). — BARENBERG. L. H., N. W. GREENSTEIN, W. LEVY and S. B. ROSEN-BLUTH: Renal thrombosis with infarction complicating diarrhoea of the newborn. Amer. J. Dis. Child. **62**, 362 (1941). — BARNEY, J. D., and E. R. MINTZ: Infarcts of the kidney. J. Amer. med. Ass. **100**, 1 (1933). — BECKMAN. O. v.: Über Thrombose der Nierenvene bei Kindern. Verh. phys.-med. Ges. Würzb. **9**, 201 (1858). — BEHR, G.: Significance of embolic glomerular lesions of subacute streptococcus. Arch. intern. Med. **27**, 262 (1921). — BEN-ASHER. S.: Hypertension caused by renal infarction. Ann. intern. Med. **23**, 431 (1945). — BIERMATH. P.: Nierenvenenthrombose unter dem Bilde eines Nierentumors. Langenbecks Arch. klin. Chir. **143**, 902 (1926). — BOURNE, W. A.: Nephrectomy in hypertension due to renal artery infarction with superimposed infection of streptococcal endocarditis by fungus. Brit. med. J. **1954 II**, 271. — BOYD. C. H., and L. G. LEWIS: Nephrectomy for arterial hypertension. Preliminary report. J. Urol. (Baltimore) **39**, 627 (1938). — CAMPBELL, M. F.: Pediatric urology. New York: Macmillan & Co. 1937. — CAMPBELL, M. F., and W. F. MATTHEWS: Renal thrombosis in infancy. J. Pediat. **20**, 604 (1942). — CARVER, G. M.: Traumatic renal infarction concurrent with massive fat embolism. J. Urol. (Baltimore) **66**, 331 (1951). — DEROW, H. A., M. J. SCHLESINGER and H. A. SAVITZ: Chronic progressive occlusion of the inferior vena cava and the renal and portal veins. Arch. intern. Med. **63**, 626 (1939). — FALCI. E.: Sur la necrose du rein. J. Urol. méd. chir. **18**, 449 (1924). — FERIOZI, D., E. C. RICE. W. BURDICK and F. J. TROENDLE: Thrombosis of inferior vena cava and renal veins with haemorrhagic infarction in infancy. J. Pediat. **38**, 235 (1951). — FISHBERG, A. M.: Hypertension due to renal embolism. J. Amer. med. Ass. **119**, 551 (1942). — FOLEY, F. E. B.: Total embolism of the renal artery. Minn. Med. **23**, 136 (1940). — FRAENKEL, G. J.: Renal infarct as a space-occupying lesion. Brit. J. Urol. **28**, 145 (1956). — GAVAN, T. L., and N. KAUFMAN: Experimental renal infarct. I. Changes in succinic dehydrogenase and cyto-chrome oxidase activity. A.M.A. Arch. Path. **60**, 580 (1955). — GOEBEL, A., L. FRIEDERICI u. H. K. FUKAS: Phosphatidneubildung in Leber und Nieren bei der CO-Vergiftung an Ratten. Z. ges. exp. Med. **123**, 51 (1954). — HALLER jr., J. A., L. R. RADIGAN and A. G. MORROW: Hypertension due to segmental infarction of the kidney. Amer. J. Med. **22**, 303 (1957). — HALPERT, B., E. E. ERICKSON and F. GYORKEY: Experimental renal infarction. II. Histo-chemical, fatty and morphologic changes. A.M.A. Arch. Path. **62**, 386 (1956). — HEPLER, A. B.: Thrombosis of the renal veins. J. Urol. (Baltimore) **31**, 527 (1934). — HIRSCHBERG, H. A., and S. N. SOLL: Renal infarction of traumatic origin. J. Amer. med. Ass. **119**, 1088 (1942). — HOWARD, J. E., M. BERTHAG, D. M. GOULD and E. R. YENDT: Hypertension resulting from unilateral renal vascular disease and its relief by nephrectomy.

Bull. Johns Hopk. Hosp. **94**, 51 (1954). — HOXIE, H. J., and C. B. COGGIN: Statistical study
of 205 cases. Arch. intern. Med. **65**, 587 (1940). — HUNTER, R. A., and W. A. MCELMOYLE:
Renal infarct with hypertension. Lancet **1956 II**, 443. — HYMAN, A.: Clinical and surgical
aspects of renal neoplasms. Surg. Gynec. Obstet. **41**, 308 (1925). — KAUFMANN, H.: Zur
traumatischen Entstehung von Aortenaneurysmen. Mschr. Unfallheilk. **36**, 182 (1929). —
KOBERNICK, S. D., J. R. MOORE and F. W. WIGGLESWORTH: Thrombosis of the renal veins
with massive haemorrhagic infarction of the kidneys in childhood. Amer. J. Path. **27**, 435
(1951). — LANGLEY, G. J., and R. PLATT: Hypertension and unilateral kidney disease.
Quart. J. Med. **16**, 143 (1947). — LIEBERTHALL, F.: Tuberculous renal infarct. J. Urol.
37, 666 (1937). — LOEB, R. F.: Diseases of the kidney. Nelson's Loose Leaf Medicine, vol. 4,
p. 703. New York: T. Nelson & Sons 1938. — MCCLELLAND, C. Q., and J. P. HUGHES:
Thrombosis of renal veins in infants. J. Pediat. **36**, 214 (1950). — MARION, G.: Necrose du
rein droit par thrombose de la veine renale, nephrectomie, guerison. J. Urol. méd. chir.
15, 455 (1923). — MARSHALL, S., and E. WHAPHAM: Case of bilateral renal infarction in a
newborn infant. Lancet **1936 I**, 428. — *Massachusetts General Hospital:* Case No 32381.
Periarteritis nodosa. Multiple infarcts of liver and kidneys. New Engl. J. Med. **235**, 441
(1946). — Case report. Renal infarct. New Engl. J. Med. **253**, 1164 (1955). — MILBURN jr..
C. L.: Haemorrhagic infarction of kidneys in infants. Report of a unilateral case in an
eight-day old male infant with survival following successful nephrectomy. J. Pediat. **41**,
133 (1952). — MELICK, W. F., and A. E. VITT: Renal thrombosis. J. Urol. (Baltimore) **51**,
587 (1944). — MORISON, J. E.: Renal vein thrombosis and infarction in the newborn. Arch.
Dis. Childh. **20**, 129 (1945). — MUNGER, H. V.: Renal thrombosis. J. Urol. (Baltimore)
71, 144 (1954). — NORDWALL, W.: Un cas de thrombose bilaterale de la veine renale chez
un nouveau-ne. Acta paediat. (Uppsala) **14**, 186 (1932). — OSLER, W., and T. MCCRAE:
Modern medicine, its theory and practice, edit. 2, vol. 3, p. 789. Philadelphia: Lea and
Febiger 1914. — PERERA, G. A., and A. W. HAELIG: Clinical characteristics of hypertension
associated with unilateral renal disease. Circulation **6**, 549 (1952). — PERRY, C. G..
and A. L. TAYLOR: Thrombotic renal hypertension. J. Path. Bact. **51**, 369 (1940). —
PONTASSE, E. F.: Occlusion of a renal artery in a case of hypertension. Circulation **13**, 31
(1956). — PRINZMETAL, M., N. HIATT and L. J. TRAGERMAN: Hypertension in patient with
bilateral renal infarction. Clinical confirmation of experiments in animals. J. Amer. med.
Ass. **118**, 44 (1942). — RAYER, P. F. O.: Traite des maladies des reins. Paris: J. B. Baillière 1937. — RECKLINGHAUSEN, V.: Quoted by P. W. ASCHNER 1922. — REGAN, F. C.,
and E. G. CRABTREE: Renal infarction. Clinical and possible surgical entity. J. Urol. (Baltimore) **59**, 981 (1948). — REXFORD, W. K., and P. J. CONNOLLY: Traumatic infarction of
kidney. Amer. J. Surg. **68**, 250 (1945). — SANDBLOM, P.: Renal thrombosis with infarction
in the newborn. Acta paediat. (Uppsala) **35**, 160 (1948). — SCHWARTZ, J.: Renal infarcts.
Amer. J. Surg. **39**, 70 (1938). — SHEA, J. D., J. W. SCHWARTZ and R. E. KOBILAK: Thrombosis
of the left renal artery with hypertension. Case report. J. Urol. (Baltemore) **59**, 302 (1948). —
STEVENS, R. C., and A. J. TOMSYKOWSKI: Thrombosis of inferior vena cava and renal veins
with haemorrhagic renal infarction. J. Urol. (Baltimore) **72**, 120 (1954). — TEPLICK, J. G..
and M. W. YARROW: Arterial infarction of the kidney. Ann. intern. Med. **42**, 1041 (1955). —
WAINWRIGHT, J.: Hypertension following infarction. Lancet **1949 II**, 62. — WEBER. H.:
Kidneys with so-called fibrinous deposits from embolism of the renal arteries. Med. Tms
and Gaz. **2**, 663 (1864). — WEISS, F.: Zur Diagnostik des Niereninfarkts. Z. klin. Med.
120, 199 (1932). — WHITE, P. D., and R. R. PORTER: A note on pain and its reference in
cases of renal infarction. New Engl. J. Med. **224**, 728 (1941). — YUILE, C. L.: Obstructive
lesions of the main renal artery in relation to hypertension. Amer. J. med. Sci. **207**, 394
(1944). — ZUELZER, W. W., S. CHARLES, R. KURNETZ, W. A. NEWTON and F. FALLON:
Circulatory diseases of the kidneys in infancy and childhood. Amer. J. Dis. Child. **81**, 1 (1951).

B. Perirenal haematoma

ASCHNER, P. W., and M. E. KLINGER: Subcapsular renal haemorrhage causing anuria
of single kidney. J. Urol. (Baltimore) **65**, 777 (1951). — BEATTY, R. A.: Hydronephrosis
with spontaneous rupture. Penn. med. J. **38**, 806 (1955). — BECK, V.: Fall von Hämatom
des Nierenlagers. Verh. dtsch. Ges. Chir. **41**, 288 (1912). — BONET: Quoted by POLKEY
and VYNALECK 1933. — CACHIN, M., P. TANRET, A. MONSAIGEON and R. LEVILLAIN: Sur
deux cas de pancréatite aigue hémorragique avec hématome périrénal. Bull. Soc. méd.
Hôp. Paris **70**, 1188 (1954). — CARVER, J.: Circumrenal haematoma. Proc. roy. Soc. Med.
32, 547 (1939). — CATHELIN, F.: Hématonephrose sous-capsulaire dans une tuberculose
rénale caverneuse; néphrectomie. Bull. Soc. anat. Paris **9**, 401 (1907). — CIBERT, J., A. VA-
CHON and H. CAVAILHER: Spontaneous perirenal haematoma J. Urol. méd. chir. **50**, 65
(1942). — COENEN, H.: Das perirenale Hämatom. Beitr. klin. Chir. **70**, 494 (1910). —
COPPRIDGE, WM. M.: Spontaneous subcapsular renal haematoma. J. Urol. (Baltimore)

39, 733 (1938).—COUNSELLER, V. S., and J. L. EMMETT: Hydronephrosis with spontaneous rupture and perinephric hematoma simulating perinephritic abscess. Proc. Mayo Clin. **11**, 44 (1936). — DARO, A. F., and M. C. TODD: Spontaneous perirenal hematoma as post-partum complication. Amer. J. Obstet. Gynec. **42**, 140 (1941). — EKMAN, H.: Displacement of kidney consequent upon spontaneous perirenal hematoma. Acta chir. scand. **93**, 531 (1946). — ELMER, R. F., and C. B. WYNGARDEN: Spontaneous perirenal hematoma. Amer J. Surg. **43**, 764 (1939). — FORT, C. A.: Spontaneous perirenal haematoma secondary to periarteritis nodosa. J. Urol. (Baltimore) **59**, 307 (1948). — HAEBLER, H.: Zur Kenntnis der subkapsulären Nierenhämatome. Dtsch. med. Wschr. **54**, 1078 (1928). — HERITAGE, K.: Spontaneous circumrenal haematoma. Proc. roy. Soc. Med. **27**, 1105 (1934). — HERMAN, L.: Massive spontaneous hemorrhage into and around parenchymal lesions of the kidney. J. Urol. (Baltimore) **59**, 544 (1948). — HINZ, R.: Massenblutung ins Nierenlager bei Myolipom. Langenbecks Arch. klin. Chir. **132**, 149 (1924). — HÜBNER, H.: Das perirenale Hämatom. Experimenteller Beitrag zur Frage des Genese. Langenbecks Arch. klin. Chir. **145**, 338 (1927). — IRWIN, F. G.: Spontaneous rupture of kidney. Case report. U.S. nav. med. Bull. **41**, 818 (1943). — JENTZER, A.: Grave hemorrhages provoked by solitary lipoma having perforated capsule of superior pole of kidney. Urol. cutan. Rev. **52**, 612 (1948). — KEEFER, C. S.: Spontaneous perirenal haematoma. J. Mt Sinai Hosp. **8**, 682 (1942). — KLINGER, M. E.: Secondary tumours of the genito-urinary tract. J. Urol. (Baltimore) **65**, 144 (1951). — LÄWEN, A.: Zur Entstehung der Massen-Blutungen ins Nierenlaer. Dtsch. Z. Chir. **118**, 374 (1912). — LENK, R.: Über Massenblutungen ins Nierenlager. Dtsch. Z. Chir. **102**, 222 (1909). — LINK, G. S.: Spontaneous bilateral perirenal hematoma. J. Urol. (Baltimore) **69**, 13 (1953). — McFARLAND, G. E., and W. R. BLISS: Haemorrhage from spontaneous rupture of a pheochromocytoma of a right adrenal gland. Ann. Surg. **133**, 404 (1951). — MACKENZIE, D. W.: Perirenal haematoma, primary, with polycythaemia. J. Urol. (Baltimore) **23**, 535 (1930). — MARTIN, K. W.: Spontaneous circumrenal haematoma. Brit. med. J. **1**, 1118 (1949). — *Massachusetts General Hospital:* Case Record No 35232. New Engl. J. Med. **240**, 933 (1949). — MILLER, J. A., and J. J. CORDONNIER: Spontaneous perirenal haematoma associated with hypertension. J. Urol. (Baltimore) **62**, 13 (1949). — MOORE, C. H.: Surgical treatment of hypertension. Sth. Surg. **7**, 353 (1938). — NICHOLSON, M. A., and M. G. GILLESPIE: Lipoma of kidney. Report of case. J. Urol. (Baltimore) **25**, 395 (1931). — NYSTROM, T. G.: Rupture of kidney tumours with hydronephrosis. Acta chir. scand. **93**, 513 (1946). — PETKOVIC, S.: Sarcome de l'uretère avec hématonéphrose et hématome périrénal. J. Urol. (Baltimore) **56**, 557 (1950). — POLKEY, H. J., and W. J. VYNALECK: Spontaneous non-traumatic perirenal and renal haematomas: experimental and clinical study. Arch. Surg. (Chicago) **26**, 196 (1933). — POWELL, T. O., and J. E. CLARK: Sarcoma: Case with spontaneous rupture of kidney. J. Urol. (Baltimore) **62**, 751 (1949). — PRICE, C. W. R.: Retroperitoneal haemorrhage from angioma of kidney. Brit. med. J. **1940 II**, 831. — RALSTON, D. E., and W. F. KVALE: Renal lesions of periarteritis nodosa. Proc. Mayo Clin. **24**, 18 (1949). — RAYER, P. F. O.: Traite des maladies des reins. Paris: J. B. Baillière & Fils 1839. — RICKER, G.: Bemerkungen zu der Abhandlung von A. LAWEN: Über das sog. perineale Hämatom und andere spontane retroperitoneale Massenblutungen. Dtsch. Z. Chir. **114**, 287 (1912). — ROBERTSON, T. D., and J. R. HAND: Primary intrarenal lipoma of surgical significance. J. Urol. (Baltimore) **46**, 458 (1941). — ROLNICK, H. C.: Practice of urology. Philadelphia: J. B. Lippincott Company 1949. — ROLNICK, H. C., and I. DAVIDSOHN: Involvement of kidney in peri-arteritis nodosa. Urol. cutan. Rev. **46**, 626 (1942). — SAGARRA, G. N. M.: Renal carbuncle and perirenal haematoma. Arch. esp. Urol. **8**, 70 (1952). — SCRIVER, W. S.: Intra-abdominal haemorrhage of renal origin complicating pregnancy. Urol. cutan. Rev. **51**, 81 (1947). — STEARNS, D. B., and H. HERSHMAN: Spontaneous perirenal haematoma, a diagnostic problem. Amer. J. Surg. **88**, 254 (1954). — UGELLI, L.: A case of spontaneous perirenal haematoma. Policlinico, Sez. chir. **44**, 162 (1937). — WEVER, G. K., and I. H. PERRY: Peri-arteritis nodosa with fatal perirenal haemorrhage. J. Amer. med. Ass. **104**, 1390 (1935). — WUNDERLICH, R. A.: Handbuch der Pathologie und Therapie, 2. Aufl. Stuttgart: Ebner & Seubert 1865.

C. Movable kidney

BELL, J. J.: Nephroptosis: its causation, symptoms and radical cure. Brit. med. J. **1923 I**, 887. — BRAASCH, W. F., and R. GOYANNA: Hypertension and its relation to nephroptosis. J. Urol. (Baltimore) **53**, 1 (1945). — BRAASCH, W. F., L. F. GREENE and R. GOYANNA: Renal ptosis and its treatment. J. Amer. med. Ass. **138**, 399 (1948). — Is nephropexy useless. J. Amer. med. Ass. **3**, 59 (1951). — BURFORD, C. E.: Nephroptosis with co-existing lesions. J. Urol. (Baltimore) **55**, 220 (1946). — DEMING, C. L.: Nephroptosis and its correction. Trans. Amer. Ass. Genito-gen.-urin. Surg. **22**, 131 (1929). — Acute and chronic symptoms and diagnosis of movable kidney: Conservative and radical treatment. Penn. med. J. **48**,

207 (1944). — DODSON, A. I.: Deming nephropexy. J. Urol. (Baltimore) 50, 515 (1943). — HAHN (1881): Quoted by A. I. DODSON, J. Urol. (Baltimore) 50, 515 (1943). — HINMAN, F.: Principles and practice of urology. London and Philadelphia: W. B. Saunders Company 1935. — KELLY, H. A., and R. M. LEWIS: Diagnosis of the particular forms of hydronephrosis due to movable kidney. Surg. Gynec. Obstet. 19, 601 (1914). — KIDD, F.: Acquired renal dystopia or movable kidney. J. Urol. (Baltimore) 26, 327 (1931). — MATHÉ, C. P.: Present-day status of nephropexy and end-results in 384 operative cases. Urol. cutan. Rev. 41, 772 (1937). — McCANN, W. S., and M. J. ROMANSKY: Effect of ptosis of kidneys on blood pressure, renal blood flow and glomerular filtration. Trans. Ass. Amer. Phycns 55, 240 (1940). — PRATHER, G. C.: Medial ptosis. New renal syndrome. New Engl. J. Med. 238, 253 (1948). — RAYER, (1841): Quoted by C. E. BURFORD, J. Urol. (Baltimore) 55, 220 (1946). — THOMSON-WALKER, J.: Surgical diseases and injuries of the genito-urinary organs, 2. edit. London: Cassell & Co. 1936. — WALKOW, u. DELITZEN: Die Wanderniere. Berlin 1899. — WOODRUFF, S. R., and R. G. SCHERER: Ptosis. J. Urol. (Baltimore) 35, 125 (1936).

D. Changes in size and position of bladder

CIBERT, J.: Ileocystoplasty for contracted bladder of tuberculosis. Brit. J. Urol. 25, 99 (1953). — CORNER, E. M., and C. W. ROWNTREE: A case of tuberculosis in a diverticulum of bladder found in an inguinal hernia. Trans. clin. Soc. Lond. 39, 21 (1906). — COUVELAIRE, R. J.: La "petite vessie" des tuberculeux genito-urinaries; essai de classification. place et variantes des cysto-intestino-plastics. J. Urol. méd. chir. 56, 381 (1950). — POTT, P.: Phil. Trans. B, 61—64 (1764). Quoted by C. P. G. WAKELEY, Brit. J. Urol. 2. 1 (1930). — PYRAH, L. N.: Use of segments of small and large intestine in urological surgery. with special reference to problem of ureterocolic anastomosis. J. Urol. (Baltimore) 78, 683 (1957). — WAKELEY, C. P. G.: Hernia of the bladder; its etiology and treatment. A report of 40 cases. Brit. J. Urol. 2, 1 (1930). — WATSON, L. F.: Hernia. London: Henry Kimpton 1924. — WILLIAMS, D. I.: Urology in Childhood. Encyclopedia of urology, vol. XV. Berlin-Göttingen-Heidelberg: Springer 1958.

E. Purpura of the bladder

DUKES, C. E.: Renal purpura. Proc. roy. Soc. Med. 32, 1322 (1939). — GAIRDNER, D.: The Schonlein-Henoch syndrome (Anaphylactoid purpura). Quart. J. Med., N. s. 17, 95 (1948). — KIDD, F.: Purpura of the urinary tract. Proc. roy. Soc. Med. 21, 1105 (1928). — LAZARUS, J. A.: Idiopathic (Schonlein) purpura associated with haematuria. J. Urol. (Baltimore) 62, 354 (1949). — MARTIN, K. W.: Purpura of the urogenital tract. Brit. J. Urol. 23, 230 (1951). — MILLER, M. M., and C. A. UHLE: Int. Clin. 3, 183 (1934). — RHODES. J.: Haematuria after use of tetanus antitoxin; case report. J. Urol. (Baltimore) 38, 410 (1937). — RICH, A. R.: Hypersensitivity to iodine as case of periarteritis nodosa. Bull. Johns Jopk. Hosp. 77, 43 (1945). — ROBSON, H. N.: Idiopathic thrombocytopenic purpura. Quart. J. Med. 18, 72, 279 (1949). — ROSENTHAL, N.: Blood picture in purpura. J. Lab. clin. Med. 13. 303 (1928).

F. Foreign body in the bladder

ASPINALL, A.: Removal of clinical thermometer (half-minute) from bladder by manual manipulation. Med. J. Aust. 2, 454 (1931). — BADENOCH, A. W.: Manual of urology, p. 402. London: W. Heinemann 1953. — BADENOCH, A. W., and R. I. CAMPBELL: Foreign bodies in urinary bladder with report of case. Brit. J. Surg. 25, 133 (1937). — BARON, S., and H. LIPSHUTZ: Unusual foreign body ulcerating into the bladder from the bowel. J. Urol. (Baltimore) 58, 112 (1947). — BINTCLIFFE, E. W.: Foreign body in bladder. Brit. med. J. 2, 543 (1942). — BOND, S. P.: Foreign bodies in bladder. J. Amer. med. Ass. 83, 1163 (1924). — BORS, E., and F. C. BOWIE: Migration of shell fragment into bladder. J. Urol. (Baltimore) 55, 358 (1946). — BRANHAM, D. W.. and H. M. RICHEY: Kirschner wire removed from bladder. J. Urol. (Baltimore) 57, 869 (1947). — BREWER, A. C., and R. MARCUS: Unusual case of foreign body in bladder. Brit. J. Surg. 35, 324 (1948). — BUTTERS, A. G.: Unusual length of plastic tubing as foreign body in bladder. Brit. med. J. 1951 I, 800. — CAPLES, B. H.: Foreign body in urinary bladder. Surg. Gynec. Obstet. 29, 315 (1919). — CHARLES, A. H.: Foreign bodies introduced in attempts to procure abortion. Brit. med. J. 1939 II, 224. — COOK, J.: Foreign bodies in the male bladder. Lancet 1935 II, 1232. — CRENSHAW, J. L.: Vesical calculus. J. Amer. med. Ass. 77, 1071 (1921). — DUNCAN, I. G.: Foreign bodies in bladder. Mississippi Doct. 15 36 (1938). — EWELL, G. H.: Chewing gum in male urinary bladder; report and treatment of case. J. Urol. (Baltimore) 24, 537 (1930). — FARNCOMBE, R.: Foreign body in bladder associated with pregnancy. Lancet 1935 II, 825. — FENGER, E. P., C. K. PETTER and G. J. THOMAS: Clinical thermometer in male urinary bladder. Minn. Med. 12, 378 (1929). — FORSHAW, H. W.: Foetal bones, unusual termination to ectopic pregnancy. Lancet 1946 II.

716. — FULLERTON, A.: Observations on bladder injury in warfare. Study of fifty-three cases. Brit. J. Surg. 6, 24 (1918). — GEYER: Blasenentzündung durch Paraffinklumpen. Dtsch. med. Wschr. 48, 1284 (1922). — GARSHWILER, W. P., A. F. WEYRBACHER and J. F. BALCH: Foreign bodies in urinary bladder. Amer. J. Surg. 22, 199 (1933). — GEYERMAN, P. T.: Medical curiosities. J. Amer. med. Ass. 108, 1409 (1937). — GILL. R. D.: Foreign body (chewing gum) in urinary bladder. Amer. J. Surg. 19, 528 (1933). — GOLDSTEIN, H. H.: Self-inserted foreign body in bladder. Urol. cutan. Rev. 45. 190 (1941). — GOODMAN, W. D.: Chewing gum in male urinary bladder. J. Urol. (Baltimore) 22, 335 (1929). — GROSS, S. D.: System of surgery, 6. edit., vol. 2. Philadelphia: Henry C. Lea 1882. — HARRIS, A.: A case of paraffin foreign body in the bladder. Med. J. Rec. 134, 372 (1931). — HOTTINGER, R.: Über Fremdkörper der Harnblase und ihre Entfernung. Korresp.-Bl. schweiz. Ärz. 49, 875 (1919). — JOELSON, J. J.: Instrumental removal of wax foreign bodies from bladder. J. Urol. (Baltimore) 64, 572 (1950). — KATZEN, P.: Foreign body in bladder. Paraffin. J. Amer. med. Ass. 105, 1422 (1935). — LATTIMER, J. K.: Late erosion of shell fragment. J. Urol. (Baltimore) 55, 483 (1946). — LINS. E. E.: Solution of incrustations in bladder by new method. J. Urol. (Baltimore) 53, 702 (1945). — LOHNSTEIN, H.: Über einen Wachsklumpen in der Blase. Entfernung desselben durch Auflösung mittels Benzin-Injection. Verh. berl. med. Ges. 38, 199 (1908). — MACKENZIE. J. H.: Huge bladder calculus with chewing-gum core. J. Urol. (Baltimore) 37, 280 (1937). — McCLINTON, J. B.: Large stone around foreign body in young female. Canad. med. Ass. 49, 204 (1943). — MOORE, T.: Chewing gum calculus in the male bladder. Brit. J. Surg. 38, 103 (1950). — NITSCHKE, P. H.: Foreign bodies in bladder. Amer. J. Surg. 40, 560 (1938). — PHILIP. P. F.: Needle calculus in the male urinary bladder. Brit. J. Urol. 27, 242 (1955). — RAVENSWAY, A. VAN: Steel pin inserted into neck of femur found in bladder. Amer. J. Surg. 31, 566 (1936). — SADEK, E.: Pathologically displaced upper femoral epiphysis as a foreign body in bladder. Brit. J. Urol. 20, 114 (1948). — SCHULTZ, W. A.: Novel way of ridding bladder of paraffin. J. Urol. (Baltimore) 34, 313 (1935). — TCHERLOCK, R.: Les corps étrangers dans les organes urogénitaux de la femme. Gynéc. et Obstét. 19, 465 (1929). — THOMSON-WALKER, J.: Quoted by BORS and BOWIE 1946. — TURNER. J. H.: Foreign body in bladder. Paraffin removed with xylol as solvent. J. Urol. (Baltimore) 33, 471 (1935). — WALKER, A. S., and D. R. KAUFMAN: Spontaneous migration of bullet into bladder after five-year interval. Urol. cutan. Rev. 46, 217 (1942). — WALLER. J. I., and F. ADNEY: Vesical calculi in children. Amer. J. Dis. Child. 79, 684 (1950). — YOUNG. H. H.: Amer. Med. (Philad.) 3, 63 (1902).

G. Priapism

AARON, G., and M. A. ROBBINS: Priapism: An unusual complication of transurethral resection. J. Urol. (Baltimore) 62, 328 (1949). — ABESHOUSE, B. S., and L. H. TANKIN: True priapism. Urol. cutan Rev. 54, 449 (1950). — ACHARD, C.: Priapisme revelateur d'une leukemie myeloide. Paris méd. 20, 539 (1930). — ALLENBACH, E.: Primäres Urethralcarcinom mit priapismusähnlichen Folgen. Dtsch. Z. Chir. 138, 152 (1916). — BAILEY, H.: Persistent priapism. Brit. J. Surg. 35, 298 (1948). — BARNEY, J. D.: Priapism complicating splenic leukaemia. N. Z. J. Med. 203, 1013 (1930). — BERKEY, H. A.: Three cases of priapism. J. Urol. (Baltimore) 22, 489 (1929). — BOUCHER, R.: Priapism in tabes dorsalis, case. Un. méd. Can. 57, 88 (1928). — BRODY, H. S., P. A. LAHR and W. A. CARROLL: Primary priapism: A new treatment. J. Urol. (Baltimore) 78, 153 (1957). — CALLOMON, F. T.: The phenomenon of priapism; its diagnostic significance; a critical review. Urol. cutan. Rev. 54, 144 (1950). — CAMPBELL, J. H., and S. D. CUMMINS: Priapism in sickle cell anaemia. J. Urol. (Baltimore) 66, 697 (1951). — COLMER, F.: Über Sarkome und Endotheliome des Penis. Beitr. path. Anat. 34, 295 (1903). — CRAVER, L. F.: Priapism in leukaemia. Surg. Clin. N. Amer. 43, 472 (1933). — CREITE: Peniscarcinom bei einem 2jährigen Kind. Dtsch. Z. Chir. 79, 299 (1905). — D'ABUNDO, E.: Priapism in neurosyphilis. Gior. ital. Derm. Sif. 74, 1331 (1933). — DAKIN, W. B.: Urological oddities. Los Angeles, Calif. 1948. — DAWSON, S. R.: Priapism, report of five cases, two with sickle cell anaemia. J. Urol. (Baltimore) 42, 821 (1939). — DEFESCHE, H. L.: Priapism; a case of traumatic origin. J. Urol. méd. chir. 47, 465 (1939). — DIGGS, L. W., and R. E. CHING: Pathology of sickle cell anaemia. Sth. med. J. (Bgham, Ala.) 27, 839 (1934). — EMODI: Urologia, No 55. 1906. — FINKLER, R. S.: Initial priapism during therapy with testosterone propionate in a eunochoid man. J. Urol. (Baltimore) 43, 866 (1940). — FOURNIER. J. A.: Leçons sur la periode pretaxique du tabes d'origine syphilitique, p. 54. Paris 1885. — FRASER, W. J.: Case of priapism. Brit. med. J. 1955 I, 419. — FREEDMAN, S. Z.: Idiopathic priapism. Urol. cutan. Rev. 50, 728 (1946). — FRONTZ, W. A., and E. P. ALYEA: Priapism of unusual etiology. J. Urol. (Baltimore) 20, 135 (1928). — GAYET, M.: A case of priapism of neoplastic origin. Lyon chir. 30, 366 (1933). — GETZOFF, P. I.: Priapism and sickle cell anaemia. J. Urol. (Baltimore) 48, 407 (1942). — GOBBIE. L.: Contribution to study of endothelial tumours. Policlinico, Sez. chir. 29, 23 (1922). — GUI-

BAL, P., et PAVI: Deux cas de metastase cancereuse rapide et massive dans l'appareil genital erectile d'une homme et d'une femme. Ann. anat. path. **6**, 1099 (1929). — GUITER-RIEZ, R.: Unusually long foreign body impacted in the urethra causing painful priapism for seven days, removal and cure by external urethrotomy. J. Urol. (Baltimore) **49**, 865 (1943). — GUTTMAN: Quoted by O. SCHEUER, Arch. Derm. Syph. (Berl.) **109**, 149 (1911). — HIN-MAN, F.: Priapism. Report of cases and a clinical study of the literature. Ann. Surg. **60**. 689 (1914). — HUGHES, B.: Persistent priapism. Brit. med. J. **2**, 571 (1932). — IKEDA, K.. F. E. B. FOLEY and J. ROSENOW: Malignant priapism; primary carcinoma of urethra with priapism. J. Urol. (Baltimore) **49**, 732 (1943). — IMBERT, M.: Case of priapism treated by incision of corpus cavernosum. Lyon méd. **138**, 535 (1926). — Case of priapism responding to surgical treatment. J. Urol. méd. chir. **50**, 227 (1942). — KAPLAN, I. I.: Persistent pria-pism of leukaemia treated by radiation therapy. Med. J. Rec. **130**, 160 (1929). — KIM-BROUGH, J. C., M. C. WORGAN and J. C. DENSLOW: Priapism, surgical treatment; case report. Urol. cutan. Rev. **54**, 53 (1950). — KLEHMET: Quoted by O. SCHEUER, Arch. Derm. Syph. (Chicago) **109**, 149 (1911). — LEWIS, B.: Case of semi-priapism, ligation of dorsal penile arteries; recovery. Med. Fortnightly (St. Louis) **12**, 332 (1897). — LEWIS, E. C., and B. E. SCHWAREZ: Priapism of unknown cause. U.S. armed Forces med. J. **8**, 271 (1957). — LEVANT, B., and R. STEPT: Priapism due to sickle cell anaemia. J. Urol. (Baltimore) **59**, 328 (1948). — LOEHE, H.: Zbl. Haut- u. Geschl.-Kr. **18**, 826 (1926). — LOWSLEY. O. S., and A. GONZALEZ: A new operation for the cure of priapism.N. Y. St. J. Med. **54**, 61 (1954). — MCKAY, R. W., and J. A. C. COLSTON: Priapism, a newmethod of treatment. J. Urol. (Baltimore) **19**, 121 (1928). — MASLOW, P.: Priapism. Sov. Vestn. venerol. i dermat. **4**, 172 (1935). Zbl. Haut- u. Geschl.-Kr. **51**, 233 (1935). — MAU-RER, M.: Quoted by JOELSON, Surg. Gynac. Obstet. **38**, 150 (1924). — MURPHY jr., R. C., and S. SHAPIRO: Sickle cell disease. Arch. intern. Med. **74**, 28 (1944). — NOE, A.: Recividating priapism of four weeks duration, cured by treating coexisting urethritis. Sem. méd. esp. **3**, 1432 (1940). — OLLINGER, P.: Case of paraplegia and priapism due to vertebral tuberculosis. Zbl. Chir. **68**, 877 (1941). — PATEL: Case of priapism treated by dorsal incision of the corpus cavernosum. Lyon chir. **26**, 231 (1929). — RICHES, E. W.: Case of priapism; new method of treatment. Brit. J. Urol. **2**, 380 (1930). — ROLNICK, D., T. L. C. COTTRELL and F. A. LLOYD: Priapism; report of two cases and discussion of treatment. Illinois med. J. **112**, 129 (1957). — ROSOKOFF, J., and E. L. BRODIE: Priapism complicating sickle celanaemia. J. Urol. (Baltimore) **56**, 544 (1946). — RUH, H. O.: Persistent priapism in splenomyelogenous leukaemia. Cleveland Med. J. **11**, 859 (1912). — SCHEUER, O.: Über Priapismus. Arch. Derm. Syph. Berl. **109**, 149 (1911). — SMITH, K. H.: Use of dicumarol in persistent priapism. J. Urol. (Baltimore) **64**, 400 (1950).: — VORSTER Zur operativen Behandlung des Priapismus. Dtsch. Z. Chir. **173** (1888). — WARTHIN, A. S.: Priapism, persistent postmortem. Internat. clin. **4**, 286 (1906). — WHITE, E. W.: Discussion of Ber-key's article, loc. cit. — WILGUS, S. D., and E. W. FELL: Priapism as an early symptom in multiple sclerosis. Arch. Neurol. Psychiat. (Chicago) **25**, 153 (1931). — WILSON, G., and J. P. MAUS:. Med. J. Rec. **251** (1922); (1925). — YAMAMOTO, K.: Operative treatment of priapism due to intravascular endothelioma of corpora cavernosa of penis; case. Jap. J. Derm. Urol. **30**, 50 (1930).

H. Peyronie's disease

AURIG, G., and H. J. SUSSE: Über neue Behandlungsmöglichkeiten der Induratio penis plastica. Strahlentherapie **89**, 433 (1952). — BODNER, H., A. H. HOWARD and J. H. KAPLAN: Peyronie's disease: Cortisone-Hyaluronidase-Hydrocortisone therapy. J. Urol. (Baltimore) **72**, 400 (1954). — BURFORD, E. H.: Fibrous cavernositis. J. Urol. (Baltimore) **43**, 208 (1940). — BURFORD, C. E., J. E. GLENN and E. H. BURFORD: Fibrous cavernositis: Further obser-vations with report of 31 additional cases. J. Urol. (Baltimore) **56**, 118 (1946). — BURFORD. E. H., J. E. GLENN and C. E. BURFORD: Therapy of Peyronie's disease. Urol. cutan. Rev. **55**, 337 (1951). — CALLOMON, F.: Induratio penis plastica; problem of its etiology and patho-genesis. Urol. cutan. Rev. **49**, 742 (1945). — CINIEWICZ, O.: Plastic induration of penis and its treatment. Urol. pol. **9**, 179 (1956). — DAHL, O.: Treatment of plastic induration of penis. Acta radiol. (Stockh.) **41**, 290 (1954). — DREYER, A.: Zur Therapie der Induratio penis plastica. Arch. Derm. Syph. (Berl.) **119**, 372 (1914). — FOGH-ANDERSEN, P.: Surgical treatment of plastic induration of penis. Acta chir. scand. **113**, 45 (1957). — FRICKE, R. E.. and J. W. OLDS: The radium treatment of Peyronie's disease. Amer. J. Roentgenol. **42**. 545 (1939). — FRICKE, R. E., and J. J. VARNEY: Peyronie's disease and its treatment with radium. J. Urol. (Baltimore) **59**, 627 (1948). — FUREY jr., C. A.: Peyronie's disease. Treat-ment by local injection of meticortelone and hydrocortisone. J. Urol. (Baltimore) **77**, 251 (1957). — HEITE, H. J., and H. H. SIEBRECHT: Beitrag zur Pathogenese der Induratio penis plastica. Derm. Wschr. **121**, 25 (1950). — KATZ-GALATZI, T.: Induratio penis plastica.

Acta med. orient. (Tel-Aviv) 8, 193 (1949). — LOWSLEY, O. S.: Surgical treatment of plastic induration of penis. N. Y. St. J. Med. 43, 2273 (1943). — LOWSLEY, O. S., and A. GENTILE: Operation for cure of certain cases of plastic induration of penis. J. Urol. (Baltimore) 57, 552 (1947). — LOWSLEY, O. S., and W. H. BOYCE: Further experiences with an operation for cure of Peyronie's disease. J. Urol. (Baltimore) 63, 888 (1950). — MARCHAND, F.: Der Progres der Wundheilung. In Deutsche Chirurgie, S. 16. Stuttgart: V. Bermann & V. Bruns 1901. — MARTENSTEIN, H.: Induratio penis plastica und Dupuytrensche Contractur. Med. Klin. 16, 205 (1920); 17, 44 (1921). — MAY, J.: Contribucion al estudio de la etiologia de la induracion esclerosa de los cuerpos cavernosos. Act. dermo-sifiliogr. (Madr.) 31, 67 (1939). — NIKOLOWSKI, W.: Die Bedeutung des Vitamins E im Rahmen der Strahlenbehandlung der Induratio penis plastica. Strahlentherapie 87, 113 (1952). — PETERSEN, O.: Plastic indura-tion of penis. Ugeskr. Laeg. 118, 487 (1956). — PEYRONIE, F. DE LA: Mem de l'acad. roy. de Chir., p. 425, 1743; new edition, p. 316, 1819. — POLKEY, H. J.: Induratio penis plastica. Urol. cutan. Rev. 32, 287 (1928). — REJKA: Quoted by F. CALLOMON, Urol. cutan. Rev. 49, 742 (1945). — RIBA, L. W.: Peyronie's disease. J. Urol. (Baltimore) 79, 114 (1958). — SCARDINO, P. L., and W. W. SCOTT: Use of tocopherols in treatment of Peyronie's disease. Ann. N. Y. Acad. Sci. 52, 390 (1949). — SCOTT, W. W., and P. L. SCARDINO: A new concept in treatment of Peyronie's disease. Sth. med. J. (Bgham, Ala.) 41, 173 (1948). — SOILAND, A.: Peyronie's disease or plastic induration. Radiology 42, 183 (1944). — STEINBERG, C. L.: New method of treatment of Dupuytren's contracture: A form of fibrositis. M. Clin. N. Amer. 30, 221 (1946). — Tocopherols in treatment of primary fibrositis. Arch. Surg. 63, 824 (1951). — TEASLEY, G. H.: Peyronie's disease. A new approach. J. Urol. (Baltimore) 71, 611 (1954). — WALLER, J. I., and W. C. DREESE: Peyronie's disease associated with Dupuytren's contracture. J. Urol. (Baltimore) 68, 623 (1952). — WESSON, M. B.: Peyronie's disease (plastic induration), cause and treatment. J. Urol. (Baltimore) 49, 350 (1948).

J. Hydrocele

ALLEN, L.: Lymphatics of the parietal tunica vaginalis propria of man. Anat. Rec. 85, 427 (1943). — BICKLE, L. W.: Abdominal or bilocular hydrocele. Brit. med. J. 1919 II. 13. — CAMPBELL, M. F.: Hydrocele of the tunica vaginalis. Surg. Gynec. Obstet. 45, 192 (1927). — CECA, B.: Eine Studie über die Genese und Funktion des Interstitiums auf Grund der Untersuchungen an senescenten Hoden. Arch. mikr. Anat. 98, 524 (1923). — DE QUER-VAIN, FR.: Clinical surgical diagnosis for students and practitioners. London: Bale & Son 1921. — DIAMOND, J. C.: Treatment and cure of 76 cases of hydrocele by one twin-injection of litium salicylate and quinine hydrochloride and urethane. Amer. J. Surg. 55, 121 (1942). — EARLE, J.: A treatise on the hydrocele, 3. edit. London 1805. — EWELL, G. H., C. R. MAR-QUARDT and J. C. SARGENT: J. Urol. (Baltimore) 44, 741 (1940). — FOOTE, R. R.: Injection treatment of hydrocele and some other conditions. Med. práct. 210, 76 (1943). — FRIES, J. W., and B. S. TALBOT: Scrotal calcification due to meconium peritonitis. J. Urol. (Balti-more) 73, 1059 (1955). — GREENE, L. B.: Hydrocele and varicocele: operative and injection treatment. Amer. J. Surg. 36, 204 (1937). — HINMAN, F.: Principles and practice of urology. London: W. B. Saunders Company 1937. — HOLMES, J. M.: A case of properitoneal hydro-cele. Brit. J. Surg. 20, 346 (1932/33). — HUBOTTER: Ein Fall von Hydrocele quadrilo-cularis intra-abdominalis. Berl. klin. Wschr. 56, 55 (1919). — HUGGINS, C. B., and F. H. ENTZ: Absorption from normal tunica vaginalis testis, hydrocele and spermatocele. J. Urol. (Baltimore) 25, 447 (1931). — JAMES, W. L.: Giant hydrocele successfully treated by the sclerosing method. Brit. med. J. 1941 I, 693. — LASBREY, F. O.: A case of abdominal or bilocular hydrocele. Brit. med. J. 1916 II, 292. — MILBERT, A. H.: Injection therapy of hydrocele and spermatocele. Amer. J. Surg. 44, 587 (1939). — NICHOLSON, J. W.: Treatment of hydrocele by injection. Brit. med. J. 2, 188 (1947). — OBNEY, C.: Hy-droceles of the testicle complicating inguinal hernias. Canad. med. Ass. 75, 753 (1956). — PATERSON, R.: Memorials of the life of James Syme. Edinburgh 1874. — RHIND, J. A.: The injection treatment of hydrocele and spermatocele. Brit. med. J. 1951 II, 711. — ROBERTSON, J. P.: Treatment of large hydroceles by injections of sodiummorrhuate. Amer. J. Surg. 53, 421 (1941). — Ross, J. M.: Treatment of hydrocele of the tunica vaginalis. Edinb. med. J. 57, 413 (1950). — SANFORD, H. L.: Quoted by H. H. YOUNG in "YOUNG's practice of urology", vol. 2, p. 538. Philadelphia: W. B. Saunders Company 1926. — SHARP, S.: A treatise on the operations of surgery, 5. edit. London 1747. — SMITH: Quoted by EWELL et al. 1940. — THOMSON-WALKER, J.: Genito-urinary surgery, edit. by K. Walker, 3. edit. London: Cassell & Co. 1948. — TORRACA, L.: Il potere di assorbimento della tunica vaginale nell'idrocele. Arch. ital. Chir. 6, 404 (1922). — WELLS, C.: Textbook of Genito-urinary surgery, edit. by H. P. Winsbury-White. Edinburgh: Livingstone Ltd. 1948. — WHITBY, M.: Non-operative treatment of chronic hydrocele of the tunica vaginalis. Brit. med. J. 2, 240 (1932). — WILLIAMS, D. I.: Handbuch der Urologie, Bd. 15, S. 253. Berlin:

Springer 1958. — WILSON, W. W.: Injection treatment of hydrocele. Lancet **1949** I, 1048. — WOLF, M., and F. R. GOMILA jr.: Spontaneous rupture of hydrocele testis. Amer. J. Surg. **89**, 703 (1955). — YOUNG, H. H.: YOUNG's practice of urology, vol. 2. Philadelphia: W. B. Saunders Company 1926.

K. Haematocele

BARRINGTON, F. J. F.: Spontaneous rupture of a haematocele of the tunica vaginalis. Brit. J. Surg. **2**, 389 (1914/15). — BLISS, P.: Case of spontaneous haematoma of the spermatic cord. J. roy. nav. med. Serv. **43**, 35 (1957). — COWDERY, J. S., E. S. GAULT and K. B. CONGER: Massive haematocele threatening life by exsanguination. J. Urol. (Baltimore) **64**, 524 (1950). — FLORIO, I.: Giant perineo-scrotal haematoma due to rupture of varicose veins in patient with left essential varicocele. Clinica (Bologna) **12**, 313 (1950). — HOWARD, R.: Some notes on scrotal operations in negroes. J. trop. Med. Hyg. **21**, 57 (1918). — KICKHAM, C. J. E.: Report of a case of calcified hydrocele simulating tumour. New Engl. J. Med. **208**, 869 (1933). — LOWSLEY, O. S., and T. J. KIRWIN: Clinical Urology, 2. edit., vol. I. Baltimore: Williams & Wilkins Company 1926. — MATIGNON, J. J., and H. VENNET: Hématome de l'epididyme. Gaz. hebd. Sci. méd. Bordeaux **25**, 102 (1904). — NELIGAN, G. E.: Traumatic orchitis and haematocele. Practitioner **136**, 496 (1936). — O'Conor, V. J.: Haematocele of epididymis. Trans. Amer. Ass. gen.-urin. Surg. **21**, 461 (1928). — PYRAH, L. N.: Properitoneal haematocele. Brit. med. J. **1933**, 734. — TAYLOR, J.: Haematocele of the epididymis. Trans. med. Soc. Lond. **53**, 103 (1930). — VALENCE, A.: Du kyste hématique epididymaire. Gaz. hebd. med. Paris **7**, 1189 (1902). — VERNON, S.: Haematoma of scrotum. Amer. J. Surg. **78**, 131 (1949). — WATSON, F. S., and J. H. CUNNINGHAM jr.: Diseases and surgery of the genito-urinary system. Philadelphia: Lea and Febiger 1908. — WEYRAUCH jr., H. M.: Calcified haematocele of the tunica vaginalis testis with spontaneous rupture. J. Urol. (Baltimore) **32**, 370 (1934). — WHITNEY, C. M.: Haematocele of the tunica vaginalis, with report of an unusual case. Brit. Med. Surg. Journ. **175**, 51 (1916). — WOOLFENDER, H. F.: On the similarity between the signs of haematocele and early malignant disease of the testicle. Med. Press **143**, 198 (1911).

L. Varicocele

BADENOCH, A. W.: Manual of urology, p. 537. London: Wm. Heinemann 1953. — BATE, J.: Symptomatic varicocele. J. Urol. (Baltimore) **18**, 649 (1927). — BENEVENTI, F. A.: Treatment of varicocele. Amer. J. Surg. **71**, 783 (1946). — BERNARDI, R.: Surgical therapy of varicocele; concepts and modifications of technique. Sem. méd. (B. Aires) **2**, 849 (1941). — BLOODGOOD, J. C.: Warning against operation for varicocele on application for enlistment. J. Tenn. med. Ass. **10**, 469 (1918). — CAMPBELL, M. F.: Varicocele. A study of 500 cases. Surg. Gynec. Obstet. **47**, 558 (1928). — Varicocele due to anomalous renal vessel; instance in thirteen-year old boy. J. Urol. (Baltimore) **52**, 502 (1944). — DAVIDSON, H. A.: Treatment of male subfertility, testicular temperature and varicoceles. Practitioner **173**, 703 (1954). — DOUGLAS, J.: Results of varicocele operation. J. Amer. med. Ass. **76**, 716 (1921). — GLOVER, T. D.: The effect of a short period of scrotal insulation on the semen of the ram. Proc. Physiol. Soc. J. Physiol. (Lond.) **128**, 22P (1953). — GUNN, R. M. C., R. N. SANDERS and W. GRANGER: Studies in fertility in sheep. 2. Seminal changes affecting fertility in rams. Bull. Coun. sci. industr. Res. Australia 148 (1942). — HANLEY. H. G.: Surgical correction of errors of testicular temperature regulation. Proc. of Second World Congr. on Fertility and Sterility, Naples, 1956. — HANSCHELL, H. M.: Injection treatment of varicocele. Brit. med. J. **1928** II, 915. — HARRISON, R. G.: Distribution of vasal and cremasteric arteries to the testis and their functional importance. J. Anat. **83**, 267 (1949). — HARRISON, R. G.: Functional importance of the vascularisation of the testis and epididymis for the maintenance of normal spermatogenesis. Fertil. and Steril. **5**, 366 (1952). — IVANISSEVICH, O.: Therapy of varicocele. Med. cir. pharm. 646 (1942). — JAVERT, C. T., and R. L. CLARK: Combined operation for varicocele and inguinal hernia. Surg. Gynec. Obstet. **79**, 644 (1944). — LEWIS, E. L.: Ivanissevich operation for varicocele. J. Urol. (Baltimore) **63**, 165 (1950). — McEVEDY, P. G.: Recent advances in injection treatment. Clin. J. **61**, 283 (1932). — OLSON, R. O., and E. P. STONE: Varicocele, symptomalogic and surgical concepts. New Engl. J. Med. **240**, 877 (1949). — PALOMO, A.: Radical cure of varicocele by new technique. Preliminary report. J. Urol. (Baltimore) **61**, 604 (1949). — PEPPER, O. H. P.: Varicocele as the first sign of sarcoma of kidney. Med. Clin. N. Amer. **10**, 29 (1926). — PHILLIPS, R. W., and F. F. McKENZIE: Thethermo-regulatory function and mechanism of the scrotum. Ref. Bull. Missouriagric. Exp. Sta. 217 (1934). — PICQUE and WORMS: Quoted by A. PALOMO: J. Urol. **61**, 604 (1949). — PORRITT, A. E.: Injection treatment of hydrocele, varicocele, bursae and naevi. Proc. roy. Soc. Med. **24**, 972 (1931). — Injection treatment of varicocele. Clin. J. **60**, 355 (1931). — PRICE, R. A.: Surgical treatment of varicocele.

Amer. J. Surg. **80**, 330 (1950). — RIBA, L. W.: Excision of internal spermatic vein for varicocele. J. Urol. (Baltimore) **57**, 889 (1947). — ROBB, W. A. T.: Operative treatment of varicocele. Brit. med. J. **1955** II, 355. — ROCHE, A. E.: Orchitis, varicocele and twisted cord. Brit. med. J. **1939** I, 27. — RUSSELL, J. K.: Varicocele in groups of fertile and subfertile males. Brit. med. J. **1954** I, 1231. — SKINNER, H. L.: Varicocele and its treatment. Ann. Surg. **113**, 123 (1941). — TULLOCH, W. S.: Varicocele in subfertility. Results of treatment. Brit. med. J. **1955** II, 356. — WHITE, S.: Acute varicocele due to pressure of a greatly distended left kidney (non-malignant). Brit. med. J. **2**, 177 (1914). — YOUNG, D.: The influence of varicocele on human spermatogenesis. Brit. J. Urol. **28**, 426 (1956).

M. Cysts of the epididymis

ABELL, I.: Cyst of testicle. Ann. Surg. **103**, 941 (1936). — ABESHOUSE, B. S.: Cysts of the epididymis. Urol. cutan. Rev. **41**, 761 (1937). — CAMPBELL, M. F.: Spermatocele. J. Urol. (Baltimore) **20**, 485 (1928). — CAVASSE, I.: Un point de l'histoire du spermatocele. Gaz. Hôp. (Paris) **33**, 378 (1860). — CORLETTE, C. E.: Two cases of polycystic disease of the epididymis. Med. J. Aust. **2**, 615 (1927). — CROSSAN, E. T.: Spermatocele. Ann. Surg. **72**, 500 (1920). — CURLING, T. B.: Practical treatise on diseases of the testis, p. 211. 1843. — EDINGTON, G. H.: Cysts of the epididymis. Postgrad. med. J. **12**, 185 (1936). — ENGLISH, J.: Über Cysten an der hinteren Blasenwand bei Männern. Med. Jb. (Wien) **2**, 127. 155 (1874). — HERBUT, P. A.: Urological pathology, vol. 2. London: Henry Kimpton 1952. — HOCHENEGG, J.: Über Cysten am Hoden und Nebenhoden. Med. Jb. (Wien) **15**, 149 (1885). — KOBELT: Quoted by H. MORRIS, Injuries and diseases of genital and urinary organs, p. 70. London 1895. — KREBS: Quoted by E. T. CROSSAN 1920. — LISTON, R.: A few observations on encysted hydrocele. Med. chir. Trans. (Lond.) **26**, 216 (1843). — LLOYD, E. A.: On the presence of spermatozoa in the fluid of hydrocele. Med. chir. Trans. (Lond.) **26**, 368 (1843). — McCRAE, E. E.: Etiology and treatment of cysts of epididymis. Brit. J. Urol. **1**, 152 (1935). — MARSAN, F.: Calcified cyst of epididymis. J. Urol. méd. chir. **8**, 157 (1919). — MATHE, C. P.: Bony-walled cyst and interstitial cell tumour of epididymis. J. Urol. (Baltimore) **41**, 188 (1939). — MILBERT, A. H.: Injection therapy of spermatocele. Amer. J. Surg. **44**, 587 (1939). — ROCHÉ: Kyste du cordon à parois calcifees. Bull. Soc. anat. Paris, ser. **3**, 347 (1889). — ROLNICK, H. C.: Etiology of spermatoceles. J. Urol. (Baltimore) **19**, 613 (1928). — ROSS, J. M.: Treatment of spermatocele by injection. Edinb. med. J. **57**, 413 (1950). — WAKELEY, C. P.: Cysts of epididymis, the so-called spermatocele. Brit. J. Surg. **31**, 165 (1943). — WHITNEY, C. M.: Spermatocele. Amer. J. Urol. **3**, 175 (1907).

N. Torsion of the testicle and of the appendages of the testis and the epididymis

ABESHOUSE, B. S.: Torsion of spermatic cord. Report of 3 cases and review of the literature. Urol. cutan Rev. **40**, 699 (1936). — ALLEN, P. D., and T. H. ANDREWS: Torsion of the spermatic cord in infancy. Amer. J. Dis. Child. **59**, 136 (1940). — AMBROSE, S. S., and J. E. SKANDALAKIS: Torsion of the appendix epididymis and testis. Report of 6 episodes. J. Urol. (Baltimore) **77**, 51 (1957). — BELLER, A. J.: Torsion of an intra-abdominal testicle. Ann. Surg. **102**, 41 (1935). — BENNETT, A. H., and W. G. SHAW: Case of undescended testis with seminomatous change and torsion. Brit. med. J. **1937** I, 256. — BIORN, C. L., and J. H. DAVIS: Torsion of the spermatic cord in the newborn. J. Amer. Ass. **145**, 1236 (1951). — BIRUKOV (1928): Quoted by L. JOSSA 1936. — BOGGON, R. H.: Polyorchidism. Brit. J. Surg. **20**, 630 (1933). — BROSTER, L. R., and R. COYTE: Torsion of appendix of testis. Brit. med. J. **1929** I, 145. — BURKE, J.: Torsion of pedicle of intra-abdominal seminoma. Zbl. Chir. **65**, 2821 (1938). — CAMPBELL, M. F.: Pediatric Urology, vol. 2. New York: Macmillan & Co. 1937. — Torsion of the spermatic cord in the newborn infant. J. Pediat. **33**, 323 (1948). — Clinical pediatric urology. Philadelphia and London: W. B. Saunders Company 1951. — CHARENDOFF, M. D., H. C. BALLON and M. A. SIMON: Torsion and rupture of an intra-abdominal seminoma. J. Urol. (Baltimore) **66**, 274 (1951). — CHATON: Rev. méd. Franche-Comte **28**, 209 (1925). — CHITTY, E. C.: Case of abdominal testis complicated by tumour formation and acute torsion of the cord. Malay. med. J. **8**, 203 (1933). — COLT, G. H.: Torsion of the hydatid of Morgagni. Brit. J. Surg. **9**, 464 (1922). — COPPRIDGE, W. M., and L. C. ROBERTS: Torsion of the appendix testis. J. Pediat. **32**, 184 (1948). — CUPLER, R. C.: Acute torsion of right intra-abdominal spermatic cord, thesymptoms of which simulated acute appendicitis. Surg. Gynec. Obstet. **21**, 250 (1915). — DELASIARVE, L. J. F.: Descente tardive du testicule gauche, prise pour une hernie étranglee; operation; gangrene du testicule; extirpation du cet organe; accidents ivers; guérison. Rev. méd. franç. et étrang. **1**, 363 (1940). — DEMING, C. L., and B. G. CLARKE: Torsion of spermatic cord. J. Amer. med. Ass. **152**, 521 (1953). — DIX, V. W.: On torsion of the appendages of the testis and epididymis. Brit. J. Urol. **3**, 245 (1931). — DONOVAN, E.

J.: Torsion of the spermatic cord in infancy. Ann. Surg. **92**, 405(1930). — Edington, G. H.: Strangulation of the fully descended testis. Lancet **1904** I, 1782. — Ewart, E. F.. and H. A. Hoffman: Torsion of the spermatic cord. J. Urol. (Baltimore) **51**, 551 (1944). — Foucault, Bull. Soc. nat. Chir. **51**, 423 (1925). — Garcia, A. L., and A. Cucullu: Sem. méd. (Paris) **36** (11), 836 (1926). — J. Chir. (Paris) **35**, 286 (1930). — Gerster, A. G.: Operation for undescended testis and inguinal hernia. Ann. Surg. **26**, 365 (1897). — Glaser, S., and H. R. E. Wallis: Torsion or spontaneous haemorrhagic infarction of testicle in newborn infant. Brit. med. J. **2**, 88 (1954). — Hamilton, G. R., C. A. De Kovessey and R. J. Getz: Torsion of the appendix epididymis and appendix testis. Three case reports. J. Urol. (Baltimore) **69**, 436 (1953). — Hamilton, G. R., A. B. Peyton and L. K. Mantell: Torsion of spermatic cord. U.S. armed Forces med. J. **1**, 733 (1950). — Handley, R. S.. and T. Crawford: Case of polyorchidism. Brit. J. Surg. **31**, 300 (1944). — Howard, R.: Discussion on diseases and displacements of the testicle. Brit. med. J. **2**, 719 (1907). — Howser, J. W., and L. P. River: Torsion and gangrene of hydatid of Morgagni. Review of literature and case report. Amer. J. Surg. **59**, 571 (1943). — Hunter, R. H.: Aetiology of congenital inguinal hernia and abnormally placed testis. Brit. J. Surg. **14**, 125 (1926). — James, T.: Torsion of the spermatic cord in the first year of life. Brit. J. Surg. **25**, 56 (1953). — Johnson, (1933): Quoted by L. Jossa 1936. — Jossa, L.: Torsion of pedicle of an abdominal testis. Beitr. klin. Chir. **163**, 45 (1936). — Key. (1914): Quoted by L. Jossa 1936. — Keyes jr., E. L., C. W. Collings and M. R. Campbell: Torsion of the spermatic cord. J. Urol. (Baltimore) **9**, 519 (1923). — Kinney, W. H.: Torsion of the spermatic cord. J. Urol. (Baltimore) **34**, 470 (1935). — Koelliker: Handbuch der Gewebelehre des Menschen, Bd. 3, S. 461. Leipzig 1902. — Kretschmer, H. L.: Torsion of spermatic cord. Report of case. J. Urol. (Baltimore) **24**, 91 (1930). — Langley, G. F.: Torsion of the spermatic cord in infancy. Lancet **1935** II, 181. — Le Conte, R. G.: Case of sarcoma of retroperitoneal undescended testis strangulated by a twist. Int. Clin., 17. s. **4**, 125 (1907). — Lopez, E. H.: Gac. méd. esp. **4**, 554 (1930). — MacLean, J. T.: Haemorrhagic infarct of testicle in the newborn. Surg. Gynec. Obstet. **76**, 319 (1943). — Meltzer, M.: Torsion of testicle. J. Urol. (Baltimore) **15**, 601 (1926). — Michel, A.: Bull. Soc. nat. Chir. **53**, 1144 (1927). — Moncalvi, L.: Torsion de l'hydatide dans un testicule ectopique. J. Urol. méd. chir. **35**, 501 (1933). — Mouchet, A.: Sur une variété d'orchite aiguë de l'enfance due à une torsionde l'hydatid de Morgagni. Presse méd. **31**, 485 (1923). — Moulder, M. K.: Bilateral torsion of the spermatic cord. Case report. Urol. cutan. Rev. **49**, 354 (1945). — Muschat, M.: Pathological anatomy of testicular torsion. Surg. Gynec. Obstet. **54**, 758 (1932). — O'Conor, V. J.: Torsion of the spermatic cord. Report of two cases and review of literature. Surg. Gynec. Obstet. **29**, 580 (1919). — Torsion of the spermatic cord. Surg. Gynec. Obstet. **57**, 242 (1933). — Ombredanne: Torsions testiculaires chez les enfants. Bull. Soc. nat. Chir. **39**, 779 (1913). — Ormond, J. K.: Torsion of intra-abdominal testis. Ann. Surg. **85**, 280 (1927). — Owen. E.: Case of axialrotation of the testis. Lancet **1903** II, 1247. — Paoli, J.: La forme tumorale de la torsion de l'hydatide de Morgagni. Marseille-méd. **67**, 209 (1930). — Patch, F. S.: Pedunculated cysts in tunica vaginalis with report of case showing torsion of the pedicle. Brit. J. Urol. **5**, 122 (1933). — Pearlman, S. J.: Malignancy in undescended abdominal testis with torsion. J. Urol. (Baltimore) **18**, 637 (1927). — Popov, K. M.: Torsion of testis in a case of intersexuality. Arch. Surg. (Chicago) **70**, 154 (1955). — Power, S.: Torsion of testis. Lancet **1933** II, 865. — Prehn, D. T.: A new sign in the differential diagnosis between torsion of spermatic cord and epididymis. J. Urol. (Baltimore) **32**, 191 (1934). — Putzu, F.: La torsione del cordone spermatico, ricerche spermentoli e considerazioni anatomo-patologiche e cliniche. Clinica chir. (Milano) **20**, 1295 (1912). — Randall, A.: Torsion of the appendix testis. J. Urol. (Baltimore) **41**, 715 (1939). — Ravich, R. A.: Haemorrhagic infarction of the testicle in the newborn. J. Urol. (Baltimore) **57**, 875 (1947). — Rhyne, J. L., F. A. Manz and J. F. Patton: Haemorrhagic infarction of testis in the newborn. Amer. J. Dis. Child. **89**, 240 (1955). — Riba, L. W., and C. J. Schmidlapp: Torsion of the spermatic cord. Surg. Gynec. Obstet. **83**, 163 (1946). — Rigby, H. M., and R. J. Howard: Torsion of the testicle. Lancet **1907** I, 1415. — Rocher, H. L., and G. Rioux: Orchite subaiguë de l'adolescence par torsion de l'hydatid sessile de Morgagni. Bull. Soc. nat. Chir. **52**, 586 (1926). — Rolnick, H. C.: Torsion of hydatid of Morgagni. J. Urol. (Baltimore) **42**, 458 (1939). — Rutolo, A.: La torsione dell'appendice del testicolo. Arch. ital. Chir. **17**, 221 (1927). — Scott, R. T.: Torsion of the appendix testis. J. Urol. (Baltimore) **44**, 755 (1940). — Scudder, C.: Strangulation of the testis by torsion of the cord. Ann. Surg. **34**, 234 (1901). — Seidel, R. F., and R. C. Yeaw: Torsion of appendix testis and appendix epididymis. J. Urol. (Baltimore) **63**, 714 (1950). — Shattock, C. E.: Case of torsion of the hydatids of Morgagni. Lancet **1922**, 693. — Solier, L. F., and P. Huard: Orchite, subaiguë par nécrobiose d'un "vas aberrans". Bull. Soc. nat. Chir. **55**, 730 (1929). — Sorrel. M. E.: A propos des torsions du testicule et de ses annexes. Bull. Soc. nat. Chir. **61**, 1270

(1935). — STILES,: Trans. med.-chir. **25**, 5 (1905). — TAYLOR, M. R.: A case of testicle strangulated at birth. Brit. med. J. **1**, 458 (1897). — THOREK, M.: Surgical errors and safeguards. Philadelphia 1932. — TRILLOT, A.: Un cas de torsion du testicule en position normale chez un nouveau-né de trois jours castration. Nourrisson **28**, 30 (1940). — VASTOLA, A. P.: Embryonaal crcinoma of abdominal testis in pseudoherma-phrodite. J. Amer. med. Ass. **101**, 111 (1933). — VERMEULEN, C. W., and C. S. HAGERTY: Torsion of appendix testis (hydatid of Morgagni). Report of 3 cases with a study of the microscopic anatomy. J. Urol. (Baltimore) **54**, 459 (1945). — WALKER, K. M.: Case of bilateral torsion of testis. Brit. J. Urol. **3**, 436 (1931). — WALTON, A. J.: Torsion of testis and its appendix. Lancet **1952 II**, 211. — WHEELER, J. S., and F. B. CLARK: Torsion of the spermatic cord. New Engl. J. Med. **247**, 973 (1952). — WHITTINGTON, C. T.: Un-descended intra-abdominal testicle with torsion of the cord and embryonic carcinoma. Amer. J. Surg. **60**, 304 (1943). — WILSON, W. A., and J. LITTLER: Polyorchidism. Report of 2 cases with torsion. Brit. J. Surg. **41**, 302 (1953). — YOUNG, H. H., and D. DAVIS: Practice of urology. 2. edit. Philadelphia: W. B. Saunders Company 1926.

Author Index

Page numbers in *italics* refer to the bibliography

Subject Index